METHODS IN MOLECULAR BIOLOGY

Series Editor
John M. Walker
School of Life and Medical Sciences
University of Hertfordshire
Hatfield, Hertfordshire, UK

For further volumes:
http://www.springer.com/series/7651

For over 35 years, biological scientists have come to rely on the research protocols and methodologies in the critically acclaimed *Methods in Molecular Biology* series. The series was the first to introduce the step-by-step protocols approach that has become the standard in all biomedical protocol publishing. Each protocol is provided in readily-reproducible step-by-step fashion, opening with an introductory overview, a list of the materials and reagents needed to complete the experiment, and followed by a detailed procedure that is supported with a helpful notes section offering tips and tricks of the trade as well as troubleshooting advice. These hallmark features were introduced by series editor Dr. John Walker and constitute the key ingredient in each and every volume of the *Methods in Molecular Biology* series. Tested and trusted, comprehensive and reliable, all protocols from the series are indexed in PubMed.

Photoswitching Proteins

Methods and Protocols

Edited by

Dominik Niopek

IPMB and BioQuant, Heidelberg University, Heidelberg, Baden-Württemberg, Germany

Humana Press

Editor
Dominik Niopek
IPMB and BioQuant
Heidelberg University
Heidelberg, Baden-Württemberg, Germany

ISSN 1064-3745 ISSN 1940-6029 (electronic)
Methods in Molecular Biology
ISBN 978-1-0716-0754-1 ISBN 978-1-0716-0755-8 (eBook)
https://doi.org/10.1007/978-1-0716-0755-8

This Humana imprint is published by the registered company Springer Science+Business Media, LLC, part of Springer Nature.
The registered company address is: 1 New York Plaza, New York, NY 10004, U.S.A.

Preface

On August 4, 2004, Edward Boyden and Karl Deisseroth made a groundbreaking discovery. They had just transplanted channelrhodopsin-2, a light-dependent cation channel from flagellate algae into neurons grown in a Petri dish, and exposed the cells to brief blue light pulses in a microscope. And suddenly, the cells responded, firing axon potentials precisely matching the pattern of the light stimulus—optogenetics was born. Since that day, the ability to remotely control neurons with optogenetics rapidly transformed the neurosciences and revolutionized our understanding of how behavior, memories, and feelings emerge from the complex interplay of cells in our brains.

Today, optogenetics is no more exclusive to the neurosciences. It is a flourishing research field and a maturing technology routinely used by many labs around the world and across various research areas in biology. The term optogenetics generally refers to controlling protein function from outside a cell or organism using light. The advantage of using light as trigger as compared to alternative triggers such as chemicals is the high spatiotemporal precision at which it can be applied. This becomes particularly relevant when studying fast and dynamic cellular processes, may it be the firing of neurons, cell migration, receptor signaling, or the transport of macromolecules within a cell, just to name a few.

At the core of optogenetics are photoswitching proteins, i.e., proteins whose activity is dependent on light. Nature presents us with a highly versatile toolbox of photoreceptors differing with respect to their cellular function, the light-switching mechanism they use as well as the wavelength of light they respond to. Generally, photoreceptors can be harnessed for optogenetic applications in two different ways. One is to transfer a photoreceptor into a non-natural context, all the while preserving its natural function. Here, the most famous example is the algae-derived channelrhodopsin-2 used to control ion flux into neurons mentioned before. A second possibility is to couple the photoreceptor's signaling state to a customized function by protein engineering. This allows adapting optogenetics to cellular functions which, in nature, are not dependent on light, but would be interesting to place under light control for basic research or biotechnological purposes.

Optogenetics has truly revolutionized the neurosciences and enabled new, fascinating insights into cell biology. There also have been considerable advances in developing optogenetics for industrial applications, such as the optimization of chemical production in microbes, as well as therapy of neurological diseases or vision restoration.

The purpose of this Methods in Molecular Biology volume is to provide cutting-edge protocols for using optogenetic techniques across a wide spectrum of applications in biology. On the one hand, we wish to offer these as a resource to newcomers, who are just about to perform their first optogenetics experiment. Therefore, we included several basic protocols from both neuronal and non-neuronal optogenetics, as well as three comprehensive overview articles highlighting emerging topics. On the other hand, the large panel of cutting-edge techniques contained in this volume should be of interest to researchers already in the field, who look for the latest updates on the methodology as well as inspiration on how to further improve their established workflows.

Chapter 1 details a method to investigate functional properties of local and long-range connectivity ex vivo by applying channelrhodopsin-2-mediated optogenetics and patch-clamp to acute brain slices. Chapter 2 introduces and discusses optogenetic and optochemical techniques to investigate G-protein-coupled receptors and their downstream signaling. In Chapter 3, the use of melanopsin, a mammalian G-protein-coupled photopigment, for timely activation of astrocyte-neuron networks in vitro and in vivo is detailed. Chapter 4 describes the use of upconversion nanoparticles for deep brain simulation of neurons by transcranial near-infrared light. Chapter 5 details a methodology for cloning, lab-scale production, and purification of recombinant adeno-associated virus (AAV) vectors encoding optogenetic switches. In Chapter 6, an AAV-mediated method and surgical procedures for cone transduction in the fovea of macaques are presented. Chapter 7 provides an overview and discussion of approaches to control proteins with light in cells and animals, with focus on those based on the light oxygen voltage 2 domain from *Avena sativa*. Chapter 8 details a method for controlling the nucleocytoplasmic transport of selected proteins in mammalian cells with blue light. Chapter 9 describes a methodology for light-inducible CRISPR labeling, which allows following in time the recruitment of CRIPSR-Cas9 to selected genomic loci. In Chapter 10, two complementary approaches facilitating light-dependent activation or inhibition of transgene expression based on cryptochrome-2 (CRY2) are detailed. Chapter 11 describes a method to control programmed cell death in mammalian cells with a dual blue light switch. Chapter 12 details a methodology for tracing reversible blue light-induced chromatin recruitment using the near-infrared fluorescent protein iRFP713. In Chapter 13, the construction and application of a multiwell platform for traceless light control of gene expression by melanopsin is detailed. Chapter 14 describes a methodology to control cAMP production with light or luciferins using an engineered, dual-activated adenylyl cyclase. Chapter 15 details a method for synthesizing an extracellular matrix with mechanical properties adjustable by red and far-red light. Chapter 16 provides fundamental considerations in the design of light-regulated receptor tyrosine kinases (Opto-RTKs) and describes protocols for their expression and characterization in mammalian cells. In Chapter 17, a co-culture assay and corresponding optical high-throughput screening method for identification of voltage-gated Ca^{2+} channel blockers is detailed. Finally, Chapter 18 reviews and discusses emerging tools at the intersection of optogenetics and CRISPR, which facilitate genome perturbations with unmet spatiotemporal precision.

In summary, this volume provides a comprehensive list of methods and tools to investigate highly dynamic pathways and behaviors in cells and animals using optogenetics. It is my hope that the protocols contained in this book will give newcomers in the field a head start and provide established labs with important updates on recent methodological advances. The optogenetics field has been extremely successful and has seen rapid growth over the past 15 years. If this book can make a humble contribution to sustaining this successful path, it has fulfilled its purpose.

I would like to thank all authors, who contributed detailed protocols for this book, thereby making their valuable expertise accessible, and for sharing those little tricks that are so critical to get a technique to work in the lab. Furthermore, I am grateful to John M. Walker, the series editor, who provided guidance and advice and without whom this volume would not exist.

I wish to devote this book to my kids, Valerie and Luisa, who were often sleeping when I worked on this book.

Heidelberg, Germany *Dominik Niopek*

Contents

Contributors

NOHELY ABREU • *Biochemistry, Cell and Molecular Biology Graduate Program, Weill Cornell Medicine, New York, NY, USA*

VIVIANA AGUS • *Department of Cell Biology, AXXAM S.p.A, Milan, Italy*

AYLA AKSOY-AKSEL • *Hertie Institute for Clinical Brain Research and Werner Reichardt Centre for Integrative Neuroscience, University of Tuebingen, Tuebingen, Germany; Department of Neurobiology, Institute of Biomaterials and Biomolecular Systems, University of Stuttgart, Stuttgart, Germany*

DANIEL L. ALTSCHULER • *Department of Pharmacology and Chemical Biology, University of Pittsburgh, Pittsburgh, PA, USA*

JULIA BAASKE • *Faculty of Biology, University of Freiburg, Freiburg, Germany*

STÉPHANE BERTIN • *Sorbonne Université, INSERM-DGOS CIC 1423, CNRS, Institut de la Vision, Paris, France; CHNO des Quinze-Vingts, INSERM-DGOS CIC 1423, Paris, France*

ELENA BRAZHNIKOVA • *Sorbonne Université, INSERM-DGOS CIC 1423, CNRS, Institut de la Vision, Paris, France*

FELIX BUBECK • *Synthetic Biology Group, BioQuant Center, University of Heidelberg, Heidelberg, Germany; Health Data Science Unit, Heidelberg University Hospital and Medical Faculty of Heidelberg University, Heidelberg, Germany*

PETER BUCHMANN • *Department of Biosystems Science and Engineering, ETH Zurich, Basel, Switzerland*

SHUO CHEN • *Helen Wills Neuroscience Institute, University of California, Berkeley, Berkeley, CA, USA*

DENIZ DALKARA • *Sorbonne Université, INSERM-DGOS CIC 1423, CNRS, Institut de la Vision, Paris, France*

JAMIE A. DAVIES • *Deanery of Biomedical Sciences, University of Edinburgh, Edinburgh, UK*

BARBARA DI VENTURA • *Signalling Research Centres BIOSS and CIBSS, University of Freiburg, Freiburg, Germany; Institute of Biology II, University of Freiburg, Freiburg, Germany*

INGRID EHRLICH • *Hertie Institute for Clinical Brain Research and Werner Reichardt Centre for Integrative Neuroscience, University of Tuebingen, Tuebingen, Germany; Department of Neurobiology, Institute of Biomaterials and Biomolecular Systems, University of Stuttgart, Stuttgart, Germany*

RAMONA EMIG • *Faculty of Biology, University of Freiburg, Freiburg im Breisgau, Germany; Institute for Experimental Cardiovascular Medicine, University Heart Center Freiburg—Bad Krozingen, Medical Center—University of Freiburg, Freiburg im Breisgau, Germany; Faculty of Medicine, University of Freiburg, Freiburg im Breisgau, Germany*

FABIAN ERDEL • *Division of Chromatin Networks, German Cancer Research Center (DKFZ) and Bioquant, Heidelberg, Germany; LBME, Centre de Biologie Intégrative (CBI), CNRS, UPS, Toulouse, France*

PATRICK FISCHBACH • *Institute of Synthetic Biology, University of Düsseldorf, Düsseldorf, Germany*

MARTIN FUSSENEGGER • *Department of Biosystems Science and Engineering, ETH Zurich, Basel, Switzerland; Faculty of Science, University of Basel, Basel, Switzerland*

Julien Genty • *Hertie Institute for Clinical Brain Research and Werner Reichardt Centre for Integrative Neuroscience, University of Tuebingen, Tuebingen, Germany*
Patrick Gonschorek • *Faculty of Biology, University of Freiburg, Freiburg, Germany; Institute of Chemical Sciences and Engineering, School of Basic Sciences, Ecole Polytechnique Fédérale de Lausanne (EPFL), Lausanne, Switzerland*
Candela González-Arias • *Department of Functional and Systems Neurobiology, Cajal Institute, CSIC, Madrid, Spain*
Dirk Grimm • *Department of Infectious Diseases/Virology, Heidelberg University Hospital, Heidelberg, Germany; German Center for Infection Research (DZIF) and German Center for Cardiovascular Research (DZHK), Heidelberg, Germany*
Janina Haar • *Department of Infectious Diseases/Virology, Heidelberg University Hospital, Heidelberg, Germany; BioQuant Center, University of Heidelberg, Heidelberg, Germany*
Klaus M. Hahn • *Department of Pharmacology, University of North Carolina, Chapel Hill, NC, USA; Lineberger Cancer Center, University of North Carolina at Chapel Hill, Chapel Hill, NC, USA*
Carmen N. Hernández-Candia • *Department of Pharmacology, University of Colorado School of Medicine, Aurora, CO, USA*
Philipp Hoess • *Faculty of Biology, University of Freiburg, Freiburg im Breisgau, Germany; Signalling Research Centres BIOSS and CIBSS, University of Freiburg, Bad Krozingen, Germany*
Mareike D. Hoffmann • *Synthetic Biology Group, BioQuant Center, University of Heidelberg, Heidelberg, Germany; Division of Chromatin Networks, German Cancer Research Center (DKFZ), Heidelberg, Germany*
Maximilian Hörner • *Faculty of Biology, University of Freiburg, Freiburg im Breisgau, Germany; Signalling Research Centres BIOSS and CIBSS, University of Freiburg, Bad Krozingen, Germany*
Céline Jaillard • *Sorbonne Université, INSERM-DGOS CIC 1423, CNRS, Institut de la Vision, Paris, France*
Harald Janovjak • *Australian Regenerative Medicine Institute (ARMI), Faculty of Medicine, Nursing and Health Sciences, Monash University, Clayton, VIC, Australia; European Molecular Biology Laboratory Australia (EMBL Australia), Monash University, Clayton, VIC, Australia*
Stephanie Kainrath • *Institute of Science and Technology Austria (IST Austria), Klosterneuburg, Austria; Australian Regenerative Medicine Institute (ARMI), Faculty of Medicine, Nursing and Health Sciences, Monash University, Clayton, VIC, Australia*
Chiara Krämer • *Department of Infectious Diseases/Virology, Heidelberg University Hospital, Heidelberg, Germany; BioQuant Center, University of Heidelberg, Heidelberg, Germany*
Joshua Levitz • *Biochemistry, Cell and Molecular Biology Graduate Program, Weill Cornell Medicine, New York, NY, USA; Department of Biochemistry, Weill Cornell Medicine, New York, NY, USA*
Samson Lichtenstein • *Department of Biosystems Science and Engineering, ETH Zurich, Basel, Switzerland*
Bei Liu • *Department of Pharmacology, University of North Carolina, Chapel Hill, NC, USA*
Maysam Mansouri • *Department of Biosystems Science and Engineering, ETH Zurich, Basel, Switzerland*
Daniel J. Marston • *Department of Pharmacology, University of North Carolina, Chapel Hill, NC, USA*

JAN MATHONY • *Synthetic Biology Group, BioQuant Center, University of Heidelberg, Heidelberg, Germany; Digital Health Center, Berlin Institute of Health (BIH) and Charité, Berlin, Germany*
SARA MEDEROS • *Department of Functional and Systems Neurobiology, Cajal Institute, CSIC, Madrid, Spain*
NYLA NAIM • *Department of Pharmacology and Chemical Biology, University of Pittsburgh, Pittsburgh, PA, USA; Molecular Pharmacology Training Program, University of Pittsburgh, Pittsburgh, PA, USA; Department of Pharmacology, Addgene, Watertown, MA, USA*
DOMINIK NIOPEK • *Synthetic Biology Group, BioQuant Center, University of Heidelberg, Heidelberg, Germany; Health Data Science Unit, Heidelberg University Hospital and Medical Faculty of Heidelberg University, Heidelberg, Germany*
GERTRUDIS PEREA • *Department of Functional and Systems Neurobiology, Cajal Institute, CSIC, Madrid, Spain*
ANNE RADEMACHER • *Division of Chromatin Networks, German Cancer Research Center (DKFZ) and Bioquant, Heidelberg, Germany*
BALDER REBMANN • *Faculty of Biology, University of Freiburg, Freiburg im Breisgau, Germany; Signalling Research Centres BIOSS and CIBSS, University of Freiburg, Bad Krozingen, Germany*
JEFF M. REECE • *Department of Pharmacology and Chemical Biology, University of Pittsburgh, Pittsburgh, PA, USA; Advanced Light Microscopy & Image Analysis Core (ALMIAC), National Institute of Diabetes and Digestive and Kidney Diseases (NIDDK), Bethesda, MD, USA*
KARSTEN RIPPE • *Division of Chromatin Networks, German Cancer Research Center (DKFZ) and Bioquant, Heidelberg, Germany*
JOSÉ-ALAIN SAHEL • *Sorbonne Université, INSERM-DGOS CIC 1423, CNRS, Institut de la Vision, Paris, France; CHNO des Quinze-Vingts, INSERM-DGOS CIC 1423, Paris, France; Foundation Ophtalmologique Adolphe de Rothschild, Paris, France; Department of Ophthalmology, The University of Pittsburgh School of Medicine, Pittsburgh, PA, USA*
TOBIAS STRITTMATTER • *Department of Biosystems Science and Engineering, ETH Zurich, Basel, Switzerland*
JORGE TROJANOWSKI • *Division of Chromatin Networks, German Cancer Research Center (DKFZ) and Bioquant, Heidelberg, Germany*
CHANDRA L. TUCKER • *Department of Pharmacology, University of Colorado School of Medicine, Aurora, CO, USA*
WILFRIED WEBER • *Faculty of Biology, University of Freiburg, Freiburg, Germany; Signalling Research Centres BIOSS and CIBSS, University of Freiburg, Freiburg, Germany*
DANIEL WEIS • *Signalling Research Centres BIOSS and CIBSS, University of Freiburg, Freiburg, Germany; Institute of Biology II, University of Freiburg, Freiburg, Germany*
MARTIN ZELLER • *Hertie Institute for Clinical Brain Research and Werner Reichardt Centre for Integrative Neuroscience, University of Tuebingen, Tuebingen, Germany*
XUEFENG ZHANG • *Department of Pharmacology and Chemical Biology, University of Pittsburgh, Pittsburgh, PA, USA*
MATIAS D. ZURBRIGGEN • *Institute of Synthetic Biology, University of Düsseldorf, Düsseldorf, Germany; CEPLAS—Cluster of Excellence on Plant Sciences, Düsseldorf, Germany*

Chapter 1

Studying Neuronal Function Ex Vivo Using Optogenetic Stimulation and Patch Clamp

Ayla Aksoy-Aksel, Julien Genty, Martin Zeller, and Ingrid Ehrlich

Abstract

Optogenetics has become a key method to interrogate the function of neural populations and circuits in the brain. This technique combines the targeted expression of light-activated proteins with subsequent manipulation of neural activity by light. Opsins such as Channelrhodopsin-2 (ChR2), which is a light-gated cation-channel, can be fused to or coexpressed with fluorescent proteins to allow for visualization and concurrent activation of neurons and their axonal projections. Via stereotaxic delivery of viral vectors, ChR2 can be constitutively or conditionally expressed in specific neurons in defined brain regions. Subsequently, identified axonal projections can be studied functionally ex vivo in combination with patch-clamp recordings in brain slices. This optogenetic mapping of neural circuitry has enabled the identification and characterization of novel synaptic connections and the detailed investigation of known anatomical connections previously not amenable with electrical stimulation techniques. Here, we describe a protocol for investigating functional properties of local and long-range connectivity in the brain using blue-light activated ChR2 variants and whole-cell patch-clamp recordings in acute brain slices.

Key words Channelrhodopsin-2, Optogenetics, Neural circuits, Synaptic connectivity, Brain slices, Ex vivo, Whole-cell patch-clamp

1 Introduction

The last decade has seen a tremendous increase in our understanding of the functional connectivity between specific cell types in neuronal microcircuits as well as the properties of long-range connections between brain areas in healthy brain function and disease states. In the past, investigation of synaptic connectivity in microcircuits entailed laborious multiple patch-clamp recordings of nearby neurons within a specific brain area. Neurons were either targeted for recording in transgenic reporter mice with labeled cell types, identified by electrophysiological properties, and/or post hoc histological processing of cell fills. The analysis of long-range

Ayla Aksoy-Aksel, Julien Genty, Martin Zeller contributed equally with all other contributors.

Dominik Niopek (ed.), *Photoswitching Proteins: Methods and Protocols*, Methods in Molecular Biology, vol. 2173,
https://doi.org/10.1007/978-1-0716-0755-8_1,

connectivity in a slice preparation of the target region had exclusively relied on the preservation of fiber tracts from the region of origin. It combines electrical stimulation of the fiber tract with recordings from neurons in the target region. However, conventional electrical stimulation has the disadvantages that it could cause tissue damage (due to placement of stimulation electrodes), recruit other fibers of passage or local nearby cells, is limited to known connections with conservation of distinct fiber tracts, and lacks cell-type and regional specificity.

The advent of optogenetics, a method using light to control the activity of neurons genetically modified to express light-gated ion channels [1, 2] dramatically increased the precision with which we can investigate the functional role of specific neurons and the properties of their synaptic connections [3–5]. The expression of Channelrhodopsin-2 (ChR2) or other activating opsins fused to fluorescent proteins enables not only rapid and temporally precise activation of neurons and their axonal trajectories in vivo and ex vivo, but also their visualization and post hoc anatomical analysis. For functional ex vivo studies of projections in acute brain slices, an important prerequisite is that ChR2-containing axons can be stimulated when severed from parent somata [6]. This allows to (1) assess inputs from brain regions inaccessible with conventional electrical stimulation, because fiber tracts are not separable or their trajectory is not known, (2) unequivocally specify the origin of inputs, and (3) investigate the functional connectivity between defined cell types in local microcircuits and long-range projections.

Therefore, optogenetic mapping of neuronal circuits in brain slices has currently become the method of choice to investigate characteristics of specific inputs [7, 8]. Expression of opsins in a specific population of neurons can be achieved with a variety of viral vectors, such as commercially available recombinant adeno-associated viral vectors (rAAVs), which encode different versions of fluorescently tagged opsins, including ChR2. Most rAAV serotypes have the advantage of being safe and reliable biosafety level 1 vectors with a good neural tropism and little or no immunogenic potential. Their packaging volume, albeit limited, is large enough for coding common opsins and a fluorescent marker (approx. 4–5 kb) [9]. Most rAAVs, unless specifically engineered, show a negligible efficacy for retrograde transport, and thus can be used for anterograde interrogation of connectivity [10, 11]. Moreover, rAAV-mediated expression of ChR2 does not appear to cause detrimental effects on neurons or labelled axons, when compared to other methods [12]. However, rAAV serotypes can differ in their transduction efficacy depending on neuronal cell type, brain area, or the rodent model used [13–15]. Viral vectors can be delivered to specific brain areas using stereotaxic injections [16], and conditional or cell-type specific expression can be achieved using Cre-dependent expression and/or specific promoters [17–

19]. After an appropriate expression time, acute brain slices are obtained for ex vivo optogenetic stimulation in combination with patch-clamp recordings.

Using this approach, several properties of synaptic transmission can be studied [8]. Firstly, the postsynaptic receptors mediating synaptic inputs onto target neurons can be identified using pharmacological blockers, as done for conventional electrical stimulation [20]. Secondly, although ChR2-activation in presynaptic terminals has been suggested to alter release probability by various mechanisms [21, 22], a number of studies including our own demonstrated that modulation of transmission by presynaptic receptors can be identified and studied [23–26]. Furthermore, with the appropriate ChR2 expression method and light stimulation technique, physiological aspects including efficacy of neurotransmitter release and fidelity of synaptic transmission can be reliably assessed [22], and stimulation protocols can be applied to induce synaptic plasticity [21, 27, 28]. Thirdly, a major advantage is that the monosynaptic nature of synaptic inputs can be directly assessed. The gold standard is to apply tetrodotoxin (TTX) during optical stimulation, which blocks action potentials in activated fibers and abolishes the synaptic response. Monosynaptic responses can be recovered selectively by addition of 4-aminopyridine (4-AP), which reenables neurotransmitter release from ChR2-activated boutons, most likely by a combination of blocking repolarizing K^+ channels and potentiating voltage-activated Ca^{2+} channels [29–32].

The field of optogenetics has seen a number of recent advances that extend and refine the ability to map neural circuits. For example, an expanded repertoire of activating opsins with different kinetics, light sensitivity, and excitation spectra was generated by identification of new opsins (e.g., Chronos and Chrimson) or directed modification of existing ones [33–36]. Thus, opsins with largely nonoverlapping activation spectra can be used to independently stimulate different populations of neurons or sets of inputs to analyze how distinct synaptic pathways interact [36–38]. Other novel approaches addressed the need for tools to map complex circuits at single-neuron resolution. This was achieved by implementing two important features, namely by confining the expression of ChR2 to the somatic domain of neurons and by restricting the light-stimulation using 2-photon excitation, to avoid activation of opsin-bearing dendrites and axons [39–41]. These methods hold promise to investigate microcircuits more rapidly compared to multiple whole-cell recordings.

In many cases, for the initial investigation of a novel connection and to characterize its general properties, a basic approach is sufficient. Here, we describe a protocol applicable in most labs using a standard slice physiology patch-clamp setup combined with fast light-emitting diode (LED)-based whole-field illumination, to study local and long-range connectivity with conventional ChR2 variants. Investigators need to take into account several aspects to meet their experimental needs: (1) brain regions of interest and choice of the rAAV serotype (2) cell type of interest and if applicable, choice of appropriate transgenic Cre-mouse line and/or promoter, (3) choice of the opsin (e.g., which version of ChR2 or other activating opsin) and type of fluorophore for visualization, and (4) light stimulation including LED wavelength and need for whole field or spatially restricted illumination.

2 Materials

Prepare all solutions with ultrapure water (e.g., from a MilliQ system) and reagent-grade chemicals in autoclaved glassware. Use dedicated separate benches for electrophysiology and histology to prevent live tissue contamination with fixatives.

2.1 Stereotaxic Surgery

1. Stereotaxic frame for mice equipped with high-precision micromanipulator arms and a gas anesthesia mask incorporated into the stereotaxic head holder.
2. Small animal gas anesthesia machine supplying oxygen–isoflurane mix during induction and surgery.
3. Drill for the skull. A basic option is a dental drill with a small drilling bit, preferable is a vertical drill with a ~0.5 mm PCB drilling bit that can be mounted on the stereotaxic micromanipulator.
4. Heating pad with closed-loop temperature control (e.g., with a rectal temperature probe) to keep body temperature during surgery.
5. Stereomicroscope with large working distance mounted on a movable arm.
6. Pressure application system (e.g., Picospritzer) connected to injection pipettes via inert tubing.
7. Sterilizer for surgical tools.
8. Surgical tools: Small scissors, blunt forceps, sterile 31G needle bent to a hook (or a dura dissector), needle holder, suture tying forceps, suture needles, silk suture thread, cotton swabs, animal shaver.
9. Disinfectant (povidone–iodine based), local anesthetic (e.g., Lidocaine), analgesic (e.g., meloxicam-based), eye ointment, wound gel, sterile phosphate buffered saline (PBS, *see* Subheading 2.5 **item 2**), 0.9% sterile NaCl solution (saline), 3% hydrogen peroxide.
10. Glass electrode puller (e.g., horizontal Sutter Instruments P-1000 equipped with a 3.3 mm trough or box filament).

11. Sharp injection micropipettes (*see* **Note 1**) pulled from borosilicate capillaries (e.g., Drummond Scientific, calibrated as 1–5 μL), and a closed micropipette storage jar.
12. Stock of rAAV vector to express light switchable opsins. Check the titer, and if necessary, dilute viral stocks with sterile phosphate buffered saline (PBS) to ~10^{12} to 10^{13} particles per mL (*see* **Note 2**).

2.2 Solutions for Electrophysiology

1. Artificial cerebrospinal fluid (ACSF) (*see* **Note 3**): 124 mM NaCl, 1.25 mM NaH_2PO_4, 1.3 mM $MgSO_4$, 2.7 mM KCl, 26 mM NaH_2CO_3, 2 mM $CaCl_2$, 18 mM D-Glucose, 4 mM ascorbic acid, gassed with carbogen (95% O_2, 5% CO_2).
2. Slicing solution: ACSF with 10 mM $MgSO_4$, gassed with carbogen.
3. Internal solutions for patch-clamp recordings (*see* **Note 4**). Cesium-based solution for voltage-clamp recordings: 115 mM Cs-methanesulfonate, 20 mM CsCl, 4 mM Mg-ATP, 0.4 mM Na-GTP, 10 mM Na_2-Phosphocreatine, 10 mM HEPES, 0.6 mM EGTA. pH adjusted to 7.2 with 1 M CsOH, osmolarity to 290–300 mOsm. Potassium gluconate–based solution for current-clamp recordings: 130 mM K-gluconate, 5 mM NaCl, 4 mM Mg-ATP, 0.4 mM Na-GTP, 10 mM Na2-Phosphocreatine, 10 mM HEPES, 0.6 mM EGTA. pH adjusted to 7.2 with 1 M KOH, 290–300 mOsm.

2.3 Acute Slice Preparation

1. Small animal isoflurane anesthesia machine.
2. Vibratome for acute live brain slices (*see* **Note 5**) with a cooling option for the slicing chamber (i.e., ice well or active cooling element).
3. Dissection tools: Large scissors for decapitation, fine scissors, forceps, fine spatula, scalpel, razor blades.
4. Supplies for slicing: Sapphire blade for the vibratome (e.g., from Delaware Diamond Knives) and appropriate blade holder (*see* **Note 6**), small block of 2% agarose, 1 mL syringes, 20G needles, glass Pasteur pipette, and rubber dropper bulb.
5. Lid of a 10 cm petri dish and equally sized circular filter paper.
6. Interface chamber for slice storage, and water bath large enough to hold the chamber. Although commercial chambers are available, we use a simple custom-made interface chamber built from a glass beaker (250 mL), a small plastic cup, the lid of a 50 mL Falcon tube, ~5 cm piece of ~0.5 cm inner diameter inert tubing, a nitrocellulose filter (diameter 5 cm, pore size 0.45 μm, ideally in a dark color), a 10 cm diameter petri dish, and a 10 cm diameter circular filter paper (*see* **Note 7**).

2.4 Acute Brain Slice Recordings

1. Electrophysiology setup enclosed by a faraday cage: A vibration isolation table holding an upright microscope with 5× air objective and a 40× or 60× water immersion objective (with high numerical aperture, e.g., 0.8–1 NA) mounted on a moving *x–y* table, a microscope camera, a recording chamber with heating unit, a platinum grid to stabilize the slices (*see* **Note 8**), and motorized precision micromanipulators for holding the headstage with pipette holder. The microscope should be equipped for oblique infrared illumination and fluorescence microscopy with appropriate filter sets and dichroic mirrors (e.g., for ChR2 excitation: excitation 472/20, dichroic 495, emission 490 longpass). Outside the faraday cage: A 19-in. electronics rack holding the patch-clamp amplifier, the analog-to-digital converter, the computer for data acquisition, optionally a digital oscilloscope, light source controllers and all other controllers, and perfusion pumps with matching tubing for circulation of ACSF.
2. LED light source (e.g., 470 nm for ChR2 excitation) with a TTL input either connected directly, or via a light guide and collimator to the fluorescence port of the microscope (*see* **Note 9**).
3. Patch electrode puller (e.g., horizontal Sutter Instruments P-1000 equipped with a FB255B box filament).
4. Patch electrodes pulled from Borosilicate glass capillaries (e.g., 0.86 mm inner and 1.5 mm outer diameter), and a closed micropipette storage jar.
5. 1 mL syringe, 0.2 μm pore size nitrocellulose syringe filter, flexible nonmetallic pipette fillers (e.g., MicroFil filling capillaries).

2.5 Post Hoc Histology and Microscopy

1. Vibratome for histology (*see* **Note 5**).
2. Phosphate buffered saline (PBS): 140 mM NaCl, 15 mM phosphate buffer, adjust the pH to 7.3–7.4 with NaOH or HCl.
3. 4% Paraformaldehyde in PBS, adjust the pH to 7.3–7.4 with NaOH or HCl.
4. 2% Agar-agar in PBS.
5. Mounting medium (e.g., Vectashield).
6. Small 35 mm petri dishes, beakers, filter papers, fine brush, forceps, 24-well plates, microscope slides and coverslips, PAP pen liquid blocker, nail polish.
7. Appropriate primary and secondary antibodies to detect fluorescent proteins (*see* **Note 10**).
8. Fluorescent microscope with appropriate filter sets or laser confocal scanning microscope with several laser lines (*see* **Note 11**).

3 Methods

Follow all regulations for working with genetically modified organisms and the regulations of handling and disposing of chemicals. All procedures with animals need to follow protocols approved by the local animal care and use committee (in accordance with the European Directive 2010/63/EU for animal experiments, if applicable). *See* Fig. 1a for experimental workflow and timeline.

3.1 Stereotaxic Surgery

1. If experimental conditions permit, mice should be ideally 6–8 weeks at time of surgery and younger than 12 weeks at time of recording (*see* **Note 12**).
2. Pull borosilicate injection pipettes with a thin taper that is long enough to cover the entire depth to the target site. Trim the tip to an inner diameter of 5–10 μm using fine scissors or forceps, and confirm under a stereoscope with a marked ocular scale. The shaft of the pipettes should be marked with a thin pen to control injected volume (e.g., with the calibrated Drummond capillaries, ticks at 1 mm spacing for a 100 nL scale).
3. Arrange surgical tools (sterile) and syringes on a clean drape next to the stereotaxic frame.
4. Thaw an aliquot of rAAV and front-fill pipettes under stereoscopic guidance. Connect the pipette to a 5 mL syringe via air-tight tubing, mount onto the stereotaxic arm or a manipulator, lower the pipette tip into the viral solution, and pull the solution into the pipette by suction with the syringe. Filled pipettes can be stored until use on the same day at 4 °C in a closed micropipette storage jar.
5. Anesthetize a mouse with 3% isoflurane in oxygen.
6. Confirm sufficient depth of anesthesia with toe or tail pinch. Pay close attention to the breathing of the mouse, which should be regular and calm. If needed, adjust concentration or flow rate of isoflurane.
7. Shave the top of the head, and apply eye ointment to prevent drying of the eyes. Inject with analgesic (e.g., 0.1 mL per 10 g body weight of 10% Metacam in sterile saline subcutaneously).
8. Mount the mouse onto the stereotaxic frame with a leveled head. Ventilate the animal with ~2% isoflurane in oxygen during all surgical procedures. Insert the rectal temperature probe and keep the mouse on the heating pad. Clean the shaved portions of the head with disinfectant and cotton swabs, and apply local anesthetic.

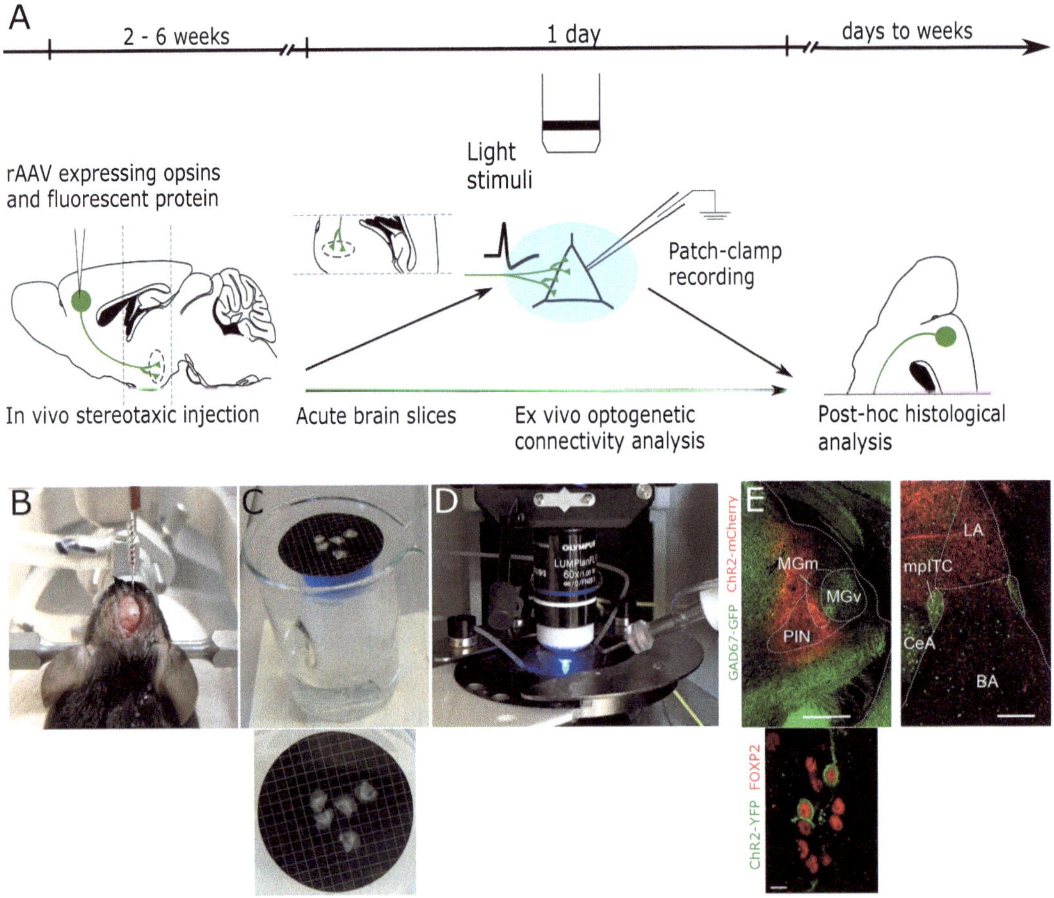

Fig. 1 Workflow and representative images of experimental steps for ex vivo optogenetic stimulation with patch-clamp recordings. (**a**) Scheme of the timeline and experimental steps. (**b**) Stereotaxic virus injection into the mouse brain. (**c**) Interface chamber in side (top) and top (bottom) view with acute brain slices containing the region for recording, here the amygdala. (**d**) Light stimulation with a 470 nm LED through the microscope objective during patch-clamp recordings. (**e**, top) Overview confocal images from post hoc histological analysis of injection and projection sites in a GAD67-GFP mouse (GABAergic neurons in green) expressing ChR2-mCherry (red). (Top left) Thalamic injection site in the medial geniculate/PIN nuclei. Scale bar: 500 μm. (Top right) Overview of the corresponding thalamic fibers at the recording site in the amygdala. Scale bar: 200 μm. (Reproduced with permission from ref. 8). (**e**, bottom) Confocal image of infected neurons in the amygdala using Cre-dependent expression of ChR2-YPF. Cells are stained with a marker (FoxP2 located in nuclei, in red), the ChR2-YPF fusion protein localizes to the cell membrane. Scale bar: 10 μm. (Reproduced with permission from ref. 23)

9. Make an incision to the skin to expose the skull. Disinfect the wound, and clean the skull with cotton swabs and a drop of hydrogen peroxide to remove the periosteum and expose a dry bone surface.
10. Identify bregma and lambda on the skull to acquire a reference for stereotaxic injection coordinates. Drill a small hole at the

desired stereotaxic coordinates until the dura is exposed. Keep holes moist with drops of sterile PBS.

11. Carefully remove the dura with minimal bleeding at the injection holes with a 31G needle bent to a hook (or dura dissector) to enable smooth penetration of the brain surface with the glass pipette.
12. Connect the glass injection pipette to the pressure application system, mount it onto the stereotaxic arm and lower it slowly to the desired depth coordinates (Fig. 1b). Place the pipette 50 μm below the target depth for 30 s to create a small pocket for the injectable, and then retract to target depth.
13. Perform injection by activating the pressure application system in a rhythmic manner (e.g., for Picospritzer we use a pressure of 20 psi and pulse length of 10–50 ms). Monitor the movement of meniscus in the pipette shaft (*see* **Note 13**) and stop once the desired volume is injected (*see* **Note 14**). Injection should proceed slowly, ideally at about 10–20 nL/min.
14. Leave the pipette in place for 5–10 min, so the injectable can diffuse. If on-track infection is a concern, pull up slowly by about 100 μm and leave in place for another 5 min. Remove the pipette at a speed of $\leq$100 μm/min for the first 500 μm.
15. Disinfect the skull with povidone–iodine and clean with cotton swabs. Close the wound with individual button sutures, disinfect, and apply wound gel.
16. Remove mouse from the stereotaxic frame and let it recover in the home cage under infrared illumination. Observe the mouse until it is fully awake, and monitor its health status in the days following the surgery.
17. Let the virus express opsins for 2–6 weeks before preparing acute brain slices, depending on the design of the experiment (*see* **Note 15**).

3.2 Acute Brain Slice Preparation

1. Prepare at least 1 L of ACSF (*see* **Note 3**) and a sufficient amount of slicing solution, that is, by adding 0.87 mL of 2 M $MgSO_4$ stock solution to 200 mL of ACSF to obtain a final concentration of 10 mM $MgSO_4$. Cool the latter on ice (<4 °C).
2. Bubble ACSF and slicing solution with carbogen (95% O_2, 5% CO_2) to ensure proper oxygenation and pH-buffering.
3. Prepare dissection tools. Make a slice transfer pipette by breaking off the sharp tip of a glass Pasteur pipette and cover with a dropper bulb, use the blunt end for slice transfer. Prepare a syringe with a 20G needle. Cut a small block of 2% agarose about the width of the brain.

4. Prepare the interface chamber (*see* **Note 7**). Place the cup with the falcon tube lid upside-down into a 250 mL glass beaker filled with ~100 mL of ACSF and bubble with carbogen using the tubing as guide. Fill the falcon tube lid with carbogenated ACSF and cover it with the nitrocellulose membrane. Cover the glass beaker with the petri dish containing the filter paper.
5. Deeply anaesthetize the mouse in 3% isoflurane in oxygen. Decapitate with large scissors and transfer the head into a small beaker filled with ice-cold slicing solution.
6. Expose the skull by cutting the scalp along the midline. Carefully open the skull with small scissors to expose the brain. Sever cranial nerves with a fine spatula, remove brain and transfer into another small beaker with ice-cold carbogenated slicing solution. The whole procedure should take less than a minute.
7. Apply a thin film of cyanoacrylate glue on the specimen holder of the vibratome (roughly the size of the brain cross-section and agarose block).
8. Transfer the brain onto a filter paper placed on the lid of a petri dish. Cut the brain with a razor blade to obtain a block containing the desired recording site, and if applicable, a separate block containing the injection site. Keep the latter in ice-cold cutting solution until acute slices are obtained.
9. Glue the brain block with recording site onto the specimen holder of the vibratome and the agar block behind it for support. Place the specimen holder into the slicing chamber of the vibratome and make sure it is fully submerged in ice-cold carbogenated slicing solution.
10. Adjust settings according to the manufacturer's instructions to cut ~320 μm thick slices of the recording site and if applicable, the nearby injection site. Ensure the blade moves slowly and vibrates at high frequency (e.g., 0.08 mm/s, 80 Hz with 0.8 mm amplitude).
11. If desired, separate the two hemispheres along the midline. Collect brain slices of the recording site with the slice transfer pipette and place them onto the nitrocellulose filter of the interface chamber (Fig. 1c). Remove excess ACSF with a syringe without touching the slices.
12. Recover the slices for recording in the interface chamber at 37 °C for 30–45 min in a water bath. Subsequently, keep the interface chamber at room temperature.
13. In the meanwhile, fix the block or slices containing the injection site in 4% PFA in PBS at 4 °C overnight (*see* **Note 16**).
14. On the next day, wash fixed samples with PBS three times for 10 min. Samples can be stably stored until post hoc analysis for

2 weeks in PBS, or for longer preservation in PBS with 0.02% azide, at 4 °C in the dark.

3.3 Recording from Acute Brain Slices and Optogenetic Stimulation

1. Prepare the electrophysiology setup for optogenetic activation of fibers and/or cells. Center the LED within the light path of the objective. Use a power meter set to the appropriate wavelength (i.e., 470 nm) to measure LED output at the back focal plane and at the end of each objective you wish to use. Repeat the procedure for different LED intensity settings (e.g., 10–100% with increment of 10%). Calculate the light power per illuminated area (mW/mm^2) from the measured values and plot LED intensity setting (%) versus power density to construct a calibration curve for each objective. For restricted aperture stimulation, close available apertures in the fluorescent light path. Assess the spatial extent of the illuminated region by taking an image with the microscope camera of the LED illumination spot on a uniform fluorescent sample (e.g., a fluorescent slide) with the desired objective. Make sure the image is not saturated. The spatial extent of the illuminated area can be quantified by determining the full width at half maximum intensity of the imaged light spot (*see* **Note 17**).
2. Prepare the electrophysiology setup for recording. Start the perfusion of carbogenated ACSF, check the flow rate (e.g., 2 mL/min), and start the heating unit to achieve the desired temperature in the recording chamber. Thaw an aliquot of the appropriate internal solution and transfer it into a 1 mL syringe equipped with a 0.2 μm syringe filter and a flexible nonmetallic pipette filler.
3. Pull patch pipettes (tip size and resistance depend on targeted cell type) and store them in a closed container.
4. Transfer a slice to the electrophysiology setup. Gently detach a slice from the nitrocellulose filter by applying ACSF from a syringe, use the slice transfer pipette to pick it up and release it into the recording chamber. Make sure the previously air-exposed surface of the slice is facing up. Stabilize the slice with the platinum grid.
5. Identify target cells by anatomical location, fluorescence from transgenic lines, or tracer injections (*see* **Note 18**) and check innervation in the target area via fluorescently labeled opsin-containing fibers, initially with a low magnification objective. Using a high magnification objective (40× or 60×), assess the density of fluorescent fibers in the target region and choose a cell for recording. Do not expose slices to the wavelength for subsequent optogenetic stimulation for long durations to avoid unnecessary activation of opsins (*see* **Note 19**).

6. Fill a pipette with the internal solution, place it into the electrode holder and apply positive pressure. Approach the cell with the micromanipulator, obtain a gigaseal and rupture the cell membrane to obtain whole-cell access (*see* **Note 20**).
7. Open or restrict the aperture of the fluorescent light path according to the design of the experiment and apply light stimulation by activating the LED (Fig. 1d, *see* **Note 17**). Stimulate with an open aperture for general assessment of long-range projections, or when the target area is poorly innervated (e.g., low density of opsin containing fibers). Restrict the aperture for more spatially selective stimulation of the fibers or cells, or when aiming at minimal stimulation of few fibers.
8. Adjust the stimulation parameters (pulse length and intensity) to record synaptic currents from the cell. Start stimulating with short pulses (on the order of milliseconds) and low LED power and increase intensity stepwise (e.g., 10% increments from 10 to 100%) or gradually increase the light-pulse duration keeping the intensity constant. Run a complete input–output curve for one or both parameters if needed. Proceed until achieving reliable and stable synaptic responses at the desired amplitude for a specific experiment (*see* **Note 21**). The timing, duration and in some cases, power modulation of the LED output can be controlled by data acquisition software and/or an external TTL device.
9. Record optogenetically evoked synaptic currents or synaptic potentials in voltage clamp or current clamp, respectively. Upon obtaining a stable baseline, pharmacological interventions can be started to address components of synaptic inputs (Fig. 2a, b) and monosynapticity (Fig. 2c, d). Depending on the design of the experiment, paired pulse stimulation can be used to address the release probability and presynaptic modulation (Fig. 2e, f), or other functional properties of synapses can be studied using more complex stimulation protocols (*see* **Note 22**).
10. After recording, fix the slices in 4% PFA in PBS overnight. To ensure a flat fixation, carefully sandwich the slices between nitrocellulose filters before submerging in fixative solution.
11. On the next day, proceed as in Subheading 3.2 **step 14**.
12. Analyze the electrophysiology data with appropriate software (*see* **Note 23**).

3.4 Post Hoc Histology

1. Resection the fixed brain tissue at ~60 μm thickness on a dedicated histology vibratome (*see* **Note 5**). Fixed brain blocks can be directly glued to the specimen holder and supported with a small agar block if necessary.
2. For fixed slices, embed them before resectioning. Place slices flat onto the bottom of a small petri dish and dry with filter

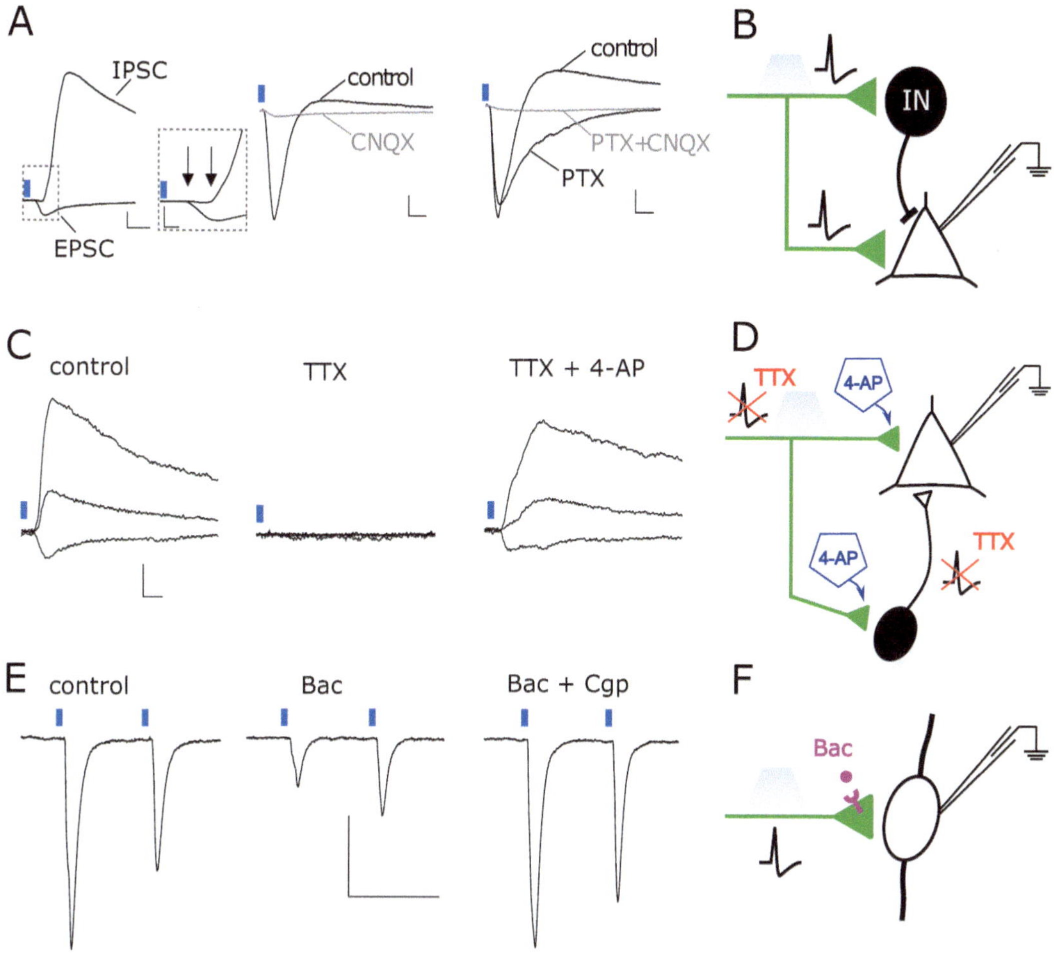

Fig. 2 Example results illustrating properties of postsynaptic currents upon optogenetic activation of long-range inputs to the mouse amygdala. (**a**, **b**) Light-evoked excitatory and inhibitory postsynaptic currents (EPSC and IPSC) from prefrontal afferents recorded in a basolateral amygdala neuron at −70 mV and 0 mV, respectively. Note the longer latency of the IPSC compared to the EPSC. Scale bars: 200 pA/10 ms and 200 pA/2 ms. The biphasic EPSC/IPSC sequence (at −50 mV) is blocked by the AMPA/Kainate-R antagonist CNQX (10 μM). Scale bar: 50 pA/5 ms. Sequential block of the EPSC/IPSC by the Cl^- channel blocker Picrotoxin (PTX, 100 μM) and PTX + CNQX (right). Scale bars: 50 pA/5 ms. (Data reproduced with permission from ref. 8). (**b**) Results from (**a**) are indicative of a feed-forward inhibitory circuit. (**c**) Light-evoked postsynaptic currents (PSCs) recorded from an amygdala neuron at different holding potentials (−70 mV, 0 mV and + 40 mV) are completely abolished in the presence of the Na^+ channel blocker tetrodotoxin (TTX, 0.5 μM) and recovered upon addition of 4-aminopyridine (4-AP, 100 μM). Scale bars: 50 pA/20 ms. (**d**) Results from (**c**) are indicative of a monosynaptic nature of the PSCs. (**e**) Light-evoked EPSCs (at −70 mV) produced by paired-pulse stimulation of thalamic afferents onto amygdala intercalated cells are reduced by application of the $GABA_B$-R agonist Baclofen (Bac, 2 μM) while paired pulse ratio is increased. Coapplication of the $GABA_B$-R antagonist CGP55845 (10 μM) recovers both amplitude and paired pulse ratio. Scale bar: 50 pA/20 ms. (Data reproduced with permission from ref. 23). (**f**) Results are indicative of a presynaptic modulation of light-activated axons by $GABA_B$-Rs

paper, so the tissue sticks to the plastic. Gently pour gelled 2% agar on top of the slice and let it harden. Cut out a small block containing the slice, glue it to the specimen holder with the slice facing upward, and proceed with slicing.

3. Use standard immunohistochemistry protocols to process injection and recording sites. Use appropriate primary and secondary antibodies directed against fluorescent proteins to visualize opsin-expressing neurons and fibers (*see* **Note 24**). Cytoarchitecture can be assessed with fluorescent Nissl stain (e.g., Neurotrace, Thermo Scientific).
4. Mount the slices on microscope slides, frame with a PAP pen, add the mounting medium and cover with coverslips. Seal with nail polish if desired.
5. Image fixed slices using a fluorescence microscope to obtain overviews of injection and recording sites. Depending on the research question, confocal microscopy may be required for detailed analysis (*see* **Notes 25** and **26**) (Fig. 1e).

4 Notes

1. Viability of neurons at the injection site is a major determinant for a successful experiment. Thin-tapered glass pipettes cause minimal tissue damage during viral injection, but have the disadvantage of being less rigid than metal needles. Bending of the flexible taper should be avoided, and removal of the dura is critical for an obstacle-free injection.
2. Although rAAV vectors are quite stable, avoid repeated freeze-thaw cycles by aliquoting the rAAV stock to small volumes required for one injection session (typically a few μL). Store aliquots at −80 °C until use. Upon thawing an aliquot of the viral vector, store it on ice until stereotaxic injection is performed.
3. ACSF can be prepared from stocks. We use a 20× stock of NaH_2CO_3, a 2 M stock of $CaCl_2$, and a 20× stock of all remaining compounds other than D-Glucose and ascorbic acid. Prepare ACSF from stocks with D-Glucose, ascorbic acid and ultrapure water at most 24 h before electrophysiological experiments and store at 4 °C until use.
4. Internal solutions with all ingredients can be prepared at a larger volume (50 or 100 mL), aliquoted, and stored in 1 mL aliquots at −20 °C until thawed on the day of the experiment. When preparing internal solution, ATP and GTP should be kept on ice, as they are easily degradable.
5. Separate dedicated vibratomes should be used for electrophysiology and histology, as cross-contamination with fixatives is a

critical concern when preparing and recording from acute brain slices.

6. Good quality acute brain slices are a prerequisite to preserve a substantial number of viable axons and recordable target neurons. We prefer sapphire blades for slicing, while other labs commonly employ zirconium ceramic blades [42].
7. Parts of the custom-built interface chamber (Fig. 1c) can be prepared beforehand. To make a holder for the slices, glue the top closed part of a 50 mL falcon tube lid to the closed bottom part of the small cup. Make an opening on one side of the cup so it does not float later. Glue a piece of inert tubing to the other side of the cup to provide a guide for the carbogen influx. To prepare a cover for the chamber, take a 10 cm petri dish and glue a similarly sized circular filter paper to its inside. The filter paper prevents dripping of condensate. The holder for the slices can be cleaned with ultrapure water and reused. The cover for the chamber can be reused as long as the filter paper remains clean.
8. The grid to stabilize slices can be built from a piece of U-shaped, flattened platinum wire on which a couple of thin nylon strings are glued. Distance between the strings should be large enough to not obstruct recording and stimulation access to the desired brain regions in the slice.
9. A wide range of LED systems covering all wavelength ranges from UV to infrared have become commercially available. They range from relatively inexpensive compact single wavelength versions with simple controllers to elaborate illumination systems with arrangements of up to 16 LEDs and more complex controllers. Compact LEDs can typically be directly mounted to the fluorescence port of the microscope with a suitable adapter, whereas more complex systems use light guides and collimators. With the currently available LEDs (especially in the blue range with peaks around 460–470 nm), sufficient light power for activation of neurons and axons in brain slices is not an issue. Adjustment of the relative power of the LED output allows in most cases to scale the synaptic response to a suitable size for a desired experimental design.
10. For GFP/YFP enhancement, we use primary antibodies directed against GFP combined with Alexa 488 conjugated secondary antibodies. For mCherry/tdTomato enhancement, we use RFP primary antibodies combined with Alexa 555 conjugated secondary antibodies.
11. For fluorescence microcopy, we use standard filtersets, for example, for GFP/Alexa 488 (ex 470/40, dichroic 495, em 525/50), for YFP (ex 500/20, dichroic 515, em 535/30) for mCherry/tdTomato/Alexa 555 (ex 550/25, dichroic 570, em

605/70). For confocal microscopy we use 488 or 514 nm laser lines for excitation of GFP/Alexa 488 or YFP, respectively, and 561 nm for excitation of mCherry/tdTomato/Alexa 555.

12. A critical factor for the viability of brain slices is the age of the animals. For adult studies, we prefer young adult mice up to 12 weeks of age at recording time. The younger the mice, the more robust is the ex vivo preparation. If a research question requires using older animals, adapted slicing protocols should be considered [42].
13. Monitor movement of the meniscus under the stereoscope when applying pressure pulses. If the meniscus does not move down, the pipette may be clogged. To alleviate this, carefully increase the pulse length until extrusion can be observed again, then revert to short pulses.
14. Adjust the delivered viral volume depending on the specifics of your experiment. Required volumes can vary from 20 to 500 nL, depending on virus type, titer, and the size of the brain structure to be infected.
15. The required expression time of the opsin to achieve effective axonal activation depends on the range of the investigated projection. Short expression times (2–3 weeks) are sufficient for local projections, but longer times (4–6 weeks) may be required for long-range projections, as opsins need to be sufficiently expressed in axons [20, 23, 25, 28, 31].
16. The precision of stereotaxic injections is important for the outcome of the slice physiology experiment. If possible, scrutinize stereotaxic targeting before proceeding with patch recordings by rapidly looking at slices that contain injection site (s) under a fluorescent microscope, and perform a detailed post hoc analysis later.
17. With the LED mounted to the fluorescence port, typically the whole field of view of a given objective is illuminated, resulting in activation of all ChR2-expressing axons or cells within that field and a certain depth. Partial restriction of the illuminated field can be achieved using adjustable apertures in the fluorescent light path [23, 43].
18. More elaborate experimental designs can be achieved by labeling specific target cell types based on molecular identity using fluorescent reporter mice combined with Cre driver lines [17, 19] or based on projection target using fluorescent retrograde tracers [20].
19. In many cases, excitation wavelengths of the opsin and fluorescent protein largely overlap (e.g., expression of ChR2 fused with GFP or YFP), leading to activation of ChR2 when screening for expressing axons in acute slices.

20. If anatomical reconstruction of recorded cells is desired, dissolve biocytin (3–5 mg/mL) in the internal solution via sonication (~30 min). When the recording is finished, gently retract the pipette and fix the slice in 4% PFA in PBS. Cellular anatomy can be visualized immunohistochemically using fluorescently labeled streptavidin [23].
21. Length of the light stimulation pulse should be kept as short as possible, for example between 0.2 and 5 ms duration. We typically use pulses <2 ms. It is important to achieve a reliable and stable response, ideally within the linear range of the input–output curve, avoiding excessive light intensity and duration.
22. For experiments requiring high frequency and/or train stimulation or repetitive stimulation protocols, be aware that depending on the opsin used, the specific on-, off-, and recovery kinetics can limit spike timing and fidelity, and thus the effective presynaptic stimulation frequency [36, 44, 45].
23. Many labs including ours acquire data in ABF format using pClamp Software (Molecular Devices). Flexible options for data analysis of this file format include NeuroMatic run in the IgorPro environment [46], or the free and open source option pyABF (https://github.com/swharden/pyABF) run in the Python environment.
24. In many cases, the expression level of the fluorescent protein in neuronal somata at the injection site is sufficient for post hoc analysis without enhancement by immunostaining. However, fluorophore bleaching may occur with longer light stimulation experiments, in particular when the excitation wavelengths of opsin and fluorescent protein overlap (e.g., ChR2 and GFP/YFP). Therefore, immunostaining against the fluorescent protein is advisable to overcome this limitation and to faithfully identify expressing cells and axons.
25. Especially in cases where connectivity between reciprocally connected brain regions is investigated, initial experiments should scrutinize the presence of opsin-labelled cell bodies in the recording area due to aberrant retrograde transport of viral vectors.
26. When opsins are expressed as fusion proteins, fluorescence localizes to the cell membrane. Therefore, for detection of opsin fusion proteins, conventional fluorescence microscopy may not be sufficient to unequivocally delineate infected cell bodies from neurites and local projections. In many cases, confocal microscopy will be required. Due to the high surface-to-volume ratio, membrane-localized fluorescence is an advantage when performing post hoc histological analysis of axonal projections.

References

1. Nagel G, Szellas T, Huhn W, Kateriya S, Adeishvili N, Berthold P, Ollig D, Hegemann P, Bamberg E (2003) Channelrhodopsin-2, a directly light-gated cation-selective membrane channel. Proc Natl Acad Sci U S A 100(24):13940–13945. https://doi.org/10.1073/pnas.1936192100
2. Boyden ES, Zhang F, Bamberg E, Nagel G, Deisseroth K (2005) Millisecond-timescale, genetically targeted optical control of neural activity. Nat Neurosci 8(9):1263–1268. https://doi.org/10.1038/nn1525
3. Tye KM, Deisseroth K (2012) Optogenetic investigation of neural circuits underlying brain disease in animal models. Nat Rev Neurosci 13(4):251–266. https://doi.org/10.1038/nrn3171
4. Yizhar O, Fenno LE, Davidson TJ, Mogri M, Deisseroth K (2011) Optogenetics in neural systems. Neuron 71(1):9–34. https://doi.org/10.1016/j.neuron.2011.06.004
5. Adamantidis A, Arber S, Bains JS, Bamberg E, Bonci A, Buzsaki G, Cardin JA, Costa RM, Dan Y, Goda Y, Graybiel AM, Hausser M, Hegemann P, Huguenard JR, Insel TR, Janak PH, Johnston D, Josselyn SA, Koch C, Kreitzer AC, Luscher C, Malenka RC, Miesenbock G, Nagel G, Roska B, Schnitzer MJ, Shenoy KV, Soltesz I, Sternson SM, Tsien RW, Tsien RY, Turrigiano GG, Tye KM, Wilson RI (2015) Optogenetics: 10 years after ChR2 in neurons—views from the community. Nat Neurosci 18(9):1202–1212. https://doi.org/10.1038/nn.4106
6. Petreanu L, Huber D, Sobczyk A, Svoboda K (2007) Channelrhodopsin-2-assisted circuit mapping of long-range callosal projections. Nat Neurosci 10(5):663–668. https://doi.org/10.1038/nn1891
7. Lim DH, Ledue J, Mohajerani MH, Vanni MP, Murphy TH (2013) Optogenetic approaches for functional mouse brain mapping. Front Neurosci 7:54–54. https://doi.org/10.3389/fnins.2013.00054
8. Bosch D, Asede D, Ehrlich I (2016) Ex vivo optogenetic dissection of fear circuits in brain slices. J Visual Exp 110:e53628. https://doi.org/10.3791/53628
9. Davidson BL, Breakefield XO (2003) Viral vectors for gene delivery to the nervous system. Nat Rev Neurosci 4(5):353–364. https://doi.org/10.1038/nrn1104
10. Tervo DG, Hwang BY, Viswanathan S, Gaj T, Lavzin M, Ritola KD, Lindo S, Michael S, Kuleshova E, Ojala D, Huang CC, Gerfen CR, Schiller J, Dudman JT, Hantman AW, Looger LL, Schaffer DV, Karpova AY (2016) A designer AAV variant permits efficient retrograde access to projection neurons. Neuron 92(2):372–382. https://doi.org/10.1016/j.neuron.2016.09.021
11. Zingg B, Chou XL, Zhang ZG, Mesik L, Liang F, Tao HW, Zhang LI (2017) AAV-mediated anterograde transsynaptic tagging: mapping corticocollicular input-defined neural pathways for defense behaviors. Neuron 93(1):33–47. https://doi.org/10.1016/j.neuron.2016.11.045
12. Miyashita T, Shao YR, Chung J, Pourzia O, Feldman DE (2013) Long-term channelrhodopsin-2 (ChR2) expression can induce abnormal axonal morphology and targeting in cerebral cortex. Front Neural Circ 7:8. https://doi.org/10.3389/fncir.2013.00008
13. Aschauer DF, Kreuz S, Rumpel S (2013) Analysis of transduction efficiency, tropism and axonal transport of AAV serotypes 1, 2, 5, 6, 8 and 9 in the mouse brain. PLoS One 8(9):e76310. https://doi.org/10.1371/journal.pone.0076310
14. Holehonnur R, Luong JA, Chaturvedi D, Ho A, Lella SK, Hosek MP, Ploski JE (2014) Adeno-associated viral serotypes produce differing titers and differentially transduce neurons within the rat basal and lateral amygdala. BMC Neurosci 15:28. https://doi.org/10.1186/1471-2202-15-28
15. McFarland NR, Lee JS, Hyman BT, McLean PJ (2009) Comparison of transduction efficiency of recombinant AAV serotypes 1, 2, 5, and 8 in the rat nigrostriatal system. J Neurochem 109(3):838–845. https://doi.org/10.1111/j.1471-4159.2009.06010.x
16. Cetin A, Komai S, Eliava M, Seeburg PH, Osten P (2006) Stereotaxic gene delivery in the rodent brain. Nat Protoc 1(6):3166–3173. https://doi.org/10.1038/nprot.2006.450
17. Madisen L, Zwingman TA, Sunkin SM, Oh SW, Zariwala HA, Gu H, Ng LL, Palmiter RD, Hawrylycz MJ, Jones AR, Lein ES, Zeng H (2010) A robust and high-throughput Cre reporting and characterization system for the whole mouse brain. Nat Neurosci 13(1):133–140. https://doi.org/10.1038/nn.2467
18. Huang ZJ, Zeng H (2013) Genetic approaches to neural circuits in the mouse. Annu Rev Neurosci 36:183–215. https://doi.org/10.1146/annurev-neuro-062012-170307

19. Daigle TL, Madisen L, Hage TA, Valley MT, Knoblich U, Larsen RS, Takeno MM, Huang L, Gu H, Larsen R, Mills M, Bosma-Moody A, Siverts LA, Walker M, Graybuck LT, Yao Z, Fong O, Nguyen TN, Garren E, Lenz GH, Chavarha M, Pendergraft J, Harrington J, Hirokawa KE, Harris JA, Nicovich PR, McGraw MJ, Ollerenshaw DR, Smith KA, Baker CA, Ting JT, Sunkin SM, Lecoq J, Lin MZ, Boyden ES, Murphy GJ, da Costa NM, Waters J, Li L, Tasic B, Zeng H (2018) A suite of transgenic driver and reporter mouse lines with enhanced brain-cell-type targeting and functionality. Cell 174(2):465–480.e422. https://doi.org/10.1016/j.cell.2018.06.035

20. Hübner C, Bosch D, Gall A, Luthi A, Ehrlich I (2014) Ex vivo dissection of optogenetically activated mPFC and hippocampal inputs to neurons in the basolateral amygdala: implications for fear and emotional memory. Front Behav Neurosci 8:64. https://doi.org/10.3389/fnbeh.2014.00064

21. Zhang YP, Oertner TG (2007) Optical induction of synaptic plasticity using a light-sensitive channel. Nat Methods 4(2):139–141. https://doi.org/10.1038/nmeth988

22. Jackman SL, Beneduce BM, Drew IR, Regehr WG (2014) Achieving high-frequency optical control of synaptic transmission. J Neurosci 34 (22):7704–7714. https://doi.org/10.1523/JNEUROSCI.4694-13.2014

23. Asede D, Bosch D, Luthi A, Ferraguti F, Ehrlich I (2015) Sensory inputs to intercalated cells provide fear-learning modulated inhibition to the basolateral amygdala. Neuron 86 (2):541–554. https://doi.org/10.1016/j.neuron.2015.03.008

24. Britt JP, Benaliouad F, McDevitt RA, Stuber GD, Wise RA, Bonci A (2012) Synaptic and behavioral profile of multiple glutamatergic inputs to the nucleus accumbens. Neuron 76 (4):790–803. https://doi.org/10.1016/j.neuron.2012.09.040

25. Chu HY, Ito W, Li J, Morozov A (2012) Target-specific suppression of GABA release from parvalbumin interneurons in the basolateral amygdala by dopamine. J Neurosci 32 (42):14815–14820. https://doi.org/10.1523/JNEUROSCI.2997-12.2012

26. Zussy C, Gomez-Santacana X, Rovira X, De Bundel D, Ferrazzo S, Bosch D, Asede D, Malhaire F, Acher F, Giraldo J, Valjent E, Ehrlich I, Ferraguti F, Pin JP, Llebaria A, Goudet C (2018) Dynamic modulation of inflammatory pain-related affective and sensory symptoms by optical control of amygdala metabotropic glutamate receptor 4. Mol Psychiatry 23(3):509–520. https://doi.org/10.1038/mp.2016.223

27. Kohl MM, Shipton OA, Deacon RM, Rawlins JN, Deisseroth K, Paulsen O (2011) Hemisphere-specific optogenetic stimulation reveals left-right asymmetry of hippocampal plasticity. Nat Neurosci 14(11):1413–1415. https://doi.org/10.1038/nn.2915

28. Morozov A, Sukato D, Ito W (2011) Selective suppression of plasticity in amygdala inputs from temporal association cortex by the external capsule. J Neurosci 31(1):339–345. https://doi.org/10.1523/JNEUROSCI.5537-10.2011

29. Petreanu L, Mao T, Sternson SM, Svoboda K (2009) The subcellular organization of neocortical excitatory connections. Nature 457 (7233):1142–1145. https://doi.org/10.1038/nature07709

30. Felix-Ortiz AC, Beyeler A, Seo C, Leppla CA, Wildes CP, Tye KM (2013) BLA to vHPC inputs modulate anxiety-related behaviors. Neuron 79(4):658–664. https://doi.org/10.1016/j.neuron.2013.06.016

31. Cho JH, Deisseroth K, Bolshakov VY (2013) Synaptic encoding of fear extinction in mPFC-amygdala circuits. Neuron 80(6):1491–1507. https://doi.org/10.1016/j.neuron.2013.09.025

32. Wu ZZ, Li DP, Chen SR, Pan HL (2009) Aminopyridines potentiate synaptic and neuromuscular transmission by targeting the voltage-activated calcium channel beta subunit. J Biol Chem 284(52):36453–36461. https://doi.org/10.1074/jbc.M109.075523

33. Zhang F, Prigge M, Beyriere F, Tsunoda SP, Mattis J, Yizhar O, Hegemann P, Deisseroth K (2008) Red-shifted optogenetic excitation: a tool for fast neural control derived from Volvox carteri. Nat Neurosci 11(6):631–633. https://doi.org/10.1038/nn.2120

34. Gradinaru V, Zhang F, Ramakrishnan C, Mattis J, Prakash R, Diester I, Goshen I, Thompson KR, Deisseroth K (2010) Molecular and cellular approaches for diversifying and extending optogenetics. Cell 141(1):154–165. https://doi.org/10.1016/j.cell.2010.02.037

35. Prigge M, Schneider F, Tsunoda SP, Shilyansky C, Wietek J, Deisseroth K, Hegemann P (2012) Color-tuned channelrhodopsins for multiwavelength optogenetics. J Biol Chem 287(38):31804–31812. https://doi.org/10.1074/jbc.M112.391185

36. Klapoetke NC, Murata Y, Kim SS, Pulver SR, Birdsey-Benson A, Cho YK, Morimoto TK, Chuong AS, Carpenter EJ, Tian Z, Wang J, Xie Y, Yan Z, Zhang Y, Chow BY, Surek B,

Melkonian M, Jayaraman V, Constantine-Paton M, Wong GK, Boyden ES (2014) Independent optical excitation of distinct neural populations. Nat Methods 11(3):338–346. https://doi.org/10.1038/nmeth.2836

37. Erbguth K, Prigge M, Schneider F, Hegemann P, Gottschalk A (2012) Bimodal activation of different neuron classes with the spectrally red-shifted channelrhodopsin chimera C1V1 in Caenorhabditis elegans. PLoS One 7(10):e46827. https://doi.org/10.1371/journal.pone.0046827

38. Yizhar O, Fenno LE, Prigge M, Schneider F, Davidson TJ, O'Shea DJ, Sohal VS, Goshen I, Finkelstein J, Paz JT, Stehfest K, Fudim R, Ramakrishnan C, Huguenard JR, Hegemann P, Deisseroth K (2011) Neocortical excitation/inhibition balance in information processing and social dysfunction. Nature 477 (7363):171–178. https://doi.org/10.1038/nature10360

39. Baker CA, Elyada YM, Parra A, Bolton MM (2016) Cellular resolution circuit mapping with temporal-focused excitation of soma-targeted channelrhodopsin. eLife 5:e14193. https://doi.org/10.7554/eLife.14193

40. Shemesh OA, Tanese D, Zampini V, Linghu C, Piatkevich K, Ronzitti E, Papagiakoumou E, Boyden ES, Emiliani V (2017) Temporally precise single-cell-resolution optogenetics. Nat Neurosci 20(12):1796–1806. https://doi.org/10.1038/s41593-017-0018-8

41. Mardinly AR, Oldenburg IA, Pégard NC, Sridharan S, Lyall EH, Chesnov K, Brohawn SG, Waller L, Adesnik H (2018) Precise multimodal optical control of neural ensemble activity. Nat Neurosci 21(6):881–893. https://doi.org/10.1038/s41593-018-0139-8

42. Ting JT, Daigle TL, Chen Q, Feng G (2014) Acute brain slice methods for adult and aging animals: application of targeted patch clamp analysis and optogenetics. Methods Mol Biol 1183:221–242. https://doi.org/10.1007/978-1-4939-1096-0_14

43. Li H, Penzo MA, Taniguchi H, Kopec CD, Huang ZJ, Li B (2013) Experience-dependent modification of a central amygdala fear circuit. Nat Neurosci 16(3):332–339. https://doi.org/10.1038/nn.3322

44. Mattis J, Tye KM, Ferenczi EA, Ramakrishnan C, O'Shea DJ, Prakash R, Gunaydin LA, Hyun M, Fenno LE, Gradinaru V, Yizhar O, Deisseroth K (2011) Principles for applying optogenetic tools derived from direct comparative analysis of microbial opsins. Nat Methods 9:159. https://doi.org/10.1038/nmeth.1808

45. Berndt A, Schoenenberger P, Mattis J, Tye KM, Deisseroth K, Hegemann P, Oertner TG (2011) High-efficiency channelrhodopsins for fast neuronal stimulation at low light levels. Proc Natl Acad Sci U S A 108 (18):7595–7600. https://doi.org/10.1073/pnas.1017210108

46. Rothman JS, Silver RA (2018) NeuroMatic: an integrated open-source software toolkit for acquisition, analysis and simulation of electrophysiological data. Front Neuroinform 12:14. https://doi.org/10.3389/fninf.2018.00014

Chapter 2

Optogenetic Techniques for Manipulating and Sensing G Protein-Coupled Receptor Signaling

Nohely Abreu and Joshua Levitz

Abstract

G protein-coupled receptors (GPCRs) form the largest class of membrane receptors in the mammalian genome with nearly 800 human genes encoding for unique subtypes. Accordingly, GPCR signaling is implicated in nearly all physiological processes. However, GPCRs have been difficult to study due in part to the complexity of their function which can lead to a plethora of converging or diverging downstream effects over different time and length scales. Classic techniques such as pharmacological control, genetic knockout and biochemical assays often lack the precision required to probe the functions of specific GPCR subtypes. Here we describe the rapidly growing set of optogenetic tools, ranging from methods for optical control of the receptor itself to optical sensing and manipulation of downstream effectors. These tools permit the quantitative measurements of GPCRs and their downstream signaling with high specificity and spatiotemporal precision.

Key words G protein-coupled receptors (GPCR), G protein, Calcium, Optogenetics, Photopharmacology, LOV domain

1 Introduction

G protein-coupled receptors (GPCRs) serve as critical membrane receptors that mediate many fundamental aspects of biological function, including sensory transduction, neuromodulation, respiration, regulation of heart rate and control of blood pressure [1]. GPCRs respond to a wide range of extracellular stimuli including neurotransmitters, peptides, hormones, ions, or lipids to initiate complex intracellular signaling pathways with precise spatial and temporal dynamics. The GPCR superfamily forms the largest class of receptors with over 800 human genes encoding distinct subtypes, and, accordingly, serves as the largest class of drug targets for the treatment of diseases [2]. Despite a long-standing appreciation of the physiological importance of GPCR signaling and recent breakthroughs at the structural level [3, 4], a relatively limited understanding of the functional mechanisms of GPCRs at the cell

Dominik Niopek (ed.), *Photoswitching Proteins: Methods and Protocols*, Methods in Molecular Biology, vol. 2173, https://doi.org/10.1007/978-1-0716-0755-8_2,

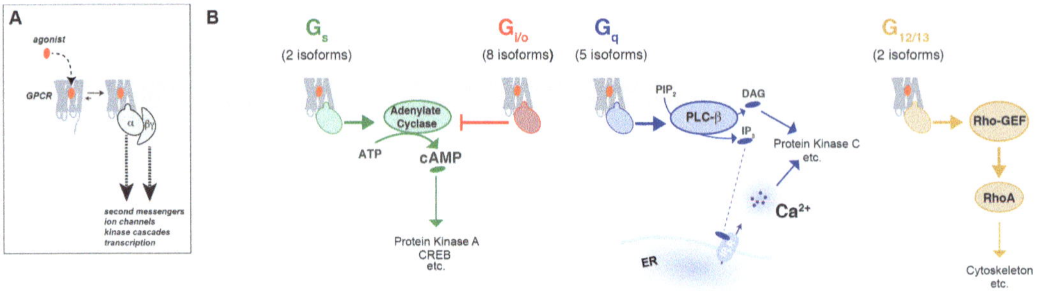

Fig. 1 Basic mechanisms of GPCR signaling. (**a**) GPCRs bind ligands which initiate a conformational change in the receptor that allows the receptor to couple to and signal through heterotrimeric G proteins. These can then activate various second messengers and effectors. (**b**) Canonical signaling pathways mediated by the four families of Gα protein subtypes

biological and systems levels currently exists, in part due to the challenge of directly manipulating and sensing receptors in physiological settings with higher spatiotemporal precision than classical pharmacological or genetic approaches. This challenge has motivated new, high resolution optical techniques, which are the topic of this review.

Following agonist binding, GPCRs undergo a series of conformational changes that allow them to activate heterotrimeric G proteins which are comprised of Gα, Gβ, and Gγ proteins (Fig. 1a). The active conformation of the receptor is recognized by the C-terminal helix of Gα [5, 6, 7], which subsequently exchanges GDP for GTP to promote dissociation of the heterotrimer into a GTP-bound Gα subunit and an active Gβγ complex. The liberated G proteins have a wide-ranging pallet of target effectors, including membrane-embedded enzymes and ion channels, depending on the G protein subtype that is engaged. In addition, G protein-mediated signals are thought to be amplified over time due to the ability of one stimulated GPCR to activate multiple heterotrimeric G proteins, as has been most clearly demonstrated for rhodopsin which can activate thousands of G proteins per second [8]. Further nonlinearity of signaling can come from the generation of second messengers which can spread with unique spatiotemporal parameters to relay the signal to different cellular locations. G protein signaling is terminated following hydrolysis of the G-α-bound GTP, which can be accelerated by regulator of G protein signaling (RGS) proteins, leading to the reassembly of the heterotrimeric G protein.

A given GPCR typically has a preferred Gα target, but in many cases can target multiple G protein families [6]. The four families of Gα proteins include the G_s, $G_{i/o}$, G_q, and $G_{12/13}$ subtypes which all contain multiple members and produce distinct, canonical responses (Fig. 1b). Generally, activated $G\alpha_s$ proteins activate adenylyl cyclase to generate cAMP from ATP which can, in turn, lead to activation of protein kinase A (PKA). PKA phosphorylates a wide range of proteins, including other GPCRs. On the other hand, activated $G\alpha_{i/o}$ proteins inhibit adenylyl cyclase which leads to reduced levels of cAMP and active PKA. Activated $G\alpha_q$ proteins activate phospholipase C-beta (PLC-ß), which produces inositol trisphosphate (IP_3) and diacylglycerol (DAG) from phosphatidylinositol 4,5-bisphosphate (PIP_2) and leads to the release of calcium from intracellular stores via IP_3 receptors of the endoplasmic reticulum. Elevations in intracellular calcium can lead to complex oscillatory responses and target myriad other signaling proteins including kinases, such as protein kinase C (PKC) and calcium-calmodulin dependent kinase II (CaMKII), and phosphatases, such as calcineurin [9]. $G\alpha_{12/13}$, the least frequently targeted family, activates Rho GTPases to regulate the actin cytoskeleton which can control cell mobility. Meanwhile, the Gβγ complex has direct targets of its own including, most prominently, the opening and closing of ion channels such as the G protein-coupled inwardly rectifying potassium (GIRK) channels and voltage-gated calcium channels (VGCCs) [10]. While no clear preference of different beta or gamma subtypes for interaction with Gα subtypes has been observed, many Gβγ-driven processes have been shown to be specific for $G_{i/o}$-coupled receptors. The mechanism for GPCR subtype specificity of Gβγ-driven signaling remains debated, but a recent study proposed that $G_{i/o}$-coupled signaling has an intrinsically higher rate of G protein activation which produces a greater local concentration of free Gβγ than other receptor subtypes [11].

The aforementioned upstream signaling pathways and effector molecules only scratch the surface of the complex cellular consequences of GPCR activation. Other direct targets of G proteins likely exist and many downstream cellular responses have been observed without clear information on the intermediate signaling events that link the response to the initial GPCR activation. For example, GPCR activation often leads to activation of the MAPK/ERK pathway which produces long-lasting effects on the cell by modulating gene transcription, cell proliferation and differentiation. Importantly, in addition to their coupling to G proteins themselves, GPCRs can couple and signal through β-arrestins [12]. Canonically, β-arrestins are recruited to the C-terminal tail of activated GPCRs following phosphorylation by G protein-coupled receptor kinases (GRKs). β-arrestin interaction with the GPCR sterically blocks interactions with G proteins while also serving as an endocytic adaptor to promote the internalization of the receptor to desensitize signaling [13]. Alternatively, following binding to the ligand-bound receptor, β-arrestin can be allosterically activated and serve as a scaffold to initiate various signaling pathways [14]. In particular, β-arrestins have been shown to initiate the MAPK/ERK pathway, independent of G proteins, either indirectly via activation of Src or through the formation of a complex with RAF-1, MEK, and ERK [15–17]. However, ongoing debate exists over the relative contributions of G proteins and β-arrestins to ERK activation by GPCRs [18–20]. Interestingly, the identity of the GPCR agonist can bias the receptor to preferentially couple and signal through either G proteins or β-arrestins, in a phenomenon termed "biased agonism" that is promising clinically but poorly understood mechanistically [21]. Finally, there is a growing body of evidence that internalized GPCRs can initiate signals from endosomes and other organelles, permitting functional diversity and an additional layer of complexity to signal transduction [22].

Ultimately, a single GPCR can initiate a mosaic of independent, converging or diverging signaling pathways that consists of a multitude of effectors and second messengers spread throughout space to achieve an assortment of short- (i.e., hundreds of milliseconds) and long-lasting (i.e., hours to days) effects on the cell and, ultimately, organism. To complicate matters, an individual receptor can serve different functions depending on its subcellular localization or the cell type in which it is expressed. Furthermore, multiple GPCR subtypes are typically expressed in any given cell, some of which can share the same endogenous ligand. Thus, signaling pathways mediated by one GPCR can crosstalk with that of a different GPCR, heightening the difficulty of studying one receptor or pathway in isolation. Finally, GPCR signaling often occurs within multicellular circuits where different cell types have distinct responses that can influence other cells within the network. This immense complexity presents a challenge to studying GPCRs: how can we achieve the precision required to thoroughly study one receptor or one component of a signaling pathway in order to dissect their direct contribution to a biological process? In this review we describe techniques that utilize optical probes to control and observe GPCR activity and signaling with high spatiotemporal precision and genetic targeting. First, we discuss existing methods for the optical manipulation (section II) and detection (section IV) of GPCRs themselves. We highlight important examples of the use of these tools to decipher mechanistic aspects about GPCR function. We then discuss methods for the optical manipulation (section III) and detection (section IV) of different intracellular molecules that contribute to GPCR signaling. This survey provides a framework for the further design and application of optogenetic tools for GPCRs.

2 Optical Control of GPCRs

One of the most successful realms of optogenetics has been the development of techniques for the optical control of GPCRs. Ideally, optical control allows one to target a specific GPCR subtype with spatial and temporal resolution in a genetically defined cell type. Such precise targeting should allow one to determine the relative contribution of a receptor subpopulation to a physiological process, including those that take place at the organismal level such as behavior, and better define the receptor's response to specific temporal patterns of activity in defined subcellular regions. A number of complementary techniques have been developed to allow for control of a range of GPCRs and recent studies have begun to exploit these tools to gain meaningful biological insight. In this section, we summarize the main categories of approaches, assess their relative advantages and caveats, and describe illustrative examples of their application.

Typically, GPCRs have been manipulated using a combination of native or synthetic pharmacology. A vast library of agonists, antagonists, and modulators exists for different receptor subtypes which represent the basis for many clinical approaches [23]. However, when applying these drugs to biological samples, especially in tissue or in vivo, many limitations emerge including lack of receptor subtype specificity, lack of cell-type specificity, and the inability to rapidly apply and remove the compound to defined locations within the tissue or cell. Taking advantage of the plethora of well-characterized ligands, photopharmacology has blossomed into a flexible and powerful field of chemical biology. The general principle of photopharmacology is to incorporate light-sensitive groups into biologically active ligands in order to control their activity with light (Fig. 2a). Early studies used "caged" ligands, most frequently caged glutamate, which are released from a chemical enclosure (i.e., 4-methoxy-7-nitroindolinyl, MNI) upon illumination [24, 25]. Various caging strategies have been developed, some of which have been applied to GPCR ligands. Banghart et al. developed caged agonists and antagonists [26–28] for opioid receptors which have been used to probe the spatial extent of enkephalin signaling in the locus coeruleus [26] and to define specific aspects of receptor kinetics [27, 29].

While applicable to many compounds and a clear enhancement over normal drugs, uncaging is irreversible, thus limiting the types of experiments that can be designed. To overcome this, over the last 15 years photoswitchable ligands have been developed for a wide range of signaling proteins, including GPCRs. Photoswitchable ligands are typically built around the azobenzene chemical photoswitch, which toggles between cis and trans configurations in response to light and can, thus, modulate the functional properties of attached compounds [30]. Dirk Trauner and others have developed azobenzene-based photoswitchable agonists for class A and B GPCRs, including mu-opioid receptors [31], muscarinic acetylcholine receptors [32], cannabinoid receptors [33], and the glucagon-like peptide receptor [34]. While agonists have been the main focus of photopharmacological development, the flexibility of the approach allows other types of ligands to be developed. For

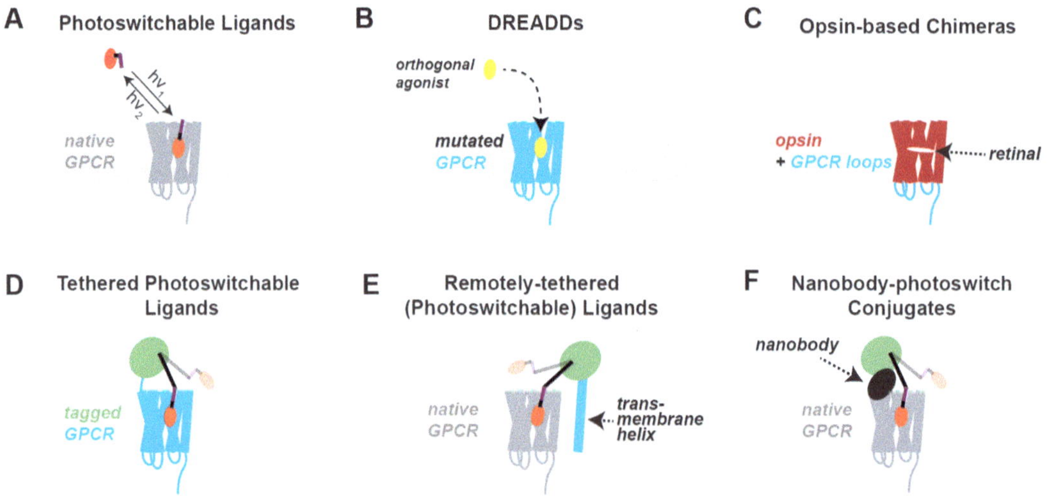

Fig. 2 Chemical and optical methods for the targeted control of GPCRs. (**a**) Photoswitchable ligands are typically composed of a well-characterized ligand (orange) and a light-sensitive moiety (purple) that either isomerizes or releases the ligand in the presence of light. This system allows the control of native GPCRs by controlling ligand availability with high time resolution. The temporal resolution of reversal is limited by ligand unbinding and spatial targeting is limited by diffusion. (**b**) DREADDs are mutated GPCRs that are engineered to respond to a synthetic agonist but to no longer respond to a native ligand. By targeting DREADD expression, one can achieve genetically-targeted chemical (i.e., "chemogenetic") control of GPCR signaling. (**c**) Opsins are naturally occurring light-activatable GPCRs that have been taken advantage of for the optical control of GPCRs. To study and manipulate signaling pathways associated with specific GPCRs, chimeras may be designed that incorporate the intracellular loops and/or the C-terminal tail of the GPCR of interest. (**d**) Photoactivation of specific GPCRs can be achieved by tethering a photoswitchable ligand to a self-labeling tag, such as SNAP (green). This system requires heterologous expression of a tagged GPCR which allows the incorporation of receptor mutants or variants but can lead to overexpression. Tethered photoswitchable ligands allow the highest temporal resolution of both ligand binding and unbinding. (**e**) Photoactivation of native GPCRs with genetic targeting can be accomplished by the expression of a tagged transmembrane domain that is tethered to a photoswitchable ligand. (**f**) Native GPCRs may also be optically controlled via photoswitchable ligands tethered via self-labeling tags (green) to nanobodies (black) that recognizes an extracellular site on the receptor. Nanobody-tethered photoswitchable ligands can either be delivered directly or can be genetically encoded for heterologous expression and cell-type targeting

example, photoswitchable allosteric modulators of metabotropic glutamate receptors (mGluRs) have been developed [35] and applied in vivo to probe the basis of mGluR4-mediated pain processing [36]. While the photopharmacological toolset is constantly expanding and in use for applications, soluble photopharmacological ligands have shortcomings due to the lack of genetic targeting, the typical lack of true subtype-specificity and limitations in both the speed and spatial precision of receptor targeting due to diffusion.

One of the most successful strategies to improve the cell-type precision of GPCR activation has been with the so-called "chemogenetic" approach. Most prominently, Bryan Roth and colleagues have developed "Designer Receptors Exclusively Activated by

Designer Drugs" (DREADDs) which are engineered muscarinic receptors that no longer retain sensitivity to their native agonist, acetylcholine, but are able to be activated by clozapine *N*-oxide (CNO), a synthetic agonist (Fig. 2b). DREADDs have been used extensively for in vivo studies in rodents to probe the cell-type and circuit basis of behavior but their application for detailed analysis of specific GPCR-driven signaling processes has been limited [37]. However, engineering of DREADDs is ongoing and recent years have seen the report of new DREADDs that allow for multiplexing [38], G protein or arrestin-biased DREADDs [39, 40], and subcellularly targeted DREADDs [41]. Overall, the genetically targeted GPCR activation enabled by chemogenetics makes this an extremely valuable approach, but optical approaches are needed for applications which require high spatiotemporal precision.

One robust approach to the optical control of GPCR activation is to employ opsins, the visual photoreceptors which use a native retinal chromophore as a light-sensing ligand [42]. A pioneering optogenetic demonstration coexpressed a G protein-coupled opsin, a Gα subunit and an arrestin from the drosophila visual system to allow for optical control of neural activity in cultured hippocampal neurons [43]. While channelrhodopsins are now typically used for direct optical excitation of neurons [44], various G protein-coupled opsins have been employed for optical control of G protein pathways via heterologous expression. This includes rhodopsin [45] or cone opsins [46] for $G_{i/o}$ activation, jellyfish opsin [47] for G_s activation and melanopsin [48, 49] for G_q activation. Notably, Karunarathne et al. [50] used spatially targeted activation of a $G_{i/o}$-coupled opsin to locally manipulate PIP_3 levels and induce neurite extension in cultured neurons and, recently, Makowka et al. [51] used jellyfish opsin to compare the effects of targeted G_s activation in the left versus right atrium of the heart. While wild-type opsins are powerful for optical control of defined G protein-mediated signaling cascades, various groups have introduced the intracellular loops and/or C-terminal tails of different GPCRs to mimic the signaling of different GPCRs while maintaining light activation (Fig. 2c) [52, 53]. This approach has been used for a variety of circuit neuroscience applications [54–56] but has not yet been employed for more precise cell biological studies.

Despite the utility of opsin-based optical control of GPCR signaling, the temporal precision of such approaches remains limited due the intrinsically slow off kinetics of opsins and the ability to probe receptor-specific function via the chimeric technique remains unclear. As a means of achieving optical control of full-length GPCRs with both spatiotemporal and genetic targeting, photoswitchable tethered ligands (PTLs) have been developed for both ion channels and GPCRs. This strategy has been most thoroughly developed for metabotropic glutamate receptors (mGluRs) and is based on the covalent attachment of a photoswitchable agonist to the receptor. Initial versions of this were based on introduction of a cysteine mutation into the ligand binding domain a full-length mGluR for attachment of a maleimide-conjugated azobenzene-glutamate compound termed "MAG" [57, 58]. The photophysical properties of the azobenzene allow rapid switching of the ligand on the millisecond time scale between inactive *trans* and active *cis* states and since the mutated receptor needs to be heterologously expressed, genetic targeting is enabled. Importantly, since the PTL is covalently attached to the receptor of interest, this system facilitates complete subtype specificity even if the ligand moiety itself is nonspecific, as is the case with MAG. Levitz et al. [59] took advantage of this high degree of control to probe the relative contribution to downstream activation of one or two agonists binding within homo and heterodimeric mGluRs. Recently, PTLs were also developed for D1 and D2 dopamine receptors [60], showing the generalizability of this approach to class A GPCRs. While maleimide-based PTLs have shown utility for biophysical and cellular applications, the cysteine-maleimide chemistry is undesirable in complex preparations due to potential off-target effects and the need to identify an optimal cysteine on the receptor is a major challenge for PTL engineering. To overcome this, "photoswitchable, orthogonal, remotely tethered ligands" (PORTLs) were developed which genetically fuse the photoswitchable ligand to a self-labelling tag (i.e., SNAP) at the N-terminus of the receptor (Fig. 2d). This method improves the orthogonality and efficiency of the system for mGluRs [61], enables multiplexing using SNAP and CLIP tags [62] and works efficiently in vivo [63]. Furthermore, branched PORTLs have recently been reported which enhance photoswitching efficiency and allow the incorporation of a fluorophore for simultaneous detection of the targeted receptor [64].

The optogenetic approaches to the manipulation of GPCRs described in this section have opened up new avenues of research and many applications remain to be pursued from molecular biophysics to cell biology and systems neuroscience. However, the challenge of many of the aforementioned techniques is that they require heterologous expression of a receptor. This provides the ability to genetically target specific cell types and the ability to introduce biologically relevant mutations or variants to the GPCR of interest, but it also raises the possibility of overexpression which can alter the fundamental biology of the system and compromise the physiological relevance of the study. One possibility is to express a genetically encoded protein that can be labeled with a tethered ligand to then target nearby native receptors (Fig. 2e). Such an approach has been demonstrated for GPCRs with a tethered antagonist for muscarinic receptors [65] or with a glutamate PORTL for mGluR2 [66]. Alternatively, receptor-targeting single-chain antibodies (i.e., nanobodies) may provide a means of delivery of a PORTL to a native GPCR (Fig. 2f). A recent proof-of-principle study showed that SNAP-tagged anti-GFP nanobodies can enable optical control of a GFP-tagged mGluR [67]. Both of these techniques have great promise for optical control of native receptors but will require further engineering and characterization to be widely applicable. Ultimately, a key aspect of future optogenetic studies of GPCRs will be the ability to apply these tools in conjunction with optical control (section III) and measurement (section IV) of other aspects of GPCR signaling to answer biological questions.

3 Optical Control of GPCR Effectors and Related Signaling Pathways

Activation of a single GPCR can lead to modulation of a diverse set of downstream effectors such as small GTPases, ion channels, kinases and other enzymes, regulators of translation and transcription factors. This divergent signaling makes it difficult to determine the relative contribution of specific GPCR effectors to subsequent physiological processes. Bypassing the receptor to control specific signaling nodes or regulatory proteins can permit the probing of the spatiotemporal properties and role of that specific component while limiting confounding effects from other signaling proteins. Although pharmacology can be used to examine the role of a particular signaling protein where available, this approach can be nonspecific and is often irreversible, especially when targeting enzymes. Recently, optogenetic tools have been developed in an effort to manipulate specific intracellular signaling proteins with high precision. While synthetic chemical photoswitches have been the typical approach to the optical control of GPCRs themselves (section II), naturally occurring light-sensitive proteins have been the core component in the design of optical control of intracellular signaling. The different configurations and mechanisms of such photoactivatable proteins have been reviewed extensively elsewhere [68, 69] but will be described where relevant below.

A limited body of work has pursued optical control of the direct effectors and regulators of GPCRs but such tools could be extremely valuable for parsing the relative roles, dynamics, and regulation of G protein or β-arrestin-mediated signaling. Notably, an optogenetic β-arrestin2 was recently reported which utilizes the cryptochrome CRY2 system [70]. CRY2 is a photoreceptor from *A. thaliana* that contains a flavin adenine dinucleotide (FAD) chromophore which is reduced following illumination with blue light (405–488 nm) which enables it to both homooligomerize and to bind its native interaction partner, CIB [71, 72]. CIB and CRY were fused to the C-terminal tail of the β2-adrenergic receptor (β_2AR) and the N-terminus of β-arrestin2, respectively [70]. Upon illumination with blue light, β-arrestin2 was translocated to the surface which was shown to be sufficient for the endocytosis of β_2AR, even in the absence of agonist. Importantly, the CRY-β-arrestin2 and β_2AR-CIB interaction is reversible; in the dark, β-arrestin2 dissociated from the receptor and redistributed throughout the cytosol over the span of ten minutes. The CRY2 system has also been used to enable photoactivation of RGS proteins [73, 74]. RGS proteins regulate GPCR signaling by accelerating the intrinsic GTPase activity of the Gα subunit to promote its reassociation with the Gβγ complex to tune the frequency, duration, and strength of signaling [75]. In one case [74], optical control was achieved by splitting RGS2 into its catalytic box and N-terminal amphipathic helix, and fusing each part to CRY2 and a truncated version of CIB (CIBN), respectively (Fig. 3a). Blue light illumination initiates the dimerization of CRY2 and CIBN, translocating the RGS catalytic box to the surface, terminating $G\alpha_q$-induced calcium signaling. This tool was used to establish the role of RGS2 in maintaining a negative feedback loop on calcium oscillations mediated by activation of the M3 muscarinic acetylcholine receptor (M3R), a G_q-coupled receptor. The CRY2-CIB system has also been used to recruit the C-terminal domain of GRKs to the membrane to bind and sequester Gβγ [73]. A similar approach has also been reported to optically recruit Gα proteins to the membrane to initiate signaling [76].

Further downstream from the receptors, optical control has been used to manipulate the levels of cAMP, a key second messenger controlled by both $G_{i/o}$ and G_q-coupled receptors (Fig. 1). Naturally occurring photoactivated adenylyl cyclases (PACs) have been identified [77], including euPAC from *Euglena gracilis* [78], bPAC from *Beggiatoa* [79], and OaPAC from *Oscillatoria acuminate* [80]. PACs contain "blue light sensor using FAD" (BLUF) domains conjugated to one or two adenylyl-cyclase domains, whereby illumination with blue light (~455 nm) induces a conformational change which activates the enzymatic activity of the adenylyl cyclase domains (Fig. 3c). Activation induces rapid and reversible generation of cAMP, and can be employed in heterologous cells or organisms to study the effects of cAMP signaling with spatiotemporal precision. Optical control of adenylyl cyclase activity has been used to manipulate cAMP concentration in a variety of contexts [81–86]. In particular, Zhou et al. [85] utilized PAC in cultured neurons to determine that transient elevations of cAMP levels increased axonal length and that this increase was dependent on the duration of increased cAMP production. Moreover, a critical caveat of utilizing PACs to control cAMP levels is that PACs exhibit some activity in the absence of the light stimulus [83], which can alter the basal cAMP levels in the cell or organism in which the tool is employed.

A more extensive toolset of photoactivatable proteins has been developed for the study of various aspects of kinase signaling [87]. For example, a number of strategies have been employed for optogenetic perturbation of the ERK/MAPK pathway, which is

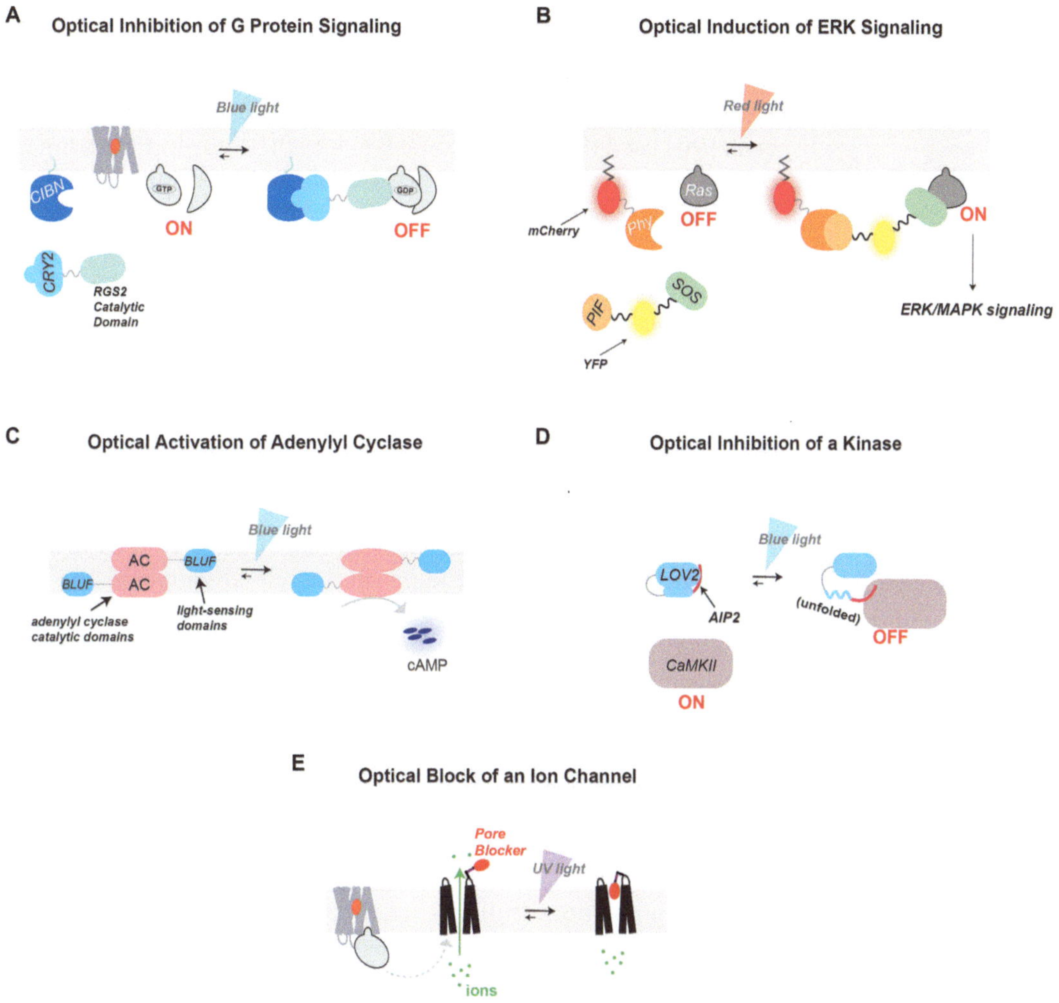

Fig. 3 Optical methods for the control of GPCR effectors. (**a**) The CRY-CIB system can be used for blue light-induced association of two proteins. For example, optical inhibition of G protein-mediated signaling has been achieved by utilizing the CRY-CIB system to induce the translocation of RGS2 to the plasma membrane with blue light. This recruitment brings RGS2 in close proximity to membrane-bound Gα-GTP which allows it to effectively inhibit G protein signaling. (**b**) The Phy-PIF system can be used for red light-induced association of two proteins. For example, photocontrol of ERK signaling can be achieved through the induction of the surface translocation of SOS using the Phy-PIF system. Phy was fused to a CAAX motif for membrane incorporation, while SOS was fused to PIF. Red light induces dimerization of Phy and PIF, translocating SOS to the membrane to promote interaction with and activation of Ras GTPase and initiate ERK signaling. Fluorescent proteins were incorporated into the constructs to observe their expression and dynamic localization. (**c**) Optical control of cAMP production can be achieved by heterologously expressing bPAC, a photoactivatable adenylyl cyclase from *Beggiatoa*, which has increased catalytic activity in the presence of blue light. (**d**) The LOV domain has been utilized to gain photocontrol of the activity of various signaling proteins. Following blue light illumination, the Jalpha helix of the LOV domain unfolds allowing it to release functional peptides or proteins for light-gated activity. In the illustrated case, LOV2 was fused to a CaMKII inhibitory peptide, AIP2, which becomes released in the presence of blue light. (**e**) Photoswitchable tethered ligands can be employed for the optical control of ion channels. For example, a photoswitchable pore blocker may be attached to an ion channel using cysteine chemistry. The pore blocker is designed such that it fits into the opening of the channel only in the presence of light to effectively block ion conductance

typically initiated following either GPCR or, more canonically, receptor tyrosine kinase (RTK) activation [88, 89]. Toettcher et al. [90] used the phytochrome B (Phy)-PIF light-induced dimerization system [91] to control plasma membrane recruitment of the small GTPase SOS with 650 nm light. Once at the plasma membrane, SOS could interact with membrane-anchored Ras GTPases, thereby triggering ERK activity (Fig. 3b). Similarly, Zhang et al. [92] used the CRY2-CIB pair to optically initiate ERK signaling by recruiting RAF-1, a vital kinase in the initiation of the ERK/MAPK cascade, which led to differentiation and neurite outgrowth in PC12 cells. Photoswitchable MEK1 and MEK2 have also been developed based on light-induced uncaging via the fluorescent protein Dronpa [93] and a photoinhibitable SRC kinase was developed using light-induced allosteric control [94]. Together these tools allow different points of the MAPK/ERK signaling cascade to be precisely perturbed depending on the relevant application.

Phosphoinositide 3-kinase (PI3K) and Akt (Protein Kinase B) signaling, additional downstream targets of GPCR signaling, and regulators of cell growth, survival, and proliferation have also been targeted for optogenetic control. Classically, PI3K phosphorylates PIP2 to generate PIP3, recruiting Akt to the cell surface, which initiates signaling to mTOR [95]. Optogenetic control of PI3K signaling was first achieved using the Phy-PIF system to recruit PI3K to the surface and induce PIP_2 phosphorylation at specific regions of the membrane with targeted 650 nm light [96]. Alternatively, manipulation of this pathway can also be achieved by using the CRY2/CIB system to optically recruit phosphatases to the plasma membrane to manipulate PIP_3 levels [97] or to optically recruit Akt itself [98–100]. Receptor tyrosine kinases, which activate many of the same effectors as GPCRs and also crosstalk with GPCRs [101], have also been targeted for optogenetic control using light-induced dimerization systems [102–105].

As an alternative to optical activation of heterologous constructs, optogenetic kinase inhibitors have also been developed to enable silencing of native kinases [87]. Most of these have been based on the LOV domain, which is a photoreceptor domain that binds a flavin chromophore and following absorption of blue light undergoes a conformational change including unfolding of the Jα helix [106]. This unfolding event can be coupled sterically or allosterically to other proteins or peptides to produce the desired output [107]. For example, an optogenetic inhibitor was recently designed for CaMKII [108], a kinase that is activated by calcium and has been shown to regulate the function of several GPCRs via phosphorylation [109–113]. The photoactivatable inhibitor of CAMKII was termed paAIP2 and is based on the fusion of a LOV domain to autocamtide inhibitory peptide 2 (AIP2), a potent CaMKII inhibitor [114]. Upon illumination with blue light, AIP2 is released and becomes available to inhibit CAMKII activity (Fig. 3d). This tool was used to inhibit long-term potentiation, a form of synaptic plasticity, in dendritic spines, and define the critical timing of CAMKII activation required for the induction of synaptic plasticity [108]. Similarly, Yi et al. [115] fused a LOV domain to an inhibitory peptide to enable optical inhibition of PKA activity and Melero-Fernandez de Mera et al. [116] fused a LOV domain to an inhibitory domain to enable the optical inhibition of JNK, a downstream target of the MAPKs. Alternatively, unnatural amino acid incorporation combined with conjugation of a tethered, photoswitchable inhibitor has been used to optically control MEK1 [117]. While optogenetic control of kinases have not been applied explicitly in the context of GPCR signaling, these tools should enable the probing of the functional roles of specific pathways following the activation of a particular GPCR.

A number of other optogenetic tools utilize the same families of light-sensitive proteins to permit manipulation of cellular processes and may have utility in the context of GPCR signaling including the optical control of Rho GTPases [118], transcription [119, 120], protein clustering and phase transition [121–123], organelle localization [124, 125], protein degradation [126, 127], protein secretion [128], and direct recruitment of proteins to membranes [129]. Further engineering efforts and new applications will be needed to advance this field and allow for these techniques to be used for studying GPCR signaling on the relatively fast time scales needed.

Finally, the GPCR effectors most commonly targeted for optogenetic control are ion channels, which have typically been manipulated via similar photopharmacological techniques to those targeting GPCRs [130, 131]. A wide range of mammalian ion channels of relevance to GPCR signaling, including voltage-gated potassium channels [132], ligand-gated cation channels [133], leak potassium channels [134], and G protein-gated GIRK channels [135] have been successfully targeted with photoswitchable ligands. A common approach has been to either use a photoswitchable pore blocker (Fig. 3e) or photoswitchable agonist conjugated to an engineered extracellular cysteine which affords the system with both optical and genetic control. The ability to directly target potential ion channel effectors with light should allow a better understanding of the coupling properties of GPCRs and ion channels. For example, Sandoz et al. [134] used a photoswitchable quaternary ammonium pore blocker to find that TREK1 channels are noncanonical downstream targets of native $GABA_B$ receptors in hippocampal neurons. While the PORTL and NB-based approaches described in section II have not yet been applied to ion channels, they hold great promise for further enhancing this efficiency and applicability of optical control of ion channels. The ongoing development of optogenetic techniques should continue to illuminate the coordinated signaling of GPCRs and ion channels.

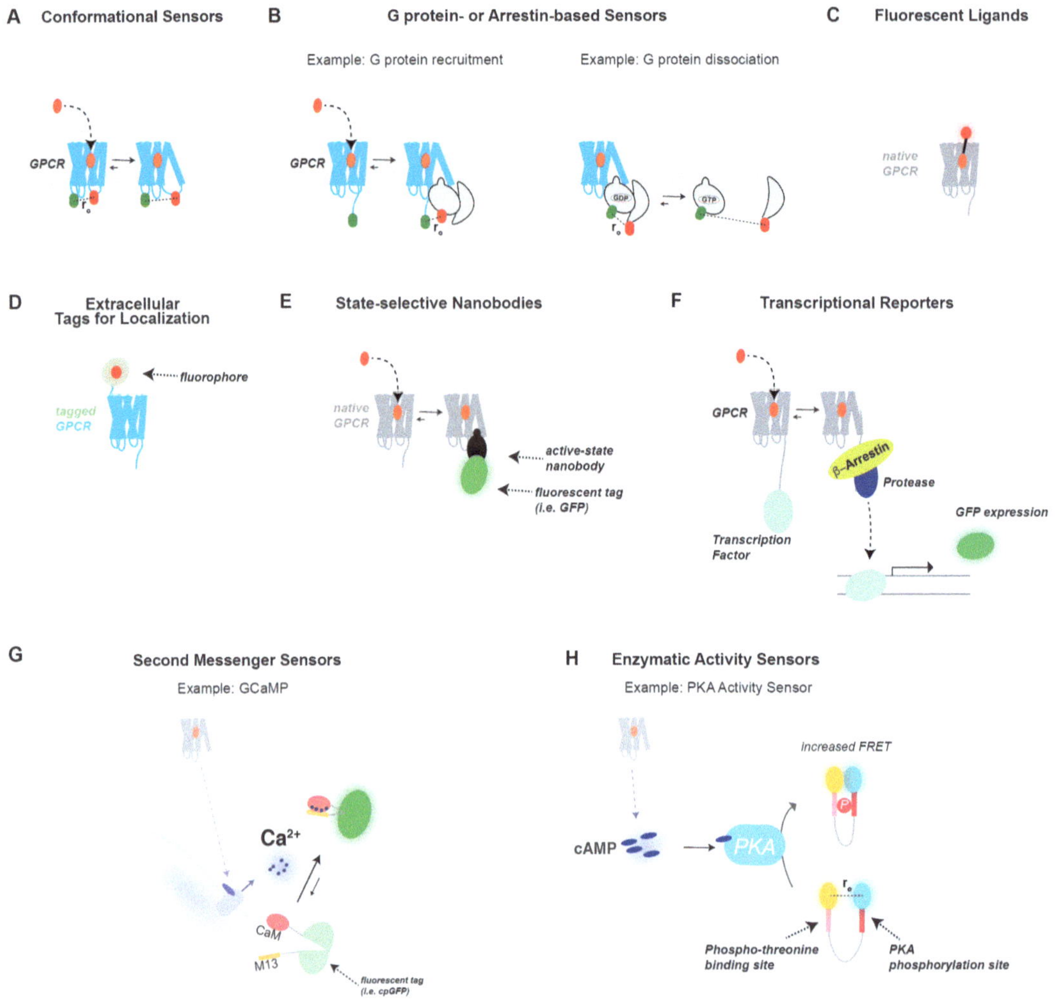

Fig. 4 Optical sensing of GPCRs and their downstream effectors. (**a**) FRET-based sensors may be used to detect the conformational changes associated with GPCR activation following ligand binding. Typically, donor and acceptor fluorophores (i.e., fluorescent proteins) are introduced into the intracellular loops of the receptor which move relative to each other during activation. Often these fluorophores disrupt coupling to G proteins making them more typically used for understanding receptor pharmacology or structure/function relationships rather than signaling properties. (**b**) FRET-based sensors have also been developed to sense G protein recruitment to the receptor (left) or dissociation of G protein heterotrimers (right). For G protein recruitment, a fluorophore is typically fused to the C-terminal tail of the receptor while the Gα subunit is tagged with a fluorophore at an internal site to avoid disrupting function and receptor specificity. To sense G protein dissociation, the Gα subunit is tagged as described above, while another fluorophore is typically placed on the N-terminus of either Gβ or Gγ. Similar sensors have also been developed to sense the recruitment or conformational changes of β-arrestins. (**c**) Ligands conjugated to fluorophores have been developed to aid in their visualization both in vitro and in vivo. (**d**) GPCRs can be tagged with a fluorophore, either a fluorescent protein or dye conjugation to a tag (i.e., SNAP, CLIP, or Halo) to permit the observation of their surface and subcellular localizations in living cells. (**e**) Fluorophore-tagged nanobodies have been developed which can be genetically encoded and used to detect the conformational state of a receptor, G protein or β-arrestin. Accumulation of fluorescence in a given location (i.e., plasma membrane or endosomes) indicates a population of receptors or effectors in a state recognized by the NB. (**f**) GPCR activation can also be sensed using a transcriptional reporter, such as the Tango assay. In this assay a transcription factor and a protease site are incorporated into the C-terminal tail of the receptor, while β-arrestin is tagged with a protease. Receptor activation

4 Optical Sensing of GPCR Signaling

While optogenetic control of GPCRs (section II) and GPCR effectors (section III) are powerful methods, the ability to observe the spatiotemporal dynamics of individual pathway components provides a complementary approach to aid in the understanding of the precise function of GPCR subtypes. In this section we will survey existing methods for detection of GPCR activation and signaling using fluorescence-based reporters. These sensors, which have been developed for nearly every node of GPCR signaling, allow for the direct observation and measurement of the intensity, duration, frequency, compartmentalization, and spread of signaling in a given stimulated pathway.

Förster resonance energy transfer (FRET) has been utilized extensively as a tool to measure the initial events of GPCR-mediated activation and signaling [136]. FRET describes the distance-dependent energy transfer between a donor fluorophore and an acceptor fluorophore that can be easily measured in live cells on standard microscopes [137]. Beginning in the mid-2000s a number of sensors have been produced to sense the ligand-induced conformational changes of GPCRs [138, 139]. Such sensors have typically been based around the introduction of fluorescent proteins into the intracellular loops of the seven helix transmembrane domains of GPCRs (Fig. 4a). More recently, organic fluorophores have also been introduced into the extracellular ligand binding domains of class C GPCRs using self-labelling tags, such as SNAP and CLIP [140], to sense the inter-subunit rearrangements that mediate activation of these dimeric receptors [141–145]. One limitation of this approach is that it is difficult to find fluorophore attachment positions that give clear signals and fluorescent proteins often disrupt the function of the fused GPCR. Unnatural amino acid incorporation has emerged as an alternative approach to fluorophore attachment without the bulk of fluorescent proteins but have so far been most valuable for biophysical studies and have been limited in their use in physiological settings [146].

Fig. 4 (continued) leads to β-arrestin-protease recruitment, leading to the cleavage and release of the transcription factor which will permit expression of a reporter gene (i.e., GFP). (**g**) A classic sensor for the detection of the second messenger, Ca^{2+}, is GCaMP. Ca^{2+} binding to CaM permits its association with the M13 peptide which induce the closure and increased fluorescence of cpGFP. Many permutations of GCaMP and other related calcium sensors exist with variable kinetics, sensitivity, subcellular targeting and spectral properties to allow the ideal construct to be used for the relevant application. (**h**) FRET-based enzymatic activity sensors are widely used to detect kinase activity. In the case of the PKA sensor, AKAR, the sensor contains N- and C-terminal donor and acceptor fluorophores, a PKA phosphorylation site, and a domain that binds to the phosphorylated residue. Following phosphorylation by active PKA, association of the two sites leads to increased FRET between the fluorophores

FRET-based approaches have also been developed to monitor the coupling of an activated GPCR with G proteins [147–150], as well as the dissociation of Gα subunit and Gβγ complex following receptor activation [150, 151] (Fig. 4b). Similar FRET-based or bioluminescence resonance energy transfer (BRET)-based approaches have also been used to measure β-arrestin recruitment to GPCRs [138] and β-arrestin conformational changes following receptor activation [152, 153]. These FRET-based tools have been used in living cells to define the millisecond time scale of receptor activation, probe structural models of receptor and effector activation, and as powerful assays for drug screening [136].

While deciphering the details of receptor activation and effector coupling in living cells remains a major goal, a parallel issue in the field of GPCR biology is to detect receptor localization dynamics within the cell. One long-standing approach is to use fluorescently labeled ligands which bind to receptors and allow their visualization (Fig. 4c). This approach is compatible with native receptors and has been used in tissue to detect receptor multimers [154]. However, the development of fluorescent agonists or antagonists for a given GPCR is challenging and the efficacy of the ligand makes it difficult to determine the effects of receptor activation on localization. An alternative has been to label the GPCR with a fluorophore directly without the need for inclusion of a pharmacophore that recognizes a binding site (Fig. 4d). A number of labeling techniques have been developed which allow one to attach fluorophores to heterologous receptors, including the SNAP, CLIP, and Halo-tags. A major advantage of this approach versus simply tagging a receptor with a fluorescent protein is that extracellular tags can be added to the N-terminus of the receptor and membrane-impermeable fluorophores may be used to limit visualization to surface receptors. For example, Thomsen et al. [155] labeled an N-terminally SNAP-tagged β_2AR with an impermeable fluorophore and used this technique to track activation-induced internalization which led to colocalization of receptors, G proteins and ß-arrestins in endosomes. In a particularly striking study, a GPCR and a Gα were labeled with SNAP- and CLIP-, respectively, and single molecule imaging was used to decipher the timing of receptor/G protein coupling in live cells [156].

The ideal optogenetic sensors for GPCRs would be able to effectively sense receptor or effector conformation and localization of native, untagged receptors. One promising avenue for this is to use nanobodies (NBs) which recognize specific states of the GPCR, G protein, or arrestin. NBs were crucial to the GPCR structural revolution, but have also been tagged with fluorophores and expressed in living cells to sense GPCR signaling [157, 158]. State-selective nanobodies that stabilize active or inactive states of GPCRs [159–161], G proteins [162, 163] or arrestins [163] exist and their accumulation in specific cellular sites can, in principle, provide information on both protein localization and conformation (Fig. 4e). In a pioneering study, Irannejad et al. used [165] an NB that recognizes the active state of the ß$_2$AR (Nb80) and a NB that recognizes the active state of Gα_s (Nb37) to show that internalized receptors can couple to G proteins from endosomes. Improved engineering methods continue to increase the rate of development of new NBs which should facilitate the widespread use of these approaches [166]. However, one caveat is that state-selective NBs themselves can serve as pharmacological agonists, antagonists or modulators and can, thus, perturb the system in a concentration-dependent manner.

As an alternative to sensing the GPCR activation process directly, transcription-coupled fluorescent sensors have also been pursued. The transcriptional activation following ß-arrestin translocation ("Tango") system has been developed and modified to allow ß-arrestin recruitment to a specific GPCR to be coupled to protease-induced release of a transcription factor to drive the transcription of luciferase or a fluorescent protein (Fig. 4f) [167, 168]. While this approach lacks temporal precision, the signal amplification and time-integration allow for detection of receptor activation over extended time periods. In an impressive study, this system was adapted for use in vivo in mice where coincident light-activation and ligand binding are necessary for protease activity and reporter gene expression. This "iTango" system was used to report on dopamine receptor activation in different cell types over a defined time period to identify reward- and locomotion-related dopamine-sensitive neurons in the striatum [169].

Downstream of receptor activation, genetically encoded fluorescent sensors have been developed for the detection of the canonical second messengers that are generated following GPCR activation including, most prominently, Ca^{2+} ions. Typically, G_q-coupled GPCR activation leads to the release of calcium from the endoplasmic reticulum (ER) (Fig. 1b). Additionally, $G_{i/o}$-coupled GPCRs can inhibit VGCCs, to reduce the influx of calcium, which is a particularly common form of presynaptic inhibition in the nervous system. While early calcium sensors were based on FRET [170], the most commonly used calcium sensors are those in the family of GCaMP [171], which has been optimized for maximal signal to noise and engineered to produce a wide range of spectral and kinetic variants [172] for simultaneous sensing of additional effectors. The general design is based on a circularly permuted GFP (cpGFP) fused to calmodulin and the M13 domain from myosin light chain kinase. Ca^{2+} binding brings calmodulin and M13 in close proximity, enclosing the GFP chromophore that is otherwise exposed to the cytosol and therefore quenched. Thus, an increase in GFP fluorescence intensity is associated with an increase in Ca^{2+} binding, enabling one-color imaging of relative calcium concentration with time resolution on the order of milliseconds (Fig. 4g). Extensive protein engineering work has also been applied to the development of FRET-based sensors for cAMP [173], a second messenger regulated by $G_{i/o}$- and G_s-coupled receptors (Fig. 1b). Most recently, a one-color sensor based on a similar design principle to GCaMP has been developed, termed cAMPr [174]. cAMPr is also based on a cpGFP, where one end is fused to the catalytic subunit of PKA (PKA-C) and the other end is tethered to a regulatory subunit of PKA (PKA-R). cAMP binding induces a conformational change which separates PKA-C and PKA-R from one another, permitting GFP to fluoresce efficiently.

Beyond the fluorescent sensors for Ca^{2+} and cAMP, genetically encoded fluorescent sensors have also been reported for voltage [174], phospholipids, pH, and other second messengers [172] and further design optimization promises to improve the sensitivity and ease with which these tools can be applied in complex systems. In a particularly exciting series of breakthroughs, the cpGFP-based sensor design has been applied to develop sensitive fluorescence reporters for a variety of GPCR-targeting neurotransmitters and neuromodulators. Patriarchi et al. [176] reported a suite of tools based on the insertion of cpGFP into intracellular loop 3 of class A GPCRs to develop sensors for dopamine, norepinephrine, melatonin, opioids, and serotonin. The dopamine sensor, "dLight1" was validated in vivo where it has been used for studies of behavior-associated dopamine dynamics in the striatum [176, 177]. Given that these constructs are unable to couple to downstream effectors their main utility will likely be as sensors for the extracellular ligand rather than GPCR-specific signaling properties. Furthermore, similarly designed sensors have also been reported for dopamine [178] and norepinephrine [179] and bacterial binding domains have been fused to cpGFP for optimized sensors of glutamate [180, 181] or GABA [182]. While extremely valuable and increasingly used throughout neuroscience, there is a major caveat to the use of fluorescent biosensors in the study of the dynamics of extracellular signals or intracellular second messengers that may limit these tools for precise, quantitative study of GPCR signaling. For example, in the case of calcium sensors, it has been reported that at high expression levels they can act as calcium buffers, making it difficult to precisely interpret the magnitude and length of signaling and requiring appropriate controls [183].

Downstream of second messengers, recent years have seen the development of improved FRET-based sensors for protein effectors including for the GPCR-relevant kinases PKA, PKC, ERK, and CaMKII [172]. Such sensors typically either monitor the activation-associated conformational changes of the kinase itself or employ a separate construct containing a specific peptide substrate to measure enzymatic activity of the kinase. For example, a large body of work has developed "Camui" sensors of CaMKII activation based on the introduction of fluorescent proteins to the N- and C-termini to report on the Ca^{2+}/Calmodulin-associated conformational changes of the CaMKII holoenzyme [184–186]. Camui has been used to assess the activation kinetics and spatial spread of CaMKII at single dendritic spines [187, 188]. However, Camui highlights one of the shortcomings of conformational biosensors in that the precise conformational changes being sensed are often unclear given the complex tertiary and quaternary structural rearrangements that are known to take place upon CaMKII activation [189, 190]. In addition, approaches such as this also require overexpression of an active enzyme, which alters the endogenous protein concentration and can lead to nonphysiological effects.

Using the enzymatic activity-based strategy, a series of FRET-based sensors, termed A kinase activity reporters ("AKAR"), have been developed to monitor PKA activity [191–196]. An example is AKAR3, which consists of a truncated CFP as the donor fluorophore, a phosphopeptide binding domain (FHA), a consensus region of PKA substrates, and a Venus as the acceptor fluorophore. Activated PKA phosphorylates the consensus region, which then can bind to FHA, bringing CFP and Venus in close proximity and thereby resulting in higher FRET (Fig. 4h). Combining this approach with 2-photon fluorescence lifetime measurements Chen et al. [192] observed increased PKA activity in hippocampal CA1 neurons following activation of β_2AR. In a more recent study, the same authors used AKAR to find that, unexpectedly, G_q-coupled receptors, including M1 muscarinic receptors, mGluR5 and a Gq-coupled DREADD, can lead to PKA activation in hippocampal neurons [197].

The peptide-phosphorylation based strategy has also been applied to develop FRET sensors for the activity of ERK ("EKAR") [195, 198], PKC ("CKAR") [199], Src [200, 201], AKT [202], and, most recently, CaMKII ("FRESCA") [203]. A recent redesign of kinase activity sensors using a cpGFP-based design to enable 1-color sensing of PKA, PKC, ERK or AKT activity with enhanced dynamics range across spectral variants [204] should facilitate multiplexing-based studies and further applications of these sensors in complex systems. Importantly, similarly to overexpression of active enzymes, overexpression of enzyme substrates can alter the cellular signaling landscape and distort the system thus requiring rigorous controls for the application of these sensors.

Given the frequent inability of peptide substrates to distinguish between related kinase subtypes and the difficulty of indirect activity sensors to provide high temporal or spatial resolution information on the enzyme itself, sensors based on the kinase of interest itself remain an important part of the tool kit. For example, Colgan et al. developed sensors to distinguish between the activities of three PKC subtypes (PKCα, PKCβ, PKCγ) in dendritic spines [205]. Two separate FRET-based sensors were designed for each isozyme, one to monitor enzyme translocation to the surface (ITRACK) and the other to monitor enzyme activity based on binding to a substrate (IDOCKS). Remarkably, these tools permitted the tracking of the activities of specific PKC isozymes during plasticity of dendritic spines of CA1 pyramidal neurons in organotypic hippocampal slices where the authors found that PKCα is the only subtype required for structural plasticity.

Crucially, the aforementioned genetically encoded biosensors can be targeted to measure signaling at specific organelles or discrete subregions of the cell, a perspective of increasing importance for understanding GPCR function [206]. For instance, GCaMP variants have been tagged with an endoplasmic reticulum (ER) signal peptide to measure the behavior of axonal ER calcium during evoked synaptic activity [207] and have been targeted to the nucleus to reveal GPCR-induced calcium responses [208]. Similarly, targeting of ERK activity sensors to the cytosol or the nucleus revealed bias between different mu-opioid receptor agonists in their potential to produce transient versus sustained ERK activation in the cytosol and nucleus [209]. Ultimately, studies of this nature will provide a crucial means of dissecting the subcellular dynamics of GPCR signaling.

The tool kit described in this review should continue to expand and enable the mechanistic study of GPCR signaling in a variety of contexts ranging from single cells to behaving animals. The continued multiplexing of optical control and optical sensing should facilitate these applications but will require continual engineering to optimize the underlying tools and uncover new design principles. As other techniques that are suited to probe GPCR function with high precision continue to advance, such as cryo-electron microscopy [210], gene editing [211, 212], tissue and in vivo imaging [213, 214], proteomics [215–217], and super-resolution imaging [218], the number of creative applications which require the high precision of optogenetic tools should only increase and allow for new aspects of GPCR biology to be revealed.

Acknowledgments

The authors thank David Simon and Margaret Stratton for helpful feedback. N.A. is supported by an NSF Graduate Research Fellowship and J.L. is supported by an R35 grant (1 R35 GM124731) from NIGMS and the Rohr Family Research Scholar Award.

References

1. Pierce KL, Premont RT, Lefkowitz RJ (2002) Seven-transmembrane receptors. Nat Rev Mol Cell Biol 3(9):639–650. https://doi.org/10.1038/nrm908
2. Hauser AS, Attwood MM, Rask-Andersen M et al (2017) Trends in GPCR drug discovery: new agents, targets and indications. Nat Rev Drug Discov 16(12):829–842. https://doi.org/10.1038/nrd.2017.178
3. Venkatakrishnan AJ, Deupi X, Lebon G et al (2013) Molecular signatures of G-protein-coupled receptors. Nature 494 (7436):185–194. https://doi.org/10.1038/nature11896

4. Erlandson SC, McMahon C, Kruse AC (2018) Structural basis for G protein signaling. Annu Rev Biophys 47(1)1–18. https://doi.org/10.1146/annurev-biophys-070317-032931
5. Rasmussen SG, DeVree BT, Zou Y et al (2011) Crystal structure of the beta2 adrenergic receptor-Gs protein complex. Nature 477(7366):549–555. https://doi.org/10.1038/nature10361
6. Flock T, Hauser AS, Lund N et al (2017) Selectivity determinants of GPCR-G-protein binding. Nature 545(7654):317–322. https://doi.org/10.1038/nature22070
7. Liu X, Xu X, Hilger D et al (2019) Structural insights into the process of GPCR-G protein complex formation. Cell 177(5):1243–1251. e1212. https://doi.org/10.1016/j.cell.2019.04.021
8. Ernst OP, Gramse V, Kolbe M et al (2007) Monomeric G protein-coupled receptor rhodopsin in solution activates its G protein transducin at the diffusion limit. Proc Natl Acad Sci U S A 104(26):10859–10864. https://doi.org/10.1073/pnas.0701967104
9. Grundmann M, Kostenis E (2017) Temporal bias: time-encoded dynamic GPCR signaling. Trends Pharmacol Sci 38(12):1110–1124. https://doi.org/10.1016/j.tips.2017.09.004
10. Khan SM, Sleno R, Gora S et al (2013) The expanding roles of Gbetagamma subunits in G protein-coupled receptor signaling and drug action. Pharmacol Rev 65(2):545–577. https://doi.org/10.1124/pr.111.005603
11. Touhara KK, MacKinnon R (2018) Molecular basis of signaling specificity between GIRK channels and GPCRs. elife 7. https://doi.org/10.7554/eLife.42908
12. DeWire SM, Ahn S, Lefkowitz RJ et al (2007) Beta-arrestins and cell signaling. Annu Rev Physiol 69:483–510. https://doi.org/10.1146/annurev.physiol.69.022405.154749
13. Rajagopal S, Shenoy SK (2018) GPCR desensitization: acute and prolonged phases. Cell Signal 41:9–16. https://doi.org/10.1016/j.cellsig.2017.01.024
14. Weis WI, Kobilka BK (2018) The molecular basis of G protein-coupled receptor activation. Annu Rev Biochem 87:897–919. https://doi.org/10.1146/annurev-biochem-060614-033910
15. Luttrell LM, Ferguson SS, Daaka Y et al (1999) Beta-arrestin-dependent formation of beta2 adrenergic receptor-Src protein kinase complexes. Science 283(5402):655–661. https://doi.org/10.1126/science.283.5402.655
16. Luttrell LM, Roudabush FL, Choy EW et al (2001) Activation and targeting of extracellular signal-regulated kinases by beta-arrestin scaffolds. Proc Natl Acad Sci U S A 98 (5):2449–2454. https://doi.org/10.1073/pnas.041604898
17. Strungs EG, Luttrell LM (2014) Arrestin-dependent activation of ERK and Src family kinases. Handb Exp Pharmacol 219:225–257. https://doi.org/10.1007/978-3-642-41199-1_12
18. O'Hayre M, Eichel K, Avino S et al (2017) Genetic evidence that beta-arrestins are dispensable for the initiation of beta2-adrenergic receptor signaling to ERK. Sci Signal 10 (484). https://doi.org/10.1126/scisignal.aal3395
19. Luttrell LM, Wang J, Plouffe B et al (2018) Manifold roles of beta-arrestins in GPCR signaling elucidated with siRNA and CRISPR/Cas9. Sci Signal 11(549). https://doi.org/10.1126/scisignal.aat7650
20. Grundmann M, Merten N, Malfacini D et al (2018) Lack of beta-arrestin signaling in the absence of active G proteins. Nat Commun 9 (1):341. https://doi.org/10.1038/s41467-017-02661-3
21. Smith JS, Lefkowitz RJ, Rajagopal S (2018) Biased signalling: from simple switches to allosteric microprocessors. Nat Rev Drug Discov 17(4):243–260. https://doi.org/10.1038/nrd.2017.229
22. Eichel K, von Zastrow M (2018) Subcellular organization of GPCR signaling. Trends Pharmacol Sci 39(2):200–208. https://doi.org/10.1016/j.tips.2017.11.009
23. Wacker D, Stevens RC, Roth BL (2017) How ligands illuminate GPCR molecular pharmacology. Cell 170(3):414–427. https://doi.org/10.1016/j.cell.2017.07.009
24. Callaway EM, Katz LC (1993) Photostimulation using caged glutamate reveals functional circuitry in living brain slices. Proc Natl Acad Sci U S A 90(16):7661–7665. https://doi.org/10.1073/pnas.90.16.7661
25. Reiner A, Levitz J, Isacoff EY (2015) Controlling ionotropic and metabotropic glutamate receptors with light: principles and potential. Curr Opin Pharmacol 20:135–143. https://doi.org/10.1016/j.coph.2014.12.008
26. Banghart MR, Sabatini BL (2012) Photoactivatable neuropeptides for spatiotemporally precise delivery of opioids in neural tissue.

Neuron 73(2):249–259. https://doi.org/10.1016/j.neuron.2011.11.016

27. Banghart MR, Williams JT, Shah RC et al (2013) Caged naloxone reveals opioid signaling deactivation kinetics. Mol Pharmacol 84 (5):687–695. https://doi.org/10.1124/mol.113.088096
28. Banghart MR, He XJ, Sabatini BL (2018) A caged enkephalin optimized for simultaneously probing mu and delta opioid receptors. ACS Chem Neurosci 9(4):684–690. https://doi.org/10.1021/acschemneuro.7b00485
29. Williams JT (2014) Desensitization of functional micro-opioid receptors increases agonist off-rate. Mol Pharmacol 86(1):52–61. https://doi.org/10.1124/mol.114.092098
30. Beharry AA, Woolley GA (2011) Azobenzene photoswitches for biomolecules. Chem Soc Rev 40(8):4422–4437. https://doi.org/10.1039/c1cs15023e
31. Schonberger M, Trauner D (2014) A photochromic agonist for mu-opioid receptors. Angew Chem Int Ed Engl 53 (12):3264–3267. https://doi.org/10.1002/anie.201309633
32. Agnetta L, Kauk M, Canizal MCA et al (2017) A photoswitchable dualsteric ligand controlling receptor efficacy. Angew Chem Int Ed Engl 56(25):7282–7287. https://doi.org/10.1002/anie.201701524
33. Westphal MV, Schafroth MA, Sarott RC et al (2017) Synthesis of photoswitchable delta(9)-tetrahydrocannabinol derivatives enables optical control of cannabinoid receptor 1 signaling. J Am Chem Soc 139(50):18206–18212. https://doi.org/10.1021/jacs.7b06456
34. Broichhagen J, Podewin T, Meyer-Berg H et al (2015) Optical control of insulin secretion using an incretin switch. Angew Chem Int Ed Engl 54(51):15565–15569. https://doi.org/10.1002/anie.201506384
35. Pittolo S, Gomez-Santacana X, Eckelt K et al (2014) An allosteric modulator to control endogenous G protein-coupled receptors with light. Nat Chem Biol 10(10):813–815. https://doi.org/10.1038/nchembio.1612
36. Zussy C, Gomez-Santacana X, Rovira X et al (2018) Dynamic modulation of inflammatory pain-related affective and sensory symptoms by optical control of amygdala metabotropic glutamate receptor 4. Mol Psychiatry 23 (3):509–520. https://doi.org/10.1038/mp.2016.223
37. Urban DJ, Roth BL (2015) DREADDs (designer receptors exclusively activated by designer drugs): chemogenetic tools with therapeutic utility. Annu Rev Pharmacol Toxicol 55:399–417. https://doi.org/10.1146/annurev-pharmtox-010814-124803
38. Vardy E, Robinson JE, Li C et al (2015) A new DREADD facilitates the multiplexed chemogenetic interrogation of behavior. Neuron 86(4):936–946. https://doi.org/10.1016/j.neuron.2015.03.065
39. Nakajima K, Wess J (2012) Design and functional characterization of a novel, arrestin-biased designer G protein-coupled receptor. Mol Pharmacol 82(4):575–582. https://doi.org/10.1124/mol.112.080358
40. Hu J, Stern M, Gimenez LE et al (2016) A G protein-biased designer G protein-coupled receptor useful for studying the physiological relevance of Gq/11-dependent signaling pathways. J Biol Chem 291(15):7809–7820. https://doi.org/10.1074/jbc.M115.702282
41. Stachniak TJ, Ghosh A, Sternson SM (2014) Chemogenetic synaptic silencing of neural circuits localizes a hypothalamus-->midbrain pathway for feeding behavior. Neuron 82 (4):797–808. https://doi.org/10.1016/j.neuron.2014.04.008
42. Palczewski K (2006) G protein-coupled receptor rhodopsin. Annu Rev Biochem 75:743–767. https://doi.org/10.1146/annurev.biochem.75.103004.142743
43. Zemelman BV, Lee GA, Ng M et al (2002) Selective photostimulation of genetically chARGed neurons. Neuron 33(1):15–22
44. Fenno L, Yizhar O, Deisseroth K (2011) The development and application of optogenetics. Annu Rev Neurosci 34:389–412. https://doi.org/10.1146/annurev-neuro-061010-113817
45. Li X, Gutierrez DV, Hanson MG et al (2005) Fast noninvasive activation and inhibition of neural and network activity by vertebrate rhodopsin and green algae channelrhodopsin. Proc Natl Acad Sci U S A 102 (49):17816–17821. https://doi.org/10.1073/pnas.0509030102
46. Masseck OA, Spoida K, Dalkara D et al (2014) Vertebrate cone opsins enable sustained and highly sensitive rapid control of Gi/o signaling in anxiety circuitry. Neuron 81(6):1263–1273. https://doi.org/10.1016/j.neuron.2014.01.041
47. Bailes HJ, Zhuang LY, Lucas RJ (2012) Reproducible and sustained regulation of Galphas signalling using a metazoan opsin as an optogenetic tool. PLoS One 7(1):e30774. https://doi.org/10.1371/journal.pone.0030774

48. Qiu X, Kumbalasiri T, Carlson SM et al (2005) Induction of photosensitivity by heterologous expression of melanopsin. Nature 433(7027):745–749. https://doi.org/10.1038/nature03345
49. Spoida K, Eickelbeck D, Karapinar R et al (2016) Melanopsin variants as intrinsic optogenetic on and off switches for transient versus sustained activation of G protein pathways. Curr Biol 26(9):1206–1212. https://doi.org/10.1016/j.cub.2016.03.007
50. Karunarathne WK, Giri L, Kalyanaraman V et al (2013) Optically triggering spatiotemporally confined GPCR activity in a cell and programming neurite initiation and extension. Proc Natl Acad Sci U S A 110(17): E1565–E1574. https://doi.org/10.1073/pnas.1220697110
51. Makowka P, Bruegmann T, Dusend V et al (2019) Optogenetic stimulation of Gs-signaling in the heart with high spatiotemporal precision. Nat Commun 10 (1):1281. https://doi.org/10.1038/s41467-019-09322-7
52. Kim JM, Hwa J, Garriga P et al (2005) Light-driven activation of beta 2-adrenergic receptor signaling by a chimeric rhodopsin containing the beta 2-adrenergic receptor cytoplasmic loops. Biochemistry 44 (7):2284–2292. https://doi.org/10.1021/bi048328i
53. Tichy AM, Gerrard EJ, Sexton PM et al (2019) Light-activated chimeric GPCRs: limitations and opportunities. Curr Opin Struct Biol 57:196–203. https://doi.org/10.1016/j.sbi.2019.05.006
54. Airan RD, Thompson KR, Fenno LE et al (2009) Temporally precise in vivo control of intracellular signalling. Nature 458 (7241):1025–1029. https://doi.org/10.1038/nature07926
55. Gunaydin LA, Grosenick L, Finkelstein JC et al (2014) Natural neural projection dynamics underlying social behavior. Cell 157 (7):1535–1551. https://doi.org/10.1016/j.cell.2014.05.017
56. Siuda ER, Copits BA, Schmidt MJ et al (2015) Spatiotemporal control of opioid signaling and behavior. Neuron 86(4):923–935. https://doi.org/10.1016/j.neuron.2015.03.066
57. Levitz J, Pantoja C, Gaub B et al (2013) Optical control of metabotropic glutamate receptors. Nat Neurosci 16(4):507–516. https://doi.org/10.1038/nn.3346
58. Carroll EC, Berlin S, Levitz J et al (2015) Two-photon brightness of azobenzene photoswitches designed for glutamate receptor optogenetics. Proc Natl Acad Sci U S A 112(7):E776–E785. https://doi.org/10.1073/pnas.1416942112
59. Levitz J, Habrian C, Bharill S et al (2016) Mechanism of assembly and cooperativity of homomeric and heteromeric metabotropic glutamate receptors. Neuron 92 (1):143–159. https://doi.org/10.1016/j.neuron.2016.08.036
60. Donthamsetti PC, Winter N, Schonberger M et al (2017) Optical control of dopamine receptors using a photoswitchable tethered inverse agonist. J Am Chem Soc 139 (51):18522–18535. https://doi.org/10.1021/jacs.7b07659
61. Broichhagen J, Damijonaitis A, Levitz J et al (2015) Orthogonal optical control of a G protein-coupled receptor with a SNAP-tethered photochromic ligand. ACS Cent Sci 1(7):383–393. https://doi.org/10.1021/acscentsci.5b00260
62. Levitz J, Broichhagen J, Leippe P et al (2017) Dual optical control and mechanistic insights into photoswitchable group II and III metabotropic glutamate receptors. Proc Natl Acad Sci U S A 114(17):E3546–E3554. https://doi.org/10.1073/pnas.1619652114
63. Berry MH, Holt A, Levitz J et al (2017) Restoration of patterned vision with an engineered photoactivatable G protein-coupled receptor. Nat Commun 8(1):1862. https://doi.org/10.1038/s41467-017-01990-7
64. Acosta-Ruiz A, Gutzeit VA, Skelly MJ, Meadows S, Lee J, Parekh P, Orr AG, Liston C, Pleil KE, Broichhagen J, Levitz J (2020) Branched photoswitchable tethered ligands enable ultra-efficient optical control and detection of G protein-coupled receptors in vivo. Neuron 105(3):446–463
65. Shields BC, Kahuno E, Kim C et al (2017) Deconstructing behavioral neuropharmacology with cellular specificity. Science 356 (6333). https://doi.org/10.1126/science.aaj2161
66. Donthamsetti PC, Broichhagen J, Vyklicky V et al (2019) Genetically targeted optical control of an endogenous G protein-coupled receptor. J Am Chem Soc 141 (29):11522–11530. https://doi.org/10.1021/jacs.9b02895
67. Farrants H, Gutzeit VA, Acosta-Ruiz A et al (2018) SNAP-tagged nanobodies enable reversible optical control of a G protein-coupled receptor via a remotely tethered photoswitchable ligand. ACS Chem Biol 13 (9):2682–2688. https://doi.org/10.1021/acschembio.8b00628
68. Repina NA, Rosenbloom A, Mukherjee A et al (2017) At light speed: advances in

optogenetic systems for regulating cell signaling and behavior. Annu Rev Chem Biomol Eng 8:13–39. https://doi.org/10.1146/annurev-chembioeng-060816-101254
69. Goglia AG, Toettcher JE (2019) A bright future: optogenetics to dissect the spatiotemporal control of cell behavior. Curr Opin Chem Biol 48:106–113. https://doi.org/10.1016/j.cbpa.2018.11.010
70. Takenouchi O, Yoshimura H, Ozawa T (2018) Unique roles of beta-arrestin in GPCR trafficking revealed by photoinducible dimerizers. Sci Rep 8(1):677. https://doi.org/10.1038/s41598-017-19130-y
71. Liu H, Yu X, Li K et al (2008) Photoexcited CRY2 interacts with CIB1 to regulate transcription and floral initiation in Arabidopsis. Science 322(5907):1535–1539. https://doi.org/10.1126/science.1163927
72. Kennedy MJ, Hughes RM, Peteya LA et al (2010) Rapid blue-light-mediated induction of protein interactions in living cells. Nat Methods 7(12):973–975. https://doi.org/10.1038/nmeth.1524
73. O'Neill PR, Gautam N (2014) Subcellular optogenetic inhibition of G proteins generates signaling gradients and cell migration. Mol Biol Cell 25(15):2305–2314. https://doi.org/10.1091/mbc.E14-04-0870
74. Hannanta-Anan P, Chow BY (2018) Optogenetic inhibition of Galphaq protein signaling reduces calcium oscillation stochasticity. ACS Synth Biol 7(6):1488–1495. https://doi.org/10.1021/acssynbio.8b00065
75. De Vries L, Zheng B, Fischer T et al (2000) The regulator of G protein signaling family. Annu Rev Pharmacol Toxicol 40:235–271. https://doi.org/10.1146/annurev.pharmtox.40.1.235
76. Yu G, Onodera H, Aono Y et al (2016) Optical manipulation of the alpha subunits of heterotrimeric G proteins using photoswitchable dimerization systems. Sci Rep 6:35777. https://doi.org/10.1038/srep35777
77. Patel N, Gold MG (2015) The genetically encoded tool set for investigating cAMP: more than the sum of its parts. Front Pharmacol 6:164. https://doi.org/10.3389/fphar.2015.00164
78. Iseki M, Matsunaga S, Murakami A et al (2002) A blue-light-activated adenylyl cyclase mediates photoavoidance in Euglena gracilis. Nature 415(6875):1047–1051. https://doi.org/10.1038/4151047a
79. Stierl M, Stumpf P, Udwari D et al (2011) Light modulation of cellular cAMP by a small bacterial photoactivated adenylyl cyclase, bPAC, of the soil bacterium Beggiatoa. J Biol Chem 286(2):1181–1188. https://doi.org/10.1074/jbc.M110.185496
80. Ohki M, Sugiyama K, Kawai F et al (2016) Structural insight into photoactivation of an adenylate cyclase from a photosynthetic cyanobacterium. Proc Natl Acad Sci U S A 113 (24):6659–6664. https://doi.org/10.1073/pnas.1517520113
81. Bellmann D, Richardt A, Freyberger R et al (2010) Optogenetically induced olfactory stimulation in Drosophila larvae reveals the neuronal basis of odor-aversion behavior. Front Behav Neurosci 4:27. https://doi.org/10.3389/fnbeh.2010.00027
82. Nagahama T, Suzuki T, Yoshikawa S et al (2007) Functional transplant of photoactivated adenylyl cyclase (PAC) into Aplysia sensory neurons. Neurosci Res 59(1):81–88. https://doi.org/10.1016/j.neures.2007.05.015
83. Schroder-Lang S, Schwarzel M, Seifert R et al (2007) Fast manipulation of cellular cAMP level by light in vivo. Nat Methods 4 (1):39–42. https://doi.org/10.1038/nmeth975
84. Weissenberger S, Schultheis C, Liewald JF et al (2011) PACalpha—an optogenetic tool for in vivo manipulation of cellular cAMP levels, neurotransmitter release, and behavior in Caenorhabditis elegans. J Neurochem 116 (4):616–625. https://doi.org/10.1111/j.1471-4159.2010.07148.x
85. Zhou Z, Tanaka KF, Matsunaga S et al (2016) Photoactivated adenylyl cyclase (PAC) reveals novel mechanisms underlying cAMP-dependent axonal morphogenesis. Sci Rep 5:19679. https://doi.org/10.1038/srep19679
86. Tsvetanova NG, von Zastrow M (2014) Spatial encoding of cyclic AMP signaling specificity by GPCR endocytosis. Nat Chem Biol 10 (12):1061–1065. https://doi.org/10.1038/nchembio.1665
87. Leopold AV, Chernov KG, Verkhusha VV (2018) Optogenetically controlled protein kinases for regulation of cellular signaling. Chem Soc Rev 47(7):2454–2484. https://doi.org/10.1039/c7cs00404d
88. Wetzker R, Bohmer FD (2003) Transactivation joins multiple tracks to the ERK/MAPK cascade. Nat Rev Mol Cell Biol 4 (8):651–657. https://doi.org/10.1038/nrm1173
89. Kolch W (2005) Coordinating ERK/MAPK signalling through scaffolds and inhibitors. Nat Rev Mol Cell Biol 6(11):827–837. https://doi.org/10.1038/nrm1743

90. Toettcher JE, Weiner OD, Lim WA (2013) Using optogenetics to interrogate the dynamic control of signal transmission by the Ras/Erk module. Cell 155(6):1422–1434. https://doi.org/10.1016/j.cell.2013.11.004
91. Levskaya A, Weiner OD, Lim WA et al (2009) Spatiotemporal control of cell signalling using a light-switchable protein interaction. Nature 461(7266):997–1001. https://doi.org/10.1038/nature08446
92. Zhang K, Duan L, Ong Q et al (2014) Light-mediated kinetic control reveals the temporal effect of the Raf/MEK/ERK pathway in PC12 cell neurite outgrowth. PLoS One 9 (3):e92917. https://doi.org/10.1371/journal.pone.0092917
93. Zhou XX, Fan LZ, Li P et al (2017) Optical control of cell signaling by single-chain photoswitchable kinases. Science 355 (6327):836–842. https://doi.org/10.1126/science.aah3605
94. Dagliyan O, Tarnawski M, Chu PH et al (2016) Engineering extrinsic disorder to control protein activity in living cells. Science 354 (6318):1441–1444. https://doi.org/10.1126/science.aah3404
95. Manning BD, Toker A (2017) AKT/PKB signaling: navigating the network. Cell 169 (3):381–405. https://doi.org/10.1016/j.cell.2017.04.001
96. Toettcher JE, Gong D, Lim WA et al (2011) Light-based feedback for controlling intracellular signaling dynamics. Nat Methods 8 (10):837–839. https://doi.org/10.1038/nmeth.1700
97. Idevall-Hagren O, Dickson EJ, Hille B et al (2012) Optogenetic control of phosphoinositide metabolism. Proc Natl Acad Sci U S A 109(35):E2316–E2323. https://doi.org/10.1073/pnas.1211305109
98. Katsura Y, Kubota H, Kunida K et al (2015) An optogenetic system for interrogating the temporal dynamics of Akt. Sci Rep 5:14589. https://doi.org/10.1038/srep14589
99. Xu Y, Nan D, Fan J et al (2016) Optogenetic activation reveals distinct roles of PIP3 and Akt in adipocyte insulin action. J Cell Sci 129(10):2085–2095. https://doi.org/10.1242/jcs.174805
100. Ong Q, Guo S, Duan L et al (2016) The timing of Raf/ERK and AKT activation in protecting PC12 cells against oxidative stress. PLoS One 11(4):e0153487. https://doi.org/10.1371/journal.pone.0153487
101. Wang W, Qiao Y, Li Z (2018) New insights into modes of GPCR activation. Trends Pharmacol Sci 39(4):367–386. https://doi.org/10.1016/j.tips.2018.01.001
102. Grusch M, Schelch K, Riedler R et al (2014) Spatio-temporally precise activation of engineered receptor tyrosine kinases by light. EMBO J 33(15):1713–1726. https://doi.org/10.15252/embj.201387695
103. Chang KY, Woo D, Jung H et al (2014) Light-inducible receptor tyrosine kinases that regulate neurotrophin signalling. Nat Commun 5:4057. https://doi.org/10.1038/ncomms5057
104. Reichhart E, Ingles-Prieto A, Tichy AM et al (2016) A phytochrome sensory domain permits receptor activation by red light. Angew Chem Int Ed Engl 55(21):6339–6342. https://doi.org/10.1002/anie.201601736
105. Kainrath S, Stadler M, Reichhart E et al (2017) Green-light-induced inactivation of receptor signaling using cobalamin-binding domains. Angew Chem Int Ed Engl 56 (16):4608–4611. https://doi.org/10.1002/anie.201611998
106. Harper SM, Neil LC, Gardner KH (2003) Structural basis of a phototropin light switch. Science 301(5639):1541–1544. https://doi.org/10.1126/science.1086810
107. Dagliyan O, Hahn KM (2019) Controlling protein conformation with light. Curr Opin Struct Biol 57:17–22. https://doi.org/10.1016/j.sbi.2019.01.012
108. Murakoshi H, Shin ME, Parra-Bueno P et al (2017) Kinetics of endogenous CaMKII required for synaptic plasticity revealed by optogenetic kinase inhibitor. Neuron 94 (1):37–47.e35. https://doi.org/10.1016/j.neuron.2017.02.036
109. Liu XY, Mao LM, Zhang GC et al (2009) Activity-dependent modulation of limbic dopamine D3 receptors by CaMKII. Neuron 61(3):425–438. https://doi.org/10.1016/j.neuron.2008.12.015
110. Guetg N, Abdel Aziz S, Holbro N et al (2010) NMDA receptor-dependent GABAB receptor internalization via CaMKII phosphorylation of serine 867 in GABAB1. Proc Natl Acad Sci U S A 107(31):13924–13929. https://doi.org/10.1073/pnas.1000909107
111. Chen YJ, Oldfield S, Butcher AJ et al (2013) Identification of phosphorylation sites in the COOH-terminal tail of the mu-opioid receptor. J Neurochem 124(2):189–199. https://doi.org/10.1111/jnc.12071
112. Jin DZ, Guo ML, Xue B et al (2013) Phosphorylation and feedback regulation of metabotropic glutamate receptor 1 by calcium/calmodulin-dependent protein kinase II. J

Neurosci 33(8):3402–3412. https://doi.org/10.1523/JNEUROSCI.3192-12.2013

113. Mockett BG, Guevremont D, Wutte M et al (2011) Calcium/calmodulin-dependent protein kinase II mediates group I metabotropic glutamate receptor-dependent protein synthesis and long-term depression in rat hippocampus. J Neurosci 31(20):7380–7391. https://doi.org/10.1523/JNEUROSCI.6656-10.2011
114. Ishida A, Shigeri Y, Tatsu Y et al (1998) Critical amino acid residues of AIP, a highly specific inhibitory peptide of calmodulin-dependent protein kinase II. FEBS Lett 427 (1):115–118. https://doi.org/10.1016/s0014-5793(98)00405-0
115. Yi JJ, Wang H, Vilela M et al (2014) Manipulation of endogenous kinase activity in living cells using photoswitchable inhibitory peptides. ACS Synth Biol 3(11):788–795. https://doi.org/10.1021/sb5001356
116. Melero-Fernandez de Mera RM, Li LL, Popinigis A et al (2017) A simple optogenetic MAPK inhibitor design reveals resonance between transcription-regulating circuitry and temporally-encoded inputs. Nat Commun 8:15017. https://doi.org/10.1038/ncomms15017
117. Tsai YH, Essig S, James JR et al (2015) Selective, rapid and optically switchable regulation of protein function in live mammalian cells. Nat Chem 7(7):554–561. https://doi.org/10.1038/nchem.2253
118. Yazawa M, Sadaghiani AM, Hsueh B et al (2009) Induction of protein-protein interactions in live cells using light. Nat Biotechnol 27(10):941–945. https://doi.org/10.1038/nbt.1569
119. Wang X, Chen X, Yang Y (2012) Spatiotemporal control of gene expression by a light-switchable transgene system. Nat Methods 9 (3):266–269. https://doi.org/10.1038/nmeth.1892
120. Motta-Mena LB, Reade A, Mallory MJ et al (2014) An optogenetic gene expression system with rapid activation and deactivation kinetics. Nat Chem Biol 10(3):196–202. https://doi.org/10.1038/nchembio.1430
121. Bugaj LJ, Choksi AT, Mesuda CK et al (2013) Optogenetic protein clustering and signaling activation in mammalian cells. Nat Methods 10(3):249–252. https://doi.org/10.1038/nmeth.2360
122. Taslimi A, Vrana JD, Chen D et al (2014) An optimized optogenetic clustering tool for probing protein interaction and function. Nat Commun 5:4925. https://doi.org/10.1038/ncomms5925
123. Shin Y, Berry J, Pannucci N et al (2017) Spatiotemporal control of intracellular phase transitions using light-activated optoDroplets. Cell 168(1–2):159–171.e114. https://doi.org/10.1016/j.cell.2016.11.054
124. van Bergeijk P, Adrian M, Hoogenraad CC et al (2015) Optogenetic control of organelle transport and positioning. Nature 518 (7537):111–114. https://doi.org/10.1038/nature14128
125. Niopek D, Benzinger D, Roensch J et al (2014) Engineering light-inducible nuclear localization signals for precise spatiotemporal control of protein dynamics in living cells. Nat Commun 5:4404. https://doi.org/10.1038/ncomms5404
126. Renicke C, Schuster D, Usherenko S et al (2013) A LOV2 domain-based optogenetic tool to control protein degradation and cellular function. Chem Biol 20(4):619–626. https://doi.org/10.1016/j.chembiol.2013.03.005
127. Bonger KM, Rakhit R, Payumo AY et al (2014) General method for regulating protein stability with light. ACS Chem Biol 9 (1):111–115. https://doi.org/10.1021/cb400755b
128. Chen D, Gibson ES, Kennedy MJ (2013) A light-triggered protein secretion system. J Cell Biol 201(4):631–640. https://doi.org/10.1083/jcb.201210119
129. Glantz ST, Berlew EE, Jaber Z et al (2018) Directly light-regulated binding of RGS-LOV photoreceptors to anionic membrane phospholipids. Proc Natl Acad Sci U S A 115 (33):E7720–E7727. https://doi.org/10.1073/pnas.1802832115
130. Hull K, Morstein J, Trauner D (2018) In vivo photopharmacology. Chem Rev 118 (21):10710–10747. https://doi.org/10.1021/acs.chemrev.8b00037
131. Paoletti P, Ellis-Davies GCR, Mourot A (2019) Optical control of neuronal ion channels and receptors. Nat Rev Neurosci 20:514–532. https://doi.org/10.1038/s41583-019-0197-2
132. Banghart M, Borges K, Isacoff E et al (2004) Light-activated ion channels for remote control of neuronal firing. Nat Neurosci 7 (12):1381–1386. https://doi.org/10.1038/nn1356
133. Volgraf M, Gorostiza P, Numano R et al (2006) Allosteric control of an ionotropic glutamate receptor with an optical switch.

Nat Chem Biol 2(1):47–52. https://doi.org/10.1038/nchembio756

134. Sandoz G, Levitz J, Kramer RH et al (2012) Optical control of endogenous proteins with a photoswitchable conditional subunit reveals a role for TREK1 in GABA(B) signaling. Neuron 74(6):1005–1014. https://doi.org/10.1016/j.neuron.2012.04.026
135. Barber DM, Schonberger M, Burgstaller J et al (2016) Optical control of neuronal activity using a light-operated GIRK channel opener (LOGO). Chem Sci 7 (3):2347–2352. https://doi.org/10.1039/C5SC04084A
136. Lohse MJ, Nuber S, Hoffmann C (2012) Fluorescence/bioluminescence resonance energy transfer techniques to study G-protein-coupled receptor activation and signaling. Pharmacol Rev 64(2):299–336. https://doi.org/10.1124/pr.110.004309
137. Miyawaki A (2011) Development of probes for cellular functions using fluorescent proteins and fluorescence resonance energy transfer. Annu Rev Biochem 80:357–373. https://doi.org/10.1146/annurev-biochem-072909-094736
138. Vilardaga JP, Bunemann M, Krasel C et al (2003) Measurement of the millisecond activation switch of G protein-coupled receptors in living cells. Nat Biotechnol 21 (7):807–812. https://doi.org/10.1038/nbt838
139. Vilardaga JP, Steinmeyer R, Harms GS et al (2005) Molecular basis of inverse agonism in a G protein-coupled receptor. Nat Chem Biol 1(1):25–28. https://doi.org/10.1038/nchembio705
140. Xue L, Karpenko IA, Hiblot J et al (2015) Imaging and manipulating proteins in live cells through covalent labeling. Nat Chem Biol 11(12):917–923. https://doi.org/10.1038/nchembio.1959
141. Doumazane E, Scholler P, Zwier JM et al (2011) A new approach to analyze cell surface protein complexes reveals specific heterodimeric metabotropic glutamate receptors. FASEB J 25(1):66–77. https://doi.org/10.1096/fj.10-163147
142. Doumazane E, Scholler P, Fabre L et al (2013) Illuminating the activation mechanisms and allosteric properties of metabotropic glutamate receptors. Proc Natl Acad Sci U S A 110(15):E1416–E1425. https://doi.org/10.1073/pnas.1215615110
143. Vafabakhsh R, Levitz J, Isacoff EY (2015) Conformational dynamics of a class C G-protein-coupled receptor. Nature 524 (7566):497–501. https://doi.org/10.1038/nature14679
144. Lecat-Guillet N, Monnier C, Rovira X et al (2017) FRET-based sensors unravel activation and allosteric modulation of the GABAB receptor. Cell Chem Biol 24 (3):360–370. https://doi.org/10.1016/j.chembiol.2017.02.011
145. Gutzeit VA, Thibado J, Stor DS et al (2019) Conformational dynamics between transmembrane domains and allosteric modulation of a metabotropic glutamate receptor. elife 8. https://doi.org/10.7554/eLife.45116
146. Tian H, Furstenberg A, Huber T (2017) Labeling and single-molecule methods to monitor G protein-coupled receptor dynamics. Chem Rev 117(1):186–245. https://doi.org/10.1021/acs.chemrev.6b00084
147. Hein P, Frank M, Hoffmann C et al (2005) Dynamics of receptor/G protein coupling in living cells. EMBO J 24(23):4106–4114. https://doi.org/10.1038/sj.emboj.7600870
148. Hein P, Rochais F, Hoffmann C et al (2006) Gs activation is time-limiting in initiating receptor-mediated signaling. J Biol Chem 281(44):33345–33351. https://doi.org/10.1074/jbc.M606713200
149. Nobles M, Benians A, Tinker A (2005) Heterotrimeric G proteins precouple with G protein-coupled receptors in living cells. Proc Natl Acad Sci U S A 102 (51):18706–18711. https://doi.org/10.1073/pnas.0504778102
150. Gales C, Rebois RV, Hogue M et al (2005) Real-time monitoring of receptor and G-protein interactions in living cells. Nat Methods 2(3):177–184. https://doi.org/10.1038/nmeth743
151. Bunemann M, Frank M, Lohse MJ (2003) Gi protein activation in intact cells involves subunit rearrangement rather than dissociation. Proc Natl Acad Sci U S A 100 (26):16077–16082. https://doi.org/10.1073/pnas.2536719100
152. Nuber S, Zabel U, Lorenz K et al (2016) beta-Arrestin biosensors reveal a rapid, receptor-dependent activation/deactivation cycle. Nature 531(7596):661–664. https://doi.org/10.1038/nature17198
153. Lee MH, Appleton KM, Strungs EG et al (2016) The conformational signature of beta-arrestin2 predicts its trafficking and signalling functions. Nature 531 (7596):665–668. https://doi.org/10.1038/nature17154

154. Albizu L, Cottet M, Kralikova M et al (2010) Time-resolved FRET between GPCR ligands reveals oligomers in native tissues. Nat Chem Biol 6(8):587–594. https://doi.org/10.1038/nchembio.396
155. Thomsen ARB, Plouffe B, Cahill TJ 3rd et al (2016) GPCR-G protein-beta-arrestin super-complex mediates sustained G protein signaling. Cell 166(4):907–919. https://doi.org/10.1016/j.cell.2016.07.004
156. Sungkaworn T, Jobin ML, Burnecki K et al (2017) Single-molecule imaging reveals receptor-G protein interactions at cell surface hot spots. Nature 550(7677):543–547. https://doi.org/10.1038/nature24264
157. Manglik A, Kobilka BK, Steyaert J (2017) Nanobodies to study G protein-coupled receptor structure and function. Annu Rev Pharmacol Toxicol 57:19–37. https://doi.org/10.1146/annurev-pharmtox-010716-104710
158. Heukers R, De Groof TWM, Smit MJ (2019) Nanobodies detecting and modulating GPCRs outside in and inside out. Curr Opin Cell Biol 57:115–122. https://doi.org/10.1016/j.ceb.2019.01.003
159. Rasmussen SG, Choi HJ, Fung JJ et al (2011) Structure of a nanobody-stabilized active state of the beta(2) adrenoceptor. Nature 469 (7329):175–180. https://doi.org/10.1038/nature09648
160. Scholler P, Nevoltris D, de Bundel D et al (2017) Allosteric nanobodies uncover a role of hippocampal mGlu2 receptor homodimers in contextual fear consolidation. Nat Commun 8(1):1967. https://doi.org/10.1038/s41467-017-01489-1
161. Che T, Majumdar S, Zaidi SA et al (2018) Structure of the nanobody-stabilized active state of the kappa opioid receptor. Cell 172 (1–2):55–67.e15. https://doi.org/10.1016/j.cell.2017.12.011
162. Westfield GH, Rasmussen SG, Su M et al (2011) Structural flexibility of the G alpha s alpha-helical domain in the beta2-adrenoceptor Gs complex. Proc Natl Acad Sci U S A 108(38):16086–16091. https://doi.org/10.1073/pnas.1113645108
163. Gulati S, Jin H, Masuho I et al (2018) Targeting G protein-coupled receptor signaling at the G protein level with a selective nanobody inhibitor. Nat Commun 9(1):1996. https://doi.org/10.1038/s41467-018-04432-0
164. Cahill TJ 3rd, Thomsen AR, Tarrasch JT et al (2017) Distinct conformations of GPCR-beta-arrestin complexes mediate desensitization, signaling, and endocytosis. Proc Natl Acad Sci U S A 114(10):2562–2567. https://doi.org/10.1073/pnas.1701529114
165. Irannejad R, Tomshine JC, Tomshine JR et al (2013) Conformational biosensors reveal GPCR signalling from endosomes. Nature 495(7442):534–538. https://doi.org/10.1038/nature12000
166. McMahon C, Baier AS, Pascolutti R et al (2018) Yeast surface display platform for rapid discovery of conformationally selective nanobodies. Nat Struct Mol Biol 25 (3):289–296. https://doi.org/10.1038/s41594-018-0028-6
167. Barnea G, Strapps W, Herrada G et al (2008) The genetic design of signaling cascades to record receptor activation. Proc Natl Acad Sci U S A 105(1):64–69. https://doi.org/10.1073/pnas.0710487105
168. Kroeze WK, Sassano MF, Huang XP et al (2015) PRESTO-Tango as an open-source resource for interrogation of the druggable human GPCRome. Nat Struct Mol Biol 22 (5):362–369. https://doi.org/10.1038/nsmb.3014
169. Lee D, Creed M, Jung K et al (2017) Temporally precise labeling and control of neuromodulatory circuits in the mammalian brain. Nat Methods 14(5):495–503. https://doi.org/10.1038/nmeth.4234
170. Miyawaki A, Llopis J, Heim R et al (1997) Fluorescent indicators for Ca2+ based on green fluorescent proteins and calmodulin. Nature 388(6645):882–887. https://doi.org/10.1038/42264
171. Nakai J, Ohkura M, Imoto K (2001) A high signal-to-noise Ca(2+) probe composed of a single green fluorescent protein. Nat Biotechnol 19(2):137–141. https://doi.org/10.1038/84397
172. Greenwald EC, Mehta S, Zhang J (2018) Genetically encoded fluorescent biosensors illuminate the spatiotemporal regulation of signaling networks. Chem Rev 118 (24):11707–11794. https://doi.org/10.1021/acs.chemrev.8b00333
173. Paramonov VM, Mamaeva V, Sahlgren C et al (2015) Genetically-encoded tools for cAMP probing and modulation in living systems. Front Pharmacol 6:196. https://doi.org/10.3389/fphar.2015.00196
174. Hackley CR, Mazzoni EO, Blau J (2018) cAMPr: a single-wavelength fluorescent sensor for cyclic AMP. Sci Signal 11(520). https://doi.org/10.1126/scisignal.aah3738
175. Platisa J, Pieribone VA (2018) Genetically encoded fluorescent voltage indicators: are

we there yet? Curr Opin Neurobiol 50:146–153. https://doi.org/10.1016/j.conb.2018.02.006

176. Patriarchi T, Cho JR, Merten K et al (2018) Ultrafast neuronal imaging of dopamine dynamics with designed genetically encoded sensors. Science 360(6396). https://doi.org/10.1126/science.aat4422

177. Mohebi A, Pettibone JR, Hamid AA et al (2019) Dissociable dopamine dynamics for learning and motivation. Nature 570 (7759):65–70. https://doi.org/10.1038/s41586-019-1235-y

178. Sun F, Zeng J, Jing M et al (2018) A genetically encoded fluorescent sensor enables rapid and specific detection of dopamine in flies, fish, and mice. Cell 174(2):481–496.e419. https://doi.org/10.1016/j.cell.2018.06.042

179. Feng J, Zhang C, Lischinsky JE et al (2019) A genetically encoded fluorescent sensor for rapid and specific in vivo detection of norepinephrine. Neuron 102(4):745–761.e748. https://doi.org/10.1016/j.neuron.2019.02.037

180. Marvin JS, Borghuis BG, Tian L et al (2013) An optimized fluorescent probe for visualizing glutamate neurotransmission. Nat Methods 10(2):162–170. https://doi.org/10.1038/nmeth.2333

181. Marvin JS, Scholl B, Wilson DE et al (2018) Stability, affinity, and chromatic variants of the glutamate sensor iGluSnFR. Nat Methods 15 (11):936–939. https://doi.org/10.1038/s41592-018-0171-3

182. Marvin JS, Shimoda Y, Magloire V et al (2019) A genetically encoded fluorescent sensor for in vivo imaging of GABA. Nat Methods 16(8):763–770. https://doi.org/10.1038/s41592-019-0471-2

183. McMahon SM, Jackson MB (2018) An inconvenient truth: calcium sensors are calcium buffers. Trends Neurosci 41 (12):880–884. https://doi.org/10.1016/j.tins.2018.09.005

184. Takao K, Okamoto K, Nakagawa T et al (2005) Visualization of synaptic Ca2+ /calmodulin-dependent protein kinase II activity in living neurons. J Neurosci 25 (12):3107–3112. https://doi.org/10.1523/JNEUROSCI.0085-05.2005

185. Kwok S, Lee C, Sanchez SA et al (2008) Genetically encoded probe for fluorescence lifetime imaging of CaMKII activity. Biochem Biophys Res Commun 369(2):519–525. https://doi.org/10.1016/j.bbrc.2008.02.070

186. Lam AJ, St-Pierre F, Gong Y et al (2012) Improving FRET dynamic range with bright green and red fluorescent proteins. Nat Methods 9(10):1005–1012. https://doi.org/10.1038/nmeth.2171

187. Lee SJ, Escobedo-Lozoya Y, Szatmari EM et al (2009) Activation of CaMKII in single dendritic spines during long-term potentiation. Nature 458(7236):299–304. https://doi.org/10.1038/nature07842

188. Chang JY, Parra-Bueno P, Laviv T et al (2017) CaMKII autophosphorylation is necessary for optimal integration of Ca(2+) signals during LTP induction, but not maintenance. Neuron 94(4):800–808.e804. https://doi.org/10.1016/j.neuron.2017.04.041

189. Chao LH, Stratton MM, Lee IH et al (2011) A mechanism for tunable autoinhibition in the structure of a human Ca2+/calmodulin-dependent kinase II holoenzyme. Cell 146 (5):732–745. https://doi.org/10.1016/j.cell.2011.07.038

190. Stratton M, Lee IH, Bhattacharyya M et al (2014) Activation-triggered subunit exchange between CaMKII holoenzymes facilitates the spread of kinase activity. elife 3: e01610. https://doi.org/10.7554/eLife.01610

191. Allen MD, Zhang J (2006) Subcellular dynamics of protein kinase A activity visualized by FRET-based reporters. Biochem Biophys Res Commun 348(2):716–721. https://doi.org/10.1016/j.bbrc.2006.07.136

192. Chen Y, Saulnier JL, Yellen G et al (2014) A PKA activity sensor for quantitative analysis of endogenous GPCR signaling via 2-photon FRET-FLIM imaging. Front Pharmacol 5:56. https://doi.org/10.3389/fphar.2014.00056

193. Barbagallo F, Xu B, Reddy GR et al (2016) Genetically encoded biosensors reveal PKA hyperphosphorylation on the myofilaments in rabbit heart failure. Circ Res 119 (8):931–943. https://doi.org/10.1161/CIRCRESAHA.116.308964

194. Demeautis C, Sipieter F, Roul J et al (2017) Multiplexing PKA and ERK1&2 kinases FRET biosensors in living cells using single excitation wavelength dual colour FLIM. Sci Rep 7:41026. https://doi.org/10.1038/srep41026

195. Tang S, Yasuda R (2017) Imaging ERK and PKA activation in single dendritic spines during structural plasticity. Neuron 93 (6):1315–1324.e1313. https://doi.org/10.1016/j.neuron.2017.02.032
196. Mo GC, Ross B, Hertel F et al (2017) Genetically encoded biosensors for visualizing live-cell biochemical activity at super-resolution. Nat Methods 14(4):427–434. https://doi.org/10.1038/nmeth.4221
197. Chen Y, Granger AJ, Tran T et al (2017) Endogenous Galphaq-coupled neuromodulator receptors activate protein kinase A. Neuron 96(5):1070–1083.e1075. https://doi.org/10.1016/j.neuron.2017.10.023
198. Harvey CD, Ehrhardt AG, Cellurale C et al (2008) A genetically encoded fluorescent sensor of ERK activity. Proc Natl Acad Sci U S A 105(49):19264–19269. https://doi.org/10.1073/pnas.0804598105
199. Violin JD, Newton AC (2003) Pathway illuminated: visualizing protein kinase C signaling. IUBMB Life 55(12):653–660. https://doi.org/10.1080/15216540310001642216
200. Wang Y, Botvinick EL, Zhao Y et al (2005) Visualizing the mechanical activation of Src. Nature 434(7036):1040–1045. https://doi.org/10.1038/nature03469
201. Ouyang M, Sun J, Chien S et al (2008) Determination of hierarchical relationship of Src and Rac at subcellular locations with FRET biosensors. Proc Natl Acad Sci U S A 105 (38):14353–14358. https://doi.org/10.1073/pnas.0807537105
202. Gao X, Zhang J (2008) Spatiotemporal analysis of differential Akt regulation in plasma membrane microdomains. Mol Biol Cell 19 (10):4366–4373. https://doi.org/10.1091/mbc.E08-05-0449
203. Ardestani G, West MC, Maresca TJ et al (2019) FRET-based sensor for CaMKII activity (FRESCA): a useful tool for assessing CaMKII activity in response to Ca(2+) oscillations in live cells. J Biol Chem 294 (31):11876–11891. https://doi.org/10.1074/jbc.RA119.009235
204. Mehta S, Zhang Y, Roth RH et al (2018) Single-fluorophore biosensors for sensitive and multiplexed detection of signalling activities. Nat Cell Biol 20(10):1215–1225. https://doi.org/10.1038/s41556-018-0200-6
205. Colgan LA, Hu M, Misler JA et al (2018) PKCalpha integrates spatiotemporally distinct Ca(2+) and autocrine BDNF signaling to facilitate synaptic plasticity. Nat Neurosci 21 (8):1027–1037. https://doi.org/10.1038/s41593-018-0184-3
206. Halls ML, Canals M (2018) Genetically encoded FRET biosensors to illuminate compartmentalised GPCR signalling. Trends Pharmacol Sci 39(2):148–157. https://doi.org/10.1016/j.tips.2017.09.005
207. de Juan-Sanz J, Holt GT, Schreiter ER et al (2017) Axonal endoplasmic reticulum Ca(2+) content controls release probability in CNS nerve terminals. Neuron 93(4):867–881.e866. https://doi.org/10.1016/j.neuron.2017.01.010
208. Vincent K, Cornea VM, Jong YI et al (2016) Intracellular mGluR5 plays a critical role in neuropathic pain. Nat Commun 7:10604. https://doi.org/10.1038/ncomms10604
209. Halls ML, Yeatman HR, Nowell CJ et al (2016) Plasma membrane localization of the mu-opioid receptor controls spatiotemporal signaling. Sci Signal 9(414):ra16. https://doi.org/10.1126/scisignal.aac9177
210. Cheng Y (2018) Single-particle cryo-EM—how did it get here and where will it go. Science 361(6405):876–880. https://doi.org/10.1126/science.aat4346
211. Nishiyama J, Mikuni T, Yasuda R (2017) Virus-mediated genome editing via homology-directed repair in mitotic and postmitotic cells in mammalian brain. Neuron 96 (4):755–768.e755. https://doi.org/10.1016/j.neuron.2017.10.004
212. Gao Y, Hisey E, Bradshaw TWA et al (2019) Plug-and-play protein modification using homology-independent universal genome engineering. Neuron 103(4):P583–597.E8. https://doi.org/10.1016/j.neuron.2019.05.047
213. Yang W, Yuste R (2017) In vivo imaging of neural activity. Nat Methods 14(4):349–359. https://doi.org/10.1038/nmeth.4230
214. Alon S, Huynh GH, Boyden ES (2019) Expansion microscopy: enabling single cell analysis in intact biological systems. FEBS J 286(8):1482–1494. https://doi.org/10.1111/febs.14597
215. Han S, Li J, Ting AY (2018) Proximity labeling: spatially resolved proteomic mapping for neurobiology. Curr Opin Neurobiol 50:17–23. https://doi.org/10.1016/j.conb.2017.10.015

216. Lobingier BT, Huttenhain R, Eichel K et al (2017) An approach to spatiotemporally resolve protein interaction networks in living cells. Cell 169(2):350–360.e312. https://doi.org/10.1016/j.cell.2017.03.022
217. Paek J, Kalocsay M, Staus DP et al (2017) Multidimensional tracking of GPCR signaling via peroxidase-catalyzed proximity labeling. Cell 169(2):338–349.e311. https://doi.org/10.1016/j.cell.2017.03.028
218. von Diezmann A, Shechtman Y, Moerner WE (2017) Three-dimensional localization of single molecules for super-resolution imaging and single-particle tracking. Chem Rev 117 (11):7244–7275. https://doi.org/10.1021/acs.chemrev.6b00629

Chapter 3

Melanopsin for Time-Controlling Activation of Astrocyte–Neuron Networks

Sara Mederos, Candela González-Arias, and Gertrudis Perea

Abstract

Melanopsin, a mammalian G-protein-coupled photopigment, is a novel optical tool which enables studying astrocyte–neuron networks. Here, we describe the required guidelines to take advantage of this promising optical tool for functional neuron–glia studies. The selective expression of melanopsin in astrocytes allows triggering astrocytic Ca^{2+} signaling, changes in synaptic transmission, and modifying behavioral responses.

Key words Astrocytes, Melanopsin, Optogenetics, Astrocyte–neuron interactions

1 Introduction

The outbreak practice of optical tools to decipher the organization of neuronal circuits and their behavioral outputs has transformed neuroscience. Indeed, the use of opsins to enhance or suppress neuronal activity has been extensively expanded [1]. Some studies have applied these approaches to astrocytes, mainly based on channelrhodopsin-2 (ChR2), to unmask the consequences of astrocyte signaling on particular brain functions [2–8]. However, photostimulation with ChR2 and other channel-based opsins has drawbacks to recapitulate the intracellular signaling observed under physiological conditions [9–12], which is mostly mediated by G-protein coupled receptor (GPCR) activation. Alternatively, the engineered G-protein coupled receptors activated by inert drug-like small molecules (DREADDs) have been also applied for astrocyte–neuron studies [13, 14]. However, the sustained activation of DREADDs by its exogenous ligand could misplace the temporal features of astrocyte signaling engaged by neuronal activity. Therefore, refined and reliable approaches involving endogenous astrocyte Ca^{2+} signaling in a time-controlled manner are required. In this spirit, melanopsin, a mammalian G-protein-coupled photopigment, has been discovered as a valuable optical tool for studying

Dominik Niopek (ed.), *Photoswitching Proteins: Methods and Protocols*, Methods in Molecular Biology, vol. 2173,
https://doi.org/10.1007/978-1-0716-0755-8_3,

astrocyte–neuron networks. The selective expression of this opsin in astrocytes is able to trigger calcium (Ca^{2+}) signaling and, additionally, induce changes in synaptic transmission and behavioral responses.

In this chapter we will describe a detailed experimental procedure to use melanopsin as optogenetic tool in astrocytes, which allows to elucidate their regulatory actions on neuronal circuits both in vitro and in vivo.

2 Materials

2.1 Viral Injections

Store Adeno-associated virus (AAV) at −80 °C until use.

1. AAV2/5-GFAP104-melanopsin-mCherry (Addgene 122630); viral titer 2.8×10^{12} (AAV5 encoding Melanopsin (Opn4-human melanopsin, AAI13559.1) fused to mCherry driven from a shortened GFAP promoter (GFAP104)).
2. AAV2/5-GFAP104-mCherry (Addgene 58909); viral titer 3.9×10^{12}.
3. AAV2/5-gfaABC1D-lck-GCaMP6f (Addgene 52924); viral titer 6.4×10^{13}.
4. AAV2/5-gfaABC1D-cyto-GCaMP6f (Addgene 52925); viral titer 6.13×10^{13}.
5. Stereotaxic apparatus.
6. Microinjector.
7. Needle/glass micropipettes.
8. Isoflurane.
9. Scalpels.
10. Surgical drill.
11. Heat blanket.
12. Mouse body temperature probe.

2.2 Optical Fiber Implantations

1. Optical power meter (e.g., PM100D coupled with sensor S120C, Thorlabs).
2. Surgical drill.
3. Miniature stainless steel screws.
4. Cement.
5. Lidocaine.
6. Antibiotics (Erythromycin).
7. Painkiller (Carprofen).

2.3 Immunohistochemistry

2.3.1 List of Primary Antibodies

1. Mouse anti-GFAP (Glial fibrillary acidic protein); dilution: 1:500; Sigma St. Louis, MO,RRID:AB_477010.
2. Rabbit anti-NeuN (Neuronal nuclei); dilution: 1:500; Millipore, RRID:AB_2571567.
3. Rabbit anti-S100; dilution: 1:100; Abcam, Cambridge, UK, RRID:AB_306716.
4. Rabbit anti-Iba1; dilution: 1:500; Wako, Huntley, IL, RRID: AB_839504.

2.3.2 List of Secondary Antibodies

1. Alexa Fluor 488 goat anti-mouse; dilution 1:200; Bioss Inc., Woburn, MA, RRID:AB_10892893.
2. Pacific Blue goat anti-rabbit; dilution 1:200; Thermo Fisher Scientific, Waltham, MA, RRID:AB_2539814.
3. Alexa Fluor 647 goat anti-rabbit; dilution 1:200; Thermo Fisher Scientific, RRID:AB_143165.
4. Alexa Fluor 405 goat anti-rabbit; dilution 1:200; Thermo Fisher Scientific, RRID:AB_10680407.

2.3.3 Additional Material

1. Vibratome (e.g., VT1000S, Leica).
2. Pentobarbital.
3. PBS.
4. Paraformaldehyde.
5. Triton X-100.
6. NGS.
7. DAPI, 1.5 μg/mL (Sigma).
8. Vectashield antifading mounting medium (Vector Laboratories, Burlingame, CA).
9. Confocal microscope.

2.4 Optogenetic Stimulation

1. Diode-pumped solid-state blue laser (473 nm, MBL-III-473, OptoEngine, LLC, Midvale, UT) (*see* **Note 1**).
2. 200 μm and 100 μm thick optic fibers ending in a ceramic ferrule (Thorlabs, Newton, NJ), which are implanted and attached to the patch cords.
3. Two-photon laser scanning microscope (e.g., Axiolmager M, LSM510, Zeiss Oberkochen, Germany).

2.5 Hippocampal Slice Preparation

1. Isoflurane.
2. Artificial Cerebrospinal Fluid (ACSF): 124 mM NaCl, 2.69 mM KCl, 1.25 mM KH_2PO_4, 2 mM $MgSO_4$, 26 mM $NaHCO_3$, 2 mM $CaCl_2$, and 10 mM glucose, gassed with 95% O_2/5% CO_2 (pH = 7.3). Daily prepared using ultrapure water and stored at room temperature (RT) (*see* **Note 2**).

3. ACSF without Mg^{2+}: 124 mM NaCl, 2.69 mM KCl, 1.25 mM KH_2PO_4, 26 mM $NaHCO_3$, 10 mM glucose, 4 mM $CaCl_2$, and 10 μM glycine, gassed with 95% O_2/5% CO_2 (pH = 7.3) (*see* **Note 3**).
4. Vibratome.
5. Agarose gel.
6. Microscope equipped with a 40× water immersion lens, infrared-DIC optics, and a CCD camera (e.g., Hamamatsu C4742-95 attached to a Olympus BX50WI or Nikon Eclipse FN1 microscope).

2.6 Calcium Imaging

1. AAV vectors (*see* Subheading 2.1).
2. Tetrodotoxin (TTX).
3. Two-photon laser scanning microscope (e.g., Axiolmager M, LSM510, Zeiss Oberkochen, Germany) equipped with a 900 nm pulsed red laser (1.80 W, 1% of power; e.g., Spectra Physics Mai-Tai, Prairie Technologies, Sioux Falls, SD).

2.7 Electrophysiological Recordings

1. Borosilicate glass pipettes.
2. Internal solution for whole cell patch-clamp in neurons: 135 mM K-gluconate, 10 mM KCl, 10 mM HEPES, 1 mM $MgCl_2$, 2 mM ATP-Na_2 (pH = 7.3, with KOH; osmolarity 280–290 mOsm/L).
3. Internal solution for whole cell patch-clamp in astrocytes: 100 mM potassium methanesulfonate, 50 mM KCl, 10 mM HEPES-K, 2 mM ATP-Na_2 (pH = 7.3, with KOH; osmolarity 280–290 mOsm/L).
4. Internal solution for intracellular loading of astrocyte network: 40 mM BAPTA-K_4, 8 mM NaCl, 1 mM $MgCl_2$, 10 mM HEPES, 0.4 mM GTP-tris salt, 2 mM ATP-Na_2 (pH = 7.3, with KOH), and 0.1% biocytin.

   Prepare internal solutions at 4 °C and store at −20 °C.
5. Amplifiers.
6. Automatic temperature controller.
7. Recording chamber.
8. Theta capillaries.
9. ACSF (*see* Subheading 2.5).
10. Stimulator.

2.8 In Vivo

1. Fiber-optic patch cord.
2. Zirconia sleeve.
3. 473 nm DPSS laser.
4. FC/PC adapter.
5. AAV vectors (*see* Subheading 2.1).

3 Methods

3.1 Virus Injection

During the whole procedure, follow biological safety guidelines: wearing lab coat, surgical mask, hairnet, and two pairs of gloves.

1. Anesthetize mice with 2% isoflurane in oxygen and place in a custom adapted stereotaxic apparatus (Fig. 1). The depth of anesthesia is checked by testing mouse's reflexes.
2. Body temperature should be continuously monitored and maintained at 37.0 ± 0.5 °C (a probe and a heat blanket can be used).
3. To target the area of interest coordinates can be taken from Paxinos atlas [15]. For hippocampus: 2 mm from Bregma, 1.4 mm lateral, and 1.2–1.4 mm dorsoventral.
4. Open the skin that protects the skull with a scalpel and leave aside, make small holes in the skull with a surgical drill (*see* **Note 4**).
5. Borosilicate glass micropipette containing virus is stereotaxically placed into selected brain region.
6. A volume of 0.3–0.4 μL of the virus is injected at 30 nL/min (*see* **Note 5**). After injection, the pipette should be held in place for 5 min prior to retraction to prevent leakage.

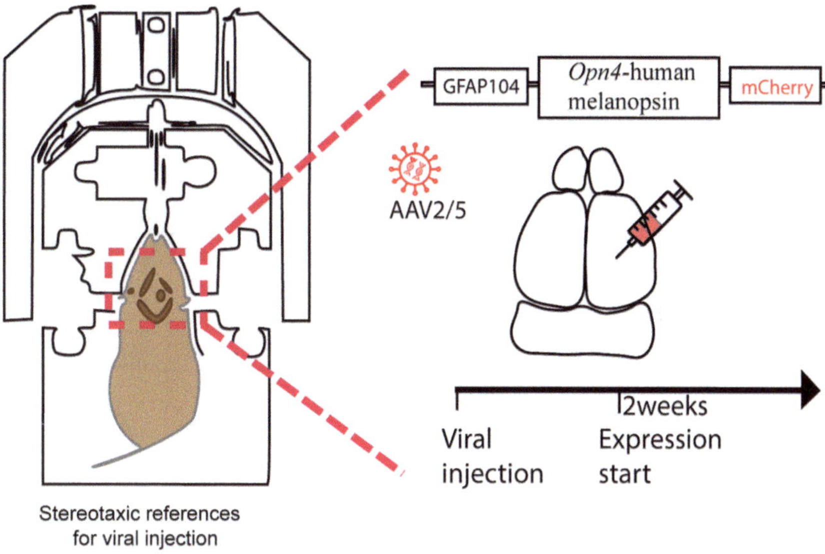

Fig. 1 Scheme for viral injection procedure and time line. Attach the needle/glass capillary to the stereotaxic holder. Place the animal under stereotaxic apparatus and perform a craniotomy following stereotaxic references for the area of interest. After confirming the mouse is fully anesthetized by assessing its response to a toe pinch, perform a midline incision placing aside the skin (bregma and lambda are visible). Clean the skull by applying NaCl 0.9%. Drill a hole above the area of interest (craniotomy). Defrost an aliquot of viral particles. Withdraw 2 μL of virus using with microinjector from a piece of Parafilm. Target area of interest with the needle/glass capillary and infuse 300–400 nL of virus

7. Remove the pipette and suture the skin.
8. Allow mice to recover from anesthesia with the help of heating pads and return to home cage once it show regular breathing and locomotion.

3.2 Optic Fibers

1. Measure the power at the tip of each optic fiber with an optical power meter by connecting it to the LED system operating in CW mode, prior to the surgery. Either commercial or custom-made optic fibers can be used [16, 17] (*see* **Note 6**).
2. Drill the skull doing one/two holes for the screws (two for more than one optic fiber). The diameter of a hole should precisely match diameter of a screw.
3. Screw in miniature stainless-steel screws. Screws should be placed in cerebellum if planning to stimulate cortex, and elsewhere if planning to stimulate cerebellum.
4. Place the optic fiber into the craniotomy previously done for viral injection.
5. Cement the screws to the ferrule ensuring that the connector is perfectly attached to enable optimum stimulation.
6. Apply lidocaine onto wound edges.
7. Inject i.p antibiotics (Erythromycin), and administer a painkiller (Carprofen) daily for 3 days after surgery.
8. Facilitate animal recovery with soft food and condensed milk (*see* Fig. 2). Proper handling of animals ensures easier connection of implants to the patch cord.

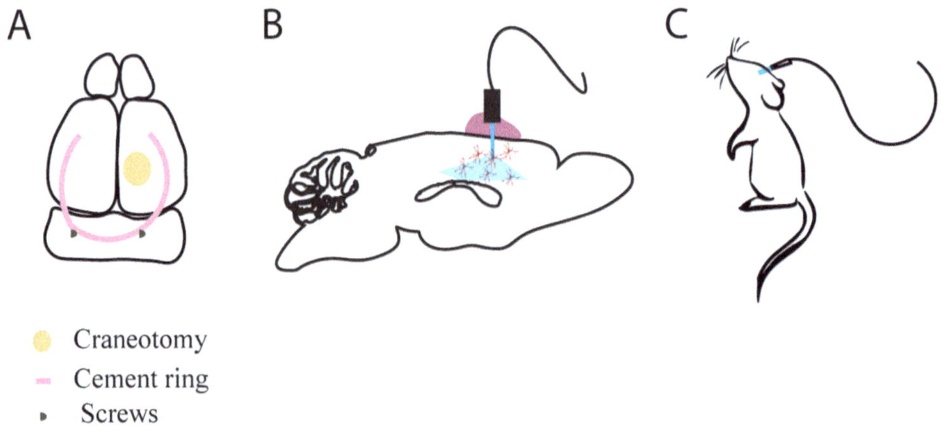

Fig. 2 Illustration of optic fibers implants. (**a**) Scheme of optic fiber implantation with miniature screws positioned on the skull. (**b**) Sagittal view of light stimulation to the targeted astrocytes. (**c**) Scheme of freely moving behaving mouse with patch cord attached for light stimulation

3.3 Immunohistochemistry

The expression of specific astrocytic constructs is confirmed by immunostaining 2–3 weeks after viral injection.

Selective labeling for GFAP, S100, and NeuN proteins can be studied by immunohistochemistry (Fig. 3). Colocalization of GFAP, S100, or NeuN labeling is analyzed with mCherry or GCaMP6f expression. Note that no colocalization should be found between NeuN labeled neurons and mCherry or GCaMP6f expressing cells (*see* **Note 7**).

1. Euthanize mice with sodium pentobarbital and transcardially perfuse with PBS followed by ice-cold 4% paraformaldehyde in PBS.
2. Remove brains and postfix overnight (o/n) at 4 °C in the same fixative solution.
3. Obtain coronal brain slices of 50 μm in a vibratome, collect as floating sections and block for 1.5 h at RT in a solution containing 0.3% Triton X-100 and 5% NGS in PBS.

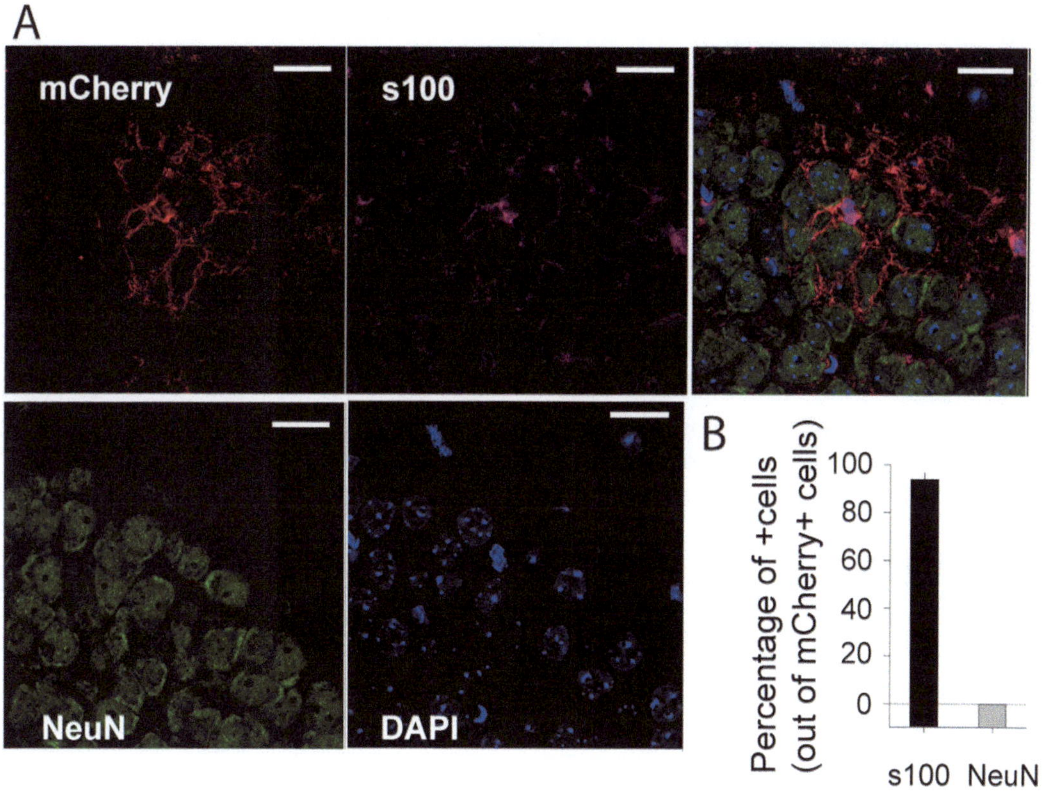

Fig. 3 Specific expression of melanopsin in astrocytes. (**a**) Magnified confocal images showing an astrocyte expressing melanopsin-mCherry (in red) intermingled with CA1 pyramidal neurons. Astrocytic s100 labeling (in pink), and the neuronal marker NeuN (in green), DAPI (in blue). Immunocytochemical localization of melanopsin-mCherry, GFAP, and NeuN is necessary from transfected mice to confirm astrocytes are correctly target in area of interest (here hippocampus). Scale bar = 20 μm. (**b**) Graphs from colocalization analysis of mCherry:s100 and mCherry:NeuN. *Figure adapted from Mederos et al. Glia 2019*

4. Incubate sections (o/n; 4 °C) with primary antibodies in a solution containing 0.3% Triton X-100 and 1% NGS in PBS.
5. Wash sections three times, 15 min each, in PBS containing 0.3% Triton X-100.
6. Incubate floating sections with the specific Alexa-conjugated antibodies in a PBS solution containing 0.3% Triton X-100 and 1% NGS for 2 h at RT.
7. Wash sections three times, 15 min each, in PBS containing 0.3% Triton X-100.
8. Incubate sections with DAPI (1.5 μg/mL) for 10 min at RT.
9. Wash sections once in PBS and mount in Vectashield antifading mounting medium.
10. Store sections at 4 °C and acquire images on a confocal microscope (e.g., Leica SP5 laser scanning confocal microscope).

Additionally, in order to check microgliosis, that is, activation of microglia in transfected area, Iba1 antibody can be used and Iba1 expression examined. *See* **Note 8** and Fig. 4.

3.4 Optogenetic Stimulation

1. Full-field photostimulation can be achieved in two ways, through a diode-pumped solid-state blue laser with analog intensity control (473 nm, i.e., MBL-III473, OptoEngine, LLC, Midvale, UT) coupled via SMA terminal to a 200 μm fiber (ThorLabs, Newton, NJ), or through the light path of microscope objective.
2. For two-photon Ca^{2+} experiments, blue light pulses at different intensities (i.e, 7 mW/mm^2) at 10 Hz (50 ms light pulses) or continuous stimulation at variable durations (1, 5, 10, and 20 s) can be delivered (*see* **Note 9**).
3. For synaptic activity recordings light pulses can follow different protocols mimicking relevant physiological processes (e.g., low vs high brain oscillations) (Fig. 5).

3.5 Hippocampal Slice Preparation

1. To obtain hippocampal slices, mice are anesthetized with isoflurane (1.5% in O_2) and decapitated.
2. Brain should be rapidly extracted and placed in ice-cold ACSF, or modified ACSF.
3. Fix the brain to the vibratome baseplate with glue (*see* **Note 10**).
4. Incubate slices of 350 μm thickness for 1 h at RT (22–24 °C) in ACSF.
5. Afterward, transfer slices to an immersion recording chamber and superfused with gassed ACSF.

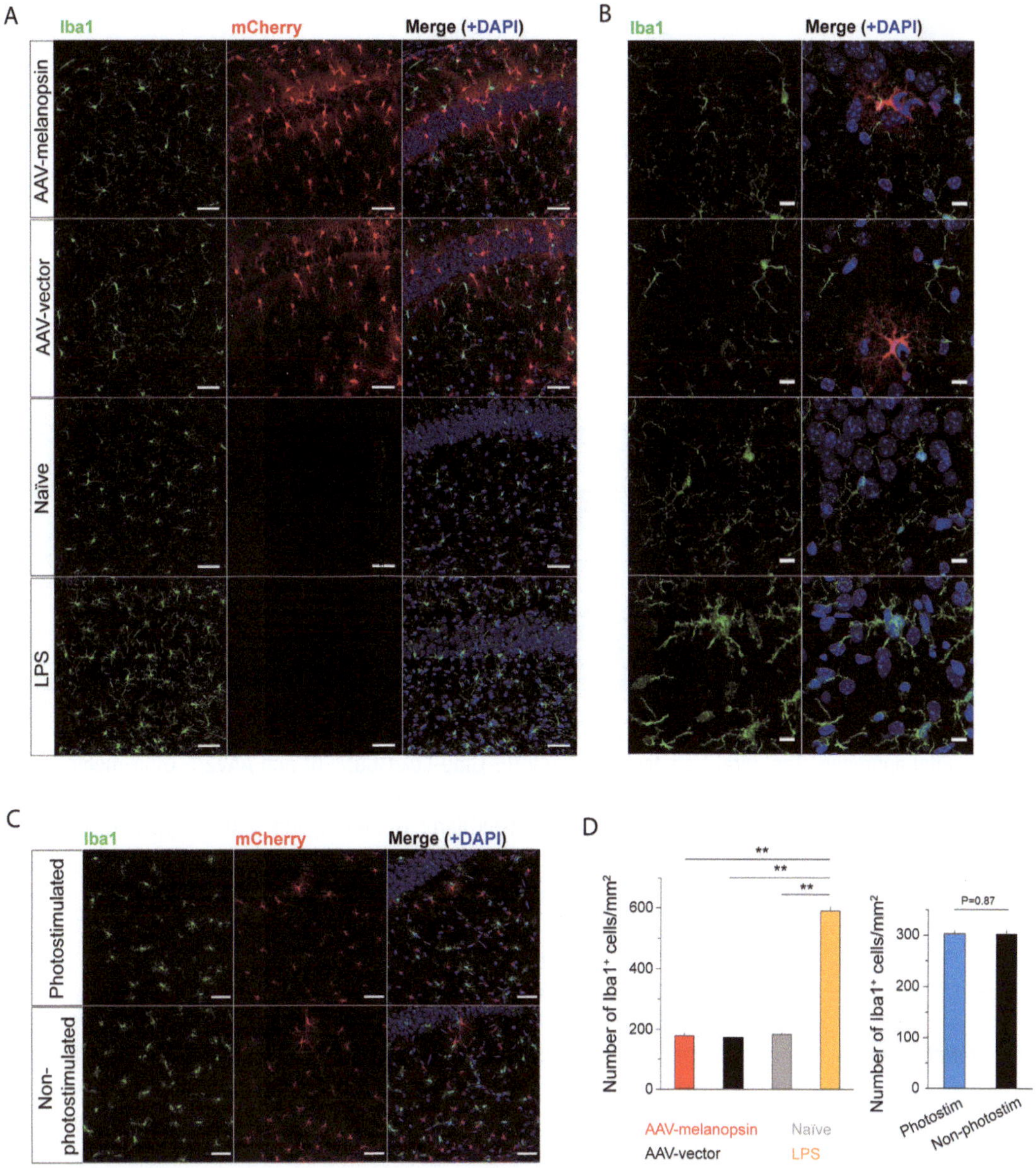

Fig. 4 Photostimulation and viral transfection do not promote microglia activation. (**a**) Iba1 labeling in melanopsin-mCherry, AAV-vector-mCherry transfected mice, nontreated animals (naïve), and LPS-treated mice (24 h after LPS stereotaxic hippocampal injection). Iba1 labeling is shown in green and mCherry in red. Scale bar = 50 μm. Blue signals correspond to DAPI labeling. Note the Iba1 labeling was performed 15 days post viral injection. (**b**) Magnified pictures of individual Iba1+ cells for the corresponding conditions in (**a**). Note the presence of gross morphological changes of Iba1+ cells in the LPS-treated animals that was absent in naïve and AAV-transfected mice. Scale bar = 10 μm. (**c**) Iba1 labeling of the melanopsin-transfected region (mCherry+ astrocytes) for the photostimulated (top row) and nonphotostimulated (bottom row) acute slices. Scale bar = 50 μm. (**d**) Left, number of Iba1+ cells/area was increased by LPS treatment, but unchanged in the rest of conditions indicating that the viral treatments used in this study did not activate microglia response. Right, Average mean number of Iba1+ cells/area after light stimulus for photostimulated slices and non-photostimulated slices. Unpaired *t*-test; $p = 0.87$. Data are shown as mean ± SEM. *Figure adapted from Mederos et al. Glia 2019*

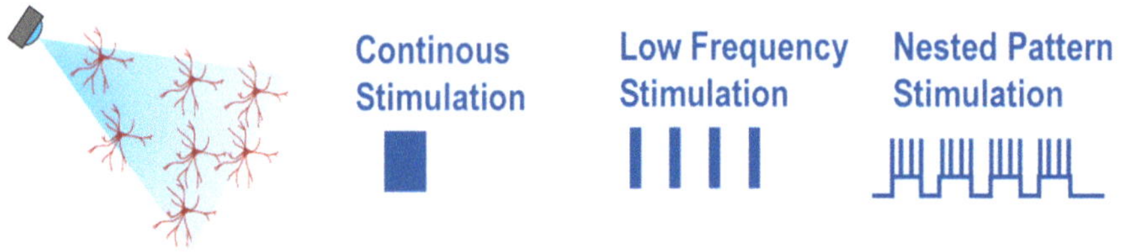

Fig. 5 Optogenetic protocols for controlling astrocyte activity. Scheme showing light application of different protocols to astrocytes. Note to select carefully your light stimulation protocol because different frequencies might differentially activate astrocytes, thereby leading to different experimental outcomes

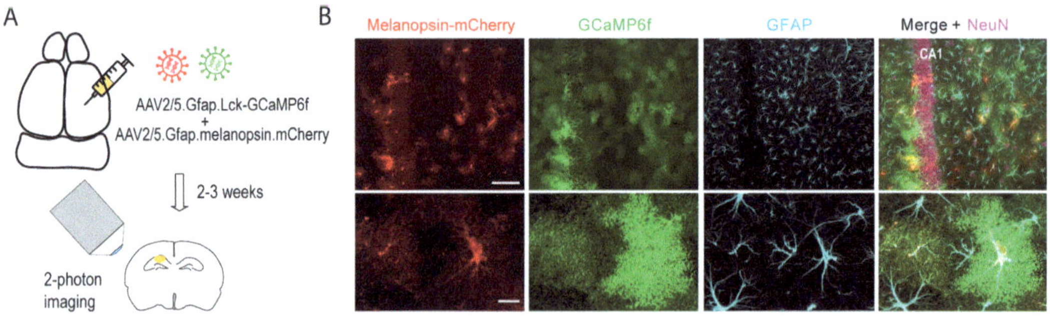

Fig. 6 Coexpression of melanopsin with GCaMP6f to study Ca^{2+} signaling in astrocytes. (**a**) Scheme of the experimental approach. Top, viral transfection of AAV2/5-Gfap-Lck.GCaMP6f and AAV2/5-Gfap-melanopsin-mCherry in hippocampus. Bottom: hippocampal slice under two-photon imaging after 2 weeks of transfection. (**b**) Top: Immunocytochemical localization of melanopsin-mCherry, GCaMP6f, and GFAP in hippocampal slices; bottom: Zoom-in image of labeled astrocytes. Merge image includes NeuN labeling (pink). Scale bar, 50 μm; inset, 10 μm. *Figure adapted from Mederos et al. Glia 2019*

This step is shared for studying calcium imaging and synaptic transmission in vitro.

3.6 Calcium Imaging

1. For Ca^{2+} imaging experiments genetically encoded calcium indicators (GECIs): Lck-GCaMP6f or cyto-GCaMP6f (*see* **Note 11**) are coinjected with melanopsin, or vector (1:1 ratio) in hippocampal astrocytes (Fig. 6). Astrocytes showing both mCherry and GCaMP6f positive labeling are selected for analysis, and regions of interest (ROI) are selected from the GCaMP6f image.
2. TTX (1 μM) is perfused for calcium imaging recordings.
3. To monitor intracellular Ca^{2+} levels in astrocytes, a two-photon microscopy equipped with a pulsed red laser tuned at 900 nm, for excitation of Lck-GCaMP6f or cyto-GCaMP6f, is needed (*see* **Note 12**). mCherry signal can be acquired in a confocal laser scanning mode at 543 nm with a HeNe laser.
4. Images are acquired by water immersion lens 40×/1.0 W Plan-Apochromat (Zeiss), a field of view between 354 × 354 μm can

be used for colocalization image, and 295 × 295 μm for acquisition.

5. Full-field photostimulation with different light pulse durations at high or low-frequency stimulation are delivered by an external laser.
6. Calcium signal analysis can be performed using available reported and free software by using ImageJ or Matlab [18–20].
7. For each ROI, different parameters can be analyzed, that is, frequency of events, amplitude, width, and onset. Onset is defined as the last point in the $\Delta F/F$ time course below 1 standard deviation (SD) before a significant peak occurred. Width value can be measured as the half prominence of each peak.
8. In parallel, experiments with control virus (expressing reported gene) should be performed to establish the responses due to melanopsin activation from those mediated by spontaneous activity or viral transfection.

3.7 Electrophysiological Recordings in Hippocampal Slices

After 2 weeks of viral injections, mice transfected with melanopsin are ready for electrophysiological analysis in situ (brain slices):

1. Fill the patch pipettes (borosilicate glass with filament) with the corresponding internal solution (*see* Subheading 2.7), input resistance between 3–5 MΩ.
2. Amplifiers (e.g., Dagan Corporation, Minneapolis, MN, USA) are used for electrical recordings, at RT (22–24 °C) or ≈32 °C, using an automatic temperature controller coupled to the recording chamber.
3. Fast and slow whole-cell capacitances must be neutralized, and series resistance has to be compensated (≈70%). Recordings will be rejected when the access resistance increased >20% during the experiment.
4. Whole-cell configuration of patch-clamp technique is applied to astrocytes to evaluate the melanopsin-induced currents (Fig. 7).
5. Synaptic stimulation can be achieved using theta capillaries (2–5 μm tip diameter) filled with ACSF. These capillaries are placed in stratum radiatum to stimulate Schaffer collateral fibers (different locations when studying other brain areas).
6. Pulses (250 μs duration) can be delivered at 0.33 Hz by stimulator.
7. Measure baseline of synaptic activity during 3–5 min before astrocyte activation (light stimulation).
8. Neurons showing *Z*-score > 2 SD of baseline EPSC amplitude (pA) are considered responding to light (Fig. 8). EPSC: excitatory postsynaptic currents.

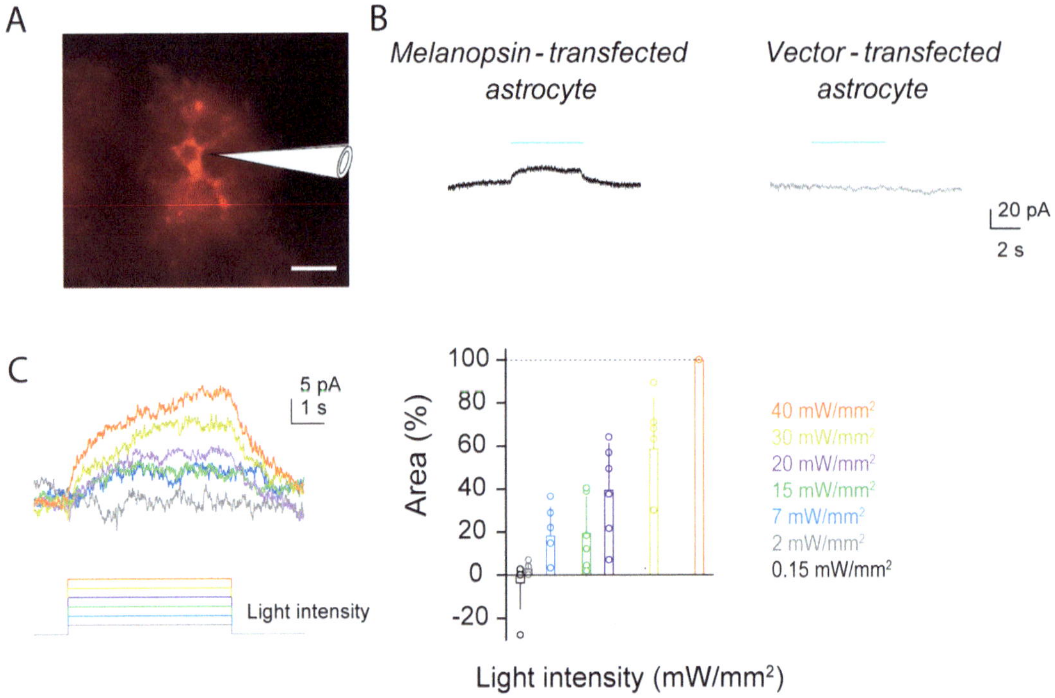

Fig. 7 Melanopsin induces astrocyte membrane currents. (**a**) Epifluorescence image of melanopsin-mCherry transfected astrocyte. Scale bar = 25 μm. (**b**) Evoked current by light stimulation (5 s, blue bar) in melanopsin-transfected astrocyte and vector-transfected astrocyte. Note light does not generate changes in membrane current in vector-transfected astrocytes. (**c**) Left, membrane currents in response to different light intensities. Right, area values of the light-evoked currents for melanopsin astrocytes. *Figure adapted from Mederos et al. Glia 2019*

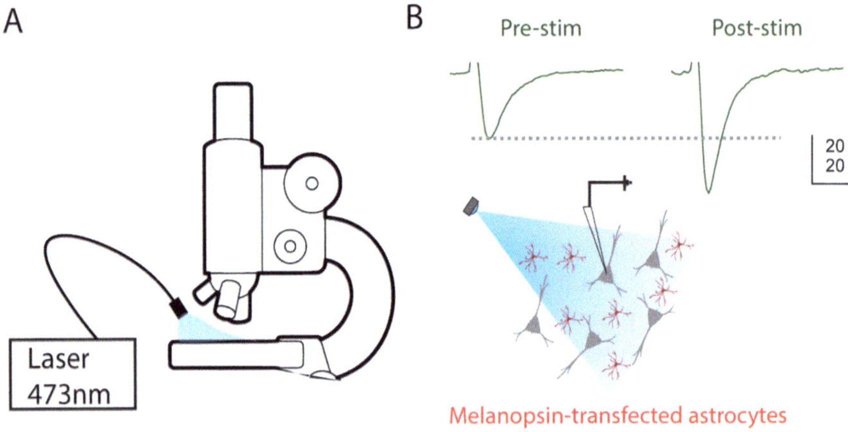

Fig. 8 Studying synaptic transmission changes in brain slices. (**a**) Scheme showing full-field stimulation with an external laser. (**b**) Assessing changes in synaptic transmission by astrocyte light activation. EPSCs can be recorded and analyzed before and after melanopsin stimulation

9. Changes in holding current levels can be monitored by analyzing Holding current index: [holding current (i) − holding current (baseline)]/absolute value [holding current (i) + holding current (baseline)]. i = holding current values at different time after light stimulus (*see* **Note 13**).
10. Astrocyte manipulation of intracellular signaling can be achieved by loading astrocytic networks with BAPTA (40 mM, Ca^{2+} chelator) or GDPβS (20 mM, a competitive blocker of G-proteins) through the recording pipette (*see* **Note 14**).
11. Pharmacological reagents can be perfused in order to isolate the receptors mediating astrocyte-induced synaptic responses (antagonists of glutamate receptors, purinergic receptors, etc.) (*see* **Note 15**).

3.8 Behavioral Experiments

After 2 weeks of viral injections, mice transfected with melanopsin are ready for behavioral analysis:

1. Connect a 1.5/3 m long fiber optic patch cord with protective tubing (Thorlabs or similar) to the chronically implanted optical fiber with a zirconia sleeve (Precision Fiber Products, Milpitas, CA, USA/ Thorlabs or similar), which will allow the mice to freely explore. The patch cord is connected to a 473 nm DPSS laser with an FC/PC adapter.
2. Control the laser output using a stimulus generator with the protocol selected to stimulate astrocytes. For control animals, light stimulation should be identical to the opsin activating stimulation, and animals should have been treated with control virus (GFAP-mCherry; *see* Subheading 2.1).
3. Use different tests to study the contribution of astrocytes to various relevant behaviors, that is, novel object recognition, passive avoidance, open field, and operant conditioning [21–24] (Fig. 9, *see* **Note 16**).

4 Notes

1. Optogenetic stimulation: LED stimulation (488 nm wavelength) through the light path of the microscope can be used to stimulate astrocytes.
2. ACSF solution can be used for brain slicing from juvenile mice ($P < 30$). For mature adult mice ($P < 50$), to improve slice viability, NMDG-HEPES solution can be used. Transcardial perfusion of 10 mL ice-cold (4 °C) NMDG and a subsequent incubation for 10 min (37 °C) NMDG-HEPES will be sufficient to preserve correctly and increase the quality of brain slices. NMDG-HEPES ACSF: 92 mM NMDG, 2.5 mM KCl,

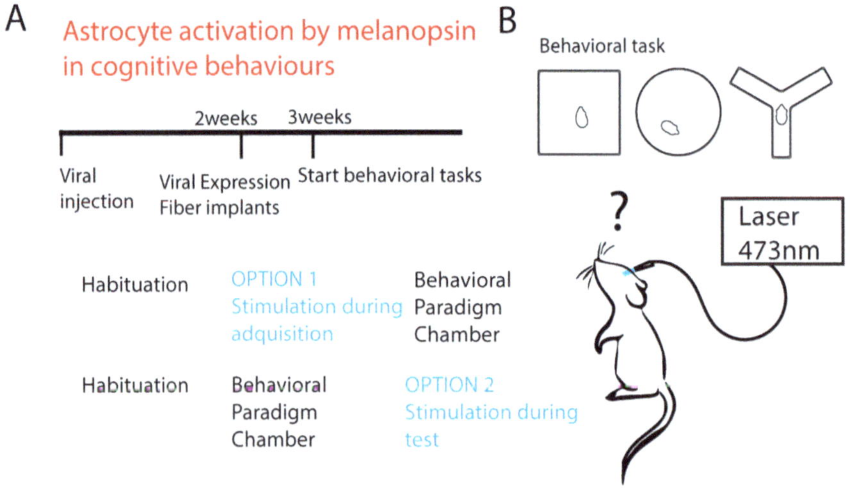

Fig. 9 Study astrocyte activation underlying cognitive behaviors. (**a**) Top: Scheme of time course since viral injection to behavioral testing. Bottom, different options to activate astrocytes with melanopsin during behavioral testing: during acquisition phase (Option 1) or during test trials (Option 2), which can modify behavioral outputs. (**b**) Top, different testing paradigms to study behaviors. Bottom, scheme of in vivo astrocyte light stimulation showing a long enough patch cord attached to a freely behaving animal. Note the requirement of melanopsin-transfected and control groups in same sessions to minimize intergroup variability by external factors. Note: All the corresponding figures are reproduced following copyright terms and conditions provided by John Wiley & Sons

1.25 mM NaH_2PO_4, 30 mM $NaHCO_3$, 20 mM HEPES, 25 mM glucose, 2 mM thiourea, 5 mM Na-ascorbate,3 mM Na-pyruvate, 0.5 mM $CaCl_2{\cdot}2H_2O$, and 10 mM $MgSO_4{\cdot}7H_2O$, gassed with 95% O_2/5% CO_2. Titrate pH to 7.3–7.4 with 17 mL ± 0.5 mL of 5 M hydrochloric acid [25].

3. Some modifications can be applied to external solutions depending on type of experiments pursued. Picrotoxin (50 μM) an antagonist of GABAa receptors, and CGP 55845 (5 μM) an antagonist of GABAb receptors, can be added to ACSF to isolate excitatory postsynaptic responses. To isolate inhibitory postsynaptic responses, CNQX (20 μM) and D-AP5 (50 μM), selective antagonists for AMPA and NMDA receptors, respectively, can be added to ACSF. For the case of miniature postsynaptic currents (mPSCs) and slow inward currents (SICs), TTX (1 μM) is added to ACSF for preventing action potential generation and propagation. Additionally, Mg^{2+}-free ACSF was used for SIC recordings. All these drugs must be applied for at least 15 min before recordings.

4. The intracranial injection of virus is different depending on mice age. For neonatal mice (P5–P8), due to the thinness of skull, it is not necessary to use a drill to make burr holes; a scalpel can be used instead. In case of neonatal surgery, the

overall duration of this procedure should be kept under 20 min to maximize the survival rate of mice.

5. A combination of different virus can be placed into the brain area by a single injection procedure. 1:1 volume ratio can be used.
6. For custom-made optical fibers the intensity of light transfer needs to be tested before implantation, optic fibers with intensity transfer >50% are used. After implantation, ferrules are secured to miniature stainless steel bone screws buried in the skull, using dental acrylic and a thin coating of a cyanoacrylate glue.
7. After 3 weeks of viral expression some damaged cells could be found and mCherry labeling might appear demeaned.
8. Light stimulation or viral transfection might induce microgliosis in acute slices. To evaluate microgliosis, illuminate acute slices for 20 s (or different light pulse durations) with blue laser (473 nm) or keep them in the dark for an equivalent amount of time. Afterward carry out the following steps:
   (a) Fix sections with ice-cold 4% paraformaldehyde in PBS (15 min).
   (b) Block sections o/n at 4 °C in a solution containing 0.5 Triton X-100 and 5% NGS.
   (c) Incubate sections (24 h; 4 °C) with rabbit anti-Iba1 (e.g., 1:500, Wako, RRID: AB_839504).
   (d) Wash sections four times × 15 min each in PBS containing 0.5% Triton X-100.
   (e) Incubate sections with secondary antibody (e.g., Alexa Fluor 488 Goat anti-rabbit).
   (f) Wash sections three times × 15 min each and mount in Vectashield antifading mounting medium (Vector Laboratories).
9. Light protocols mimicking neuronal activity can be applied to melanopsin, which improves the chances to get meaningful astrocyte calcium responses and synaptic-mediated effects. Different frequencies (theta, gamma oscillations) and durations can be used to stimulate melanopsin-transfected astroctyes.
10. Small block of agarose gel can be used to reduce brain movement during cutting.
11. Highly branched nature of astrocytes can be visualized and fine processes monitored by Lck-GCaMP6f. Alternatively, cyto-GCaMP6f can be used to monitor Ca^{2+} signals in proximal processes and astrocyte somas.
12. Because GCaMP and melanopsin are stimulated at the same wavelength, it is necessary to use a two-photon microscope. In

this way, GCaMP can be visualized without practically activating the opsin.

13. It is important that during the stimulation of melanopsin the holding current is monitored to verify that no significant changes occur. Conversely, other opsins, such as ChR2, induce sustained depolarizations in neuronal membrane potential due to extracellular enhancement of K^+ [10].
14. For astrocyte network loading, the holding potential should be −70 mV. Intracellular solution containing GDPβS (20 mM), BAPTA (40 mM) and biocytin (0.1%) is used for astrocyte filling for 20–30 min. Slices are then fixed and biocytin revealed by Alexa 488-Streptavidin, showing the wide area covered by the intracellular biocytin loading, and confirming the broad downregulation of Ca^{2+} signals by BAPTA intracellular filling astrocytes.
15. Drugs are bath applied for at least 15 min before recordings.
16. Astrocyte optogenetic stimulation during acquisition periods or during testing trials might change the final outcome of behavioral experiments [26]. Bilateral light stimulation is generally more efficient; however, unilateral optical stimulation might be sufficient considering the brain area studied.

References

1. Bernstein JG, Boyden ES (2011) Optogenetic tools for analyzing the neural circuits of behavior. Trends Cogn Sci 15(12):592–600. https://doi.org/10.1016/j.tics.2011.10.003
2. Perea G, Yang A, Boyden ES et al (2014) Optogenetic astrocyte activation modulates response selectivity of visual cortex neurons in vivo. Nat Commun 5:3262. https://doi.org/10.1038/ncomms4262
3. Sasaki T, Beppu K, Tanaka KF et al (2012) Application of an optogenetic byway for perturbing neuronal activity via glial photostimulation. Proc Natl Acad Sci U S A 109 (50):20720–20725. https://doi.org/10.1073/pnas.1213458109
4. Masamoto K, Unekawa M, Watanabe T et al (2015) Unveiling astrocytic control of cerebral blood flow with optogenetics. Sci Rep 5:11455. https://doi.org/10.1038/srep11455
5. Gourine AV, Kasymov V, Marina N et al (2010) Astrocytes control breathing through pH-dependent release of ATP. Science 329 (5991):571–575. https://doi.org/10.1126/science.1190721
6. Pelluru D, Konadhode RR, Bhat NR et al (2016) Optogenetic stimulation of astrocytes in the posterior hypothalamus increases sleep at night in C57BL/6J mice. Eur J Neurosci 43 (10):1298–1306. https://doi.org/10.1111/ejn.13074
7. Yamashita A, Hamada A, Suhara Y et al (2014) Astrocytic activation in the anterior cingulate cortex is critical for sleep disorder under neuropathic pain. Synapse 68(6):235–247. https://doi.org/10.1002/syn.21733
8. Tang F, Lane S, Korsak A et al (2014) Lactate-mediated glia-neuronal signalling in the mammalian brain. Nat Commun 5:3284. https://doi.org/10.1038/ncomms4284
9. Mederos S, Hernandez-Vivanco A, Ramirez-Franco J et al (2019) Melanopsin for precise optogenetic activation of astrocyte-neuron networks. Glia 67(5):915–934. https://doi.org/10.1002/glia.23580
10. Octeau JC, Gangwani MR, Allam SL et al (2019) Transient, consequential increases in extracellular potassium ions accompany channelrhodopsin2 excitation. Cell Rep 27 (8):2249–2261.e2247. https://doi.org/10.1016/j.celrep.2019.04.078
11. Losi G, Mariotti L, Sessolo M et al (2017) New tools to study astrocyte ca(2+) signal dynamics in brain networks in vivo. Front Cell Neurosci

11:134. https://doi.org/10.3389/fncel.2017.00134

12. Xie AX, Petravicz J, McCarthy KD (2015) Molecular approaches for manipulating astrocytic signaling in vivo. Front Cell Neurosci 9:144. https://doi.org/10.3389/fncel.2015.00144
13. Yang L, Qi Y, Yang Y (2015) Astrocytes control food intake by inhibiting AGRP neuron activity via adenosine A1 receptors. Cell Rep 11 (5):798–807. https://doi.org/10.1016/j.celrep.2015.04.002
14. Chen N, Sugihara H, Kim J et al (2016) Direct modulation of GFAP-expressing glia in the arcuate nucleus bi-directionally regulates feeding. elife 5. https://doi.org/10.7554/eLife.18716
15. Franklin K, Paxinos G (2012) Paxinos and Franklin's the mouse brain in stereotaxic coordinates. Elsevier, Amsterdam
16. Ung K, Arenkiel BR (2012) Fiber-optic implantation for chronic optogenetic stimulation of brain tissue. J Vis Exp 68:e50004. https://doi.org/10.3791/50004
17. Bender F, Korotkova T, Ponomarenko A (2018) Optogenetic entrainment of hippocampal theta oscillations in behaving mice. J Vis Exp 136. https://doi.org/10.3791/57349
18. Srinivasan R, Huang BS, Venugopal S et al (2015) Ca(2+) signaling in astrocytes from Ip3r2(−/−) mice in brain slices and during startle responses in vivo. Nat Neurosci 18 (5):708–717. https://doi.org/10.1038/nn.4001
19. Bindocci E, Savtchouk I, Liaudet N et al (2017) Three-dimensional Ca(2+) imaging advances understanding of astrocyte biology. Science 356(6339). https://doi.org/10.1126/science.aai8185
20. Agarwal A, Wu PH, Hughes EG et al (2017) Transient opening of the mitochondrial permeability transition pore induces microdomain calcium transients in astrocyte processes. Neuron 93(3):587–605.e587. https://doi.org/10.1016/j.neuron.2016.12.034
21. Adamsky A, Kol A, Kreisel T et al (2018) Astrocytic activation generates de novo neuronal potentiation and memory enhancement. Cell 174(1):59–71.e14. https://doi.org/10.1016/j.cell.2018.05.002
22. Agulhon C, Boyt KM, Xie AX et al (2013) Modulation of the autonomic nervous system and behaviour by acute glial cell Gq protein-coupled receptor activation in vivo. J Physiol 591(22):5599–5609. https://doi.org/10.1113/jphysiol.2013.261289
23. Martin-Fernandez M, Jamison S, Robin LM et al (2017) Synapse-specific astrocyte gating of amygdala-related behavior. Nat Neurosci 20:1540–1548. https://doi.org/10.1038/nn.4649
24. Oliveira JF, Sardinha VM, Guerra-Gomes S et al (2015) Do stars govern our actions? Astrocyte involvement in rodent behavior. Trends Neurosci 38(9):535–549. https://doi.org/10.1016/j.tins.2015.07.006
25. Ting JT, Lee BR, Chong P et al (2018) Preparation of acute brain slices using an optimized n-methyl-D-glucamine protective recovery method. J Vis Exp 132. https://doi.org/10.3791/53825
26. Allen BD, Singer AC, Boyden ES (2015) Principles of designing interpretable optogenetic behavior experiments. Learn Mem 22 (4):232–238. https://doi.org/10.1101/lm.038026.114

Chapter 4

Near-infrared Deep Brain Stimulation in Living Mice

Shuo Chen

Abstract

Optogenetics has revolutionized the experimental interrogation of neural circuits in the past decade and holds potential for the treatment of neurological disorders. However, optogenetic stimulation of deep brain neurons requires the insertion of invasive optical fibers because the activating blue-green light cannot penetrate deep inside brain tissue. Here we describe a minimally invasive technique for the stimulation of deep brain neurons by transcranial near-infrared light (NIR), where upconversion nanoparticles (UCNPs) are used as optogenetic actuators to locally convert NIR into visible light. We detail the protocol to use locally injected UCNPs to stimulate dopamine neurons in the ventral tegmental area (VTA) of anesthetized mice by transcranial NIR.

Key words Optogenetics, Near-infrared, Deep brain stimulation, Upconversion, Nanoparticles, Transcranial

1 Introduction

Optogenetics, a recently developed optical approach to modulate the activity of neurons, harnesses genetically encoded light-gated ion channels, the rhodopsins, to achieve unprecedented precision in stimulating target neurons [1]. While having revolutionized the experimental interrogation of neural systems, optogenetics has hitherto required the insertion of invasive optical fibers for deep brain applications, because the activating blue-green light is strongly scattered and absorbed by endogenous chromophores in brain tissue [2]. To make optogenetics noninvasive, an obvious option is to use near-infrared light (NIR, 650–1350 nm), which can efficiently penetrate biological tissue and reach deep brain regions [3]. However, the development of NIR-responsive rhodopsin variants has proved difficult: the optimal activation wavelengths of recently developed red-shifted rhodopsins all fall short of 650 nm [3–8].

We developed a novel approach to minimally invasive NIR optogenetics, in which tissue-penetrating NIR light is locally

Dominik Niopek (ed.), *Photoswitching Proteins: Methods and Protocols*, Methods in Molecular Biology, vol. 2173, https://doi.org/10.1007/978-1-0716-0755-8_4,

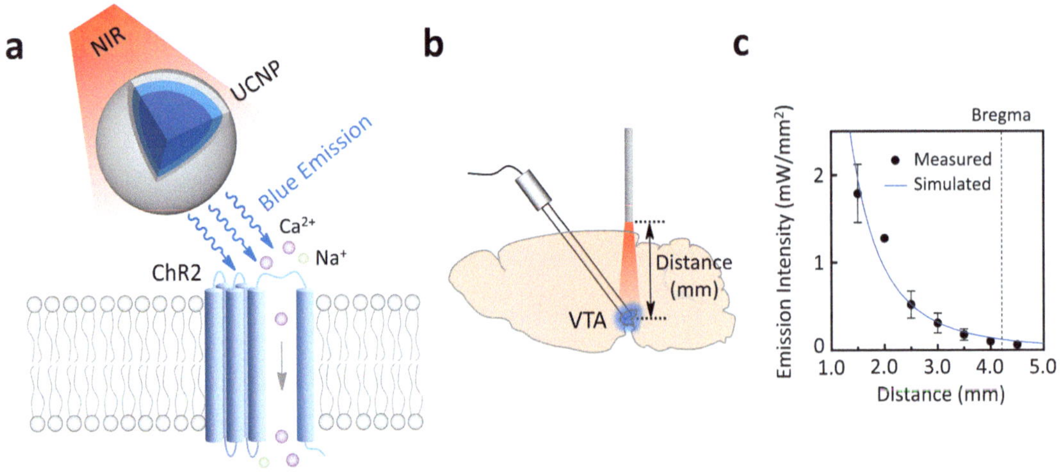

Fig. 1 UCNP-mediated NIR upconversion optogenetics for deep brain stimulation. (**a**) Schematic principle of UCNP-mediated NIR upconversion optogenetics. (**b**) Scheme of in vivo fiber photometry for measuring UCNP-mediated NIR upconversion in deep brain tissue. The tip of an optic fiber transmitting NIR excitation light is positioned at various distances from the VTA where UCNPs are injected, and their emission is recorded by a second optic fiber. (**c**) Measured ($n = 4$ mice) and simulated intensity of upconversion emission at the VTA as a function of the distance from the 980-nm NIR irradiation source (25-ms pulses at 20 Hz, 2.0-W peak power). Data are presented as mean ± s.e.m

converted to visible light in deep brain to activate rhodopsin-expressing neurons (Fig. 1) [9]. To achieve this requires an optically unique material, the upconversion nanoparticle (UCNP), which can convert low-energy NIR photons to high-energy visible emission [10, 11]. UCNPs have an upconversion efficiency orders of magnitude greater than that of multiphoton processes. As a result, a continuous-wave NIR laser diode of low power can result in intense upconversion emission by UCNPs.

To optimize their biocompatibility and long-term stability, we coated UCNPs with silica capable of chemically stabilizing the nanoparticles and preventing direct contact of their lanthanide-doped core with the tissue [9]. The resulting monodispersed blue-emitting UCNPs ($NaYF_4$:Yb/Tm@SiO_2) of ~90-nm diameter showed both minimum cytotoxicity and long-term stability: one month after injection UCNPs still remained at the target site in the brain [9].

We used UCNPs to realize optogenetic stimulation of the ventral tegmental area (VTA) of the mouse brain by transcranial NIR delivery, because the VTA has significant medial implications: it is a well-established node in the brain's reward system and the dysregulation of dopamine (DA) release by VTA neurons is casually linked to many neurological disorders, such as major depression [12]. We expressed channelrhodposin-2 (ChR2) [13] in the VTA of tyrosine hydroxylase (TH)-driven Cre recombinase (TH-Cre) transgenic mice and injected blue-emitting UCNPs to the same site (Figs. 2 and 3). Before delivering NIR transcranially, we evaluated

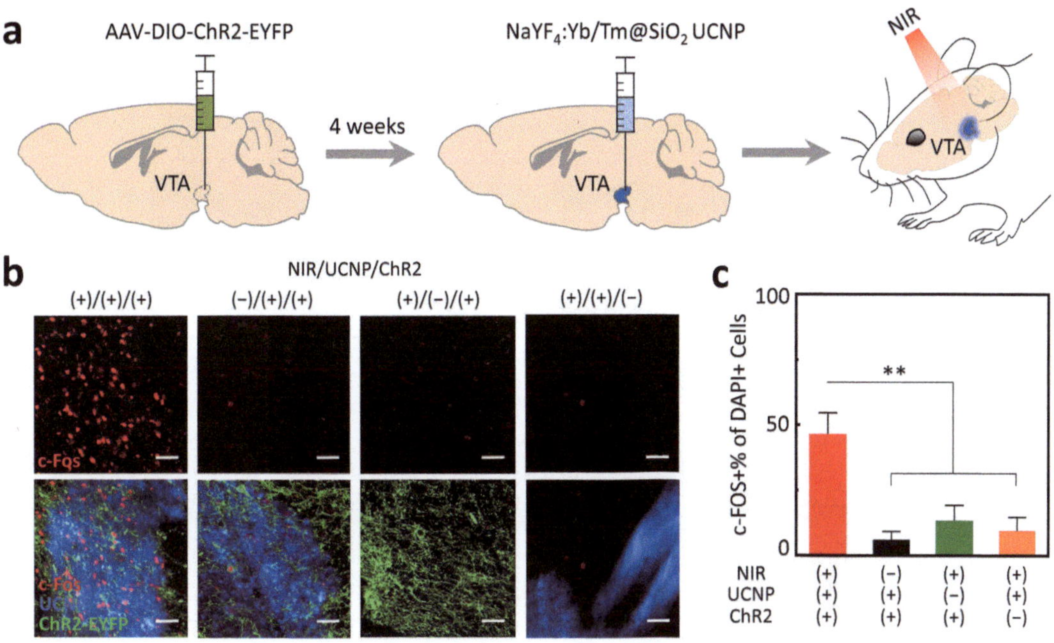

Fig. 2 Transcranial NIR stimulation of VTA dopamine neurons in vivo. (**a**) In vivo experimental scheme for transcranial NIR stimulation of the VTA in anesthetized mice. (**b**) Confocal images of the VTA after transcranial NIR stimulation under different conditions. Extensive NIR-driven c-Fos (red) expression was observed only in the presence of both UCNPs (blue) and ChR2 expression (labeled with EYFP, green). Scale bars: 100 μm. (**c**) Percentage of c-Fos–positive neurons within cell population indicated by DAPI, corresponding to the four conditions presented in (**b**) ($n = 3$ mice each, $P < 0.01$)

the irradiation safety by in vivo temperature measurement (Fig. 4) and examined the upconversion emission intensity by in vivo fiber photometry (Fig. 1) [14]. The activation of the VTA by transcranial NIR was confirmed by immunohistochemistry to map the activated neurons as well as in vivo fast scan cyclic voltammetry to examine DA release (Figs. 2 and 3). These experiments established UCNP-mediated NIR optogenetics as a viable minimally invasive method to modulate the activity of deep brain structures with high spatio-temporal and cell-type specificity [9]. Here we detail the protocols of these experiments.

2 Materials

2.1 Subjects and Reagents

1. TH-Cre mouse line (European Mutant Mouse Archives, B6.129X1-Thtm1(cre)Te/Kieg, Stock#00254). Mice were aged between 1.5 and 6 months prior to the commencement of experiments and were maintained on a 12-h light–dark cycle with ad libitum access to food and water.
2. $NaYF_4$:Yb/Tm@SiO_2 UCNP solution (200 mg/mL).

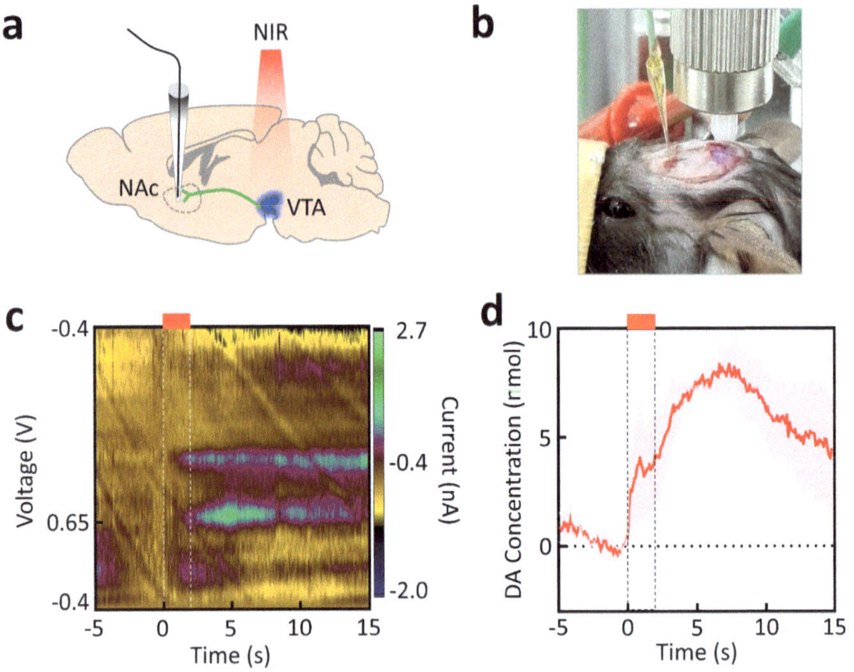

Fig. 3 In vivo fast-scan cyclic voltammetry to measure dopamine transients in ventral striatum during NIR stimulation of the VTA. (**a**) Experimental scheme. (**b**) A typical experimental setup. (**c**) A trace of background-subtracted current measured in the ventral striatum of a nomifensine-pretreated mouse in response to transcranial NIR stimulation of the VTA (15-ms pulses at 20 Hz, 700-mW peak power). Vertical dashed lines marked by a horizontal orange line in between indicate the start and end of 2-s transcranial NIR stimulation. (**d**) Transient dopamine concentration in ventral striatum in response to transcranial VTA stimulation. Significant dopamine release temporally locked to NIR stimulation was detected

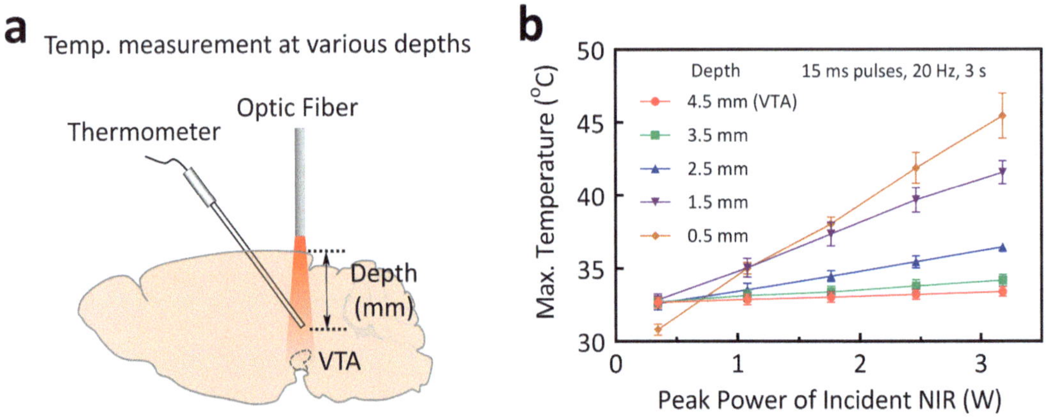

Fig. 4 Evaluation of NIR heating effect on brain tissue at various depths. (**a**) Schematic illustration of the measurement of temperature change at various depths in the brain upon transcranial 980 nm NIR irradiation. (**b**) Local temperature change as a function of the depth in brain under transcranial NIR irradiation at various intensities. Only the maximum temperature recorded was presented under each condition. Data are presented as mean $\pm$ s.e.m. ($n = 3$ for all experiments)

3. Adeno-associated virus: AAV.EF1a.DIO.hChR2(H134R). EYFP (3×10^8 vg).
4. Avertin.
5. Isoflurane.
6. Saline.
7. Nomifensine (dopamine transporter blocker, Sigma, St. Louis MO).

2.2 Stereotactic Injection

1. Stereotaxic frame (Narishige, Tokyo).
2. 10 μL Hamilton microsyringe (701LT, Hamilton).
3. Beveled 33 gauge needle (NF33BL, WPI).
4. Microsyringe pump (UMP3, WPI).
5. Pump controller (Micro4, WPI).

2.3 In Vivo Fiber Photometry

1. 980 nm laser resource (MDL-980, Changchun New Industries CNI Laser) under control of a function generator.
2. Optic fibers (*see* **Note 1**, 200 μm in diameter, one for NIR-transmitting, the other for emission collection).
3. Fiber photometric system (Olympus Engineering), including a dichroic mirror (DM515YFP, Olympus), a photomultiplier tube (H10722-210, Hamamatsu Photonics), a data acquisition module (NI USB-6211, National Instruments) with a custom-made LabVIEW program (National Instruments).
4. Blue light emitting diode (LED) irradiating 0.13 mW/mm^2 light for calibrating the detected blue light intensity.

2.4 In Vivo NIR Stimulation of VTA and Fast-Scan Cyclic Voltammetry

1. Stereotaxic frame (Narishige, Tokyo).
2. 980 nm laser resource (MDL-980, Changchun New Industries CNI Laser) under control of a function generator.
3. Optic fiber for light-transmitting (200 μm in diameter, *see* **Note 1**).
4. Carbon fiber electrode.
5. Ag wire (auxiliary electrode).
6. Ag/AgCl wire (reference electrode).
7. Counter electrode-grounded type potentiostat (Model HECS-972E, Huso Electrochemical Systems, Kawasaki, Japan).
8. Control/recording system (TH-1, ESA Biosciences, Inc., MA, USA) with two multifunction boards (NI-PCI-6221, National Instruments, TX, USA).

Table 1
List of antibodies

Primary antibody	Secondary antibody
Rabbit anti-c-Fos (1:1000, SYSY)	Donkey anti-rabbit, Alexa Fluor 594 (1:500, Life Technologies)
Rabbit anti-GFAP (1:1000, abcam)	Donkey anti-rabbit, Alexa Fluor 594 (1:500, Life Technologies)
Rabbit anti-Iba1 (1:2000, Wako)	Donkey anti-rabbit, Alexa Fluor 594 (1:500, Life Technologies)

2.5 Immunohistochemistry, Cytotoxicity Assay, and Evaluation of NIR Heating Effect

1. 0.1 M PBS.
2. 4% PFA in 0.1 M PBS.
3. PBST: PBS containing 0.3% Triton X-100.
4. PBST supplemented with 3% normal donkey serum.
5. Antibodies (Table 1).
6. Mounting medium for fluorescence with DAPI (4′,6-diamidino-2-phenylindole, Vectashield).
7. Microscope slides, coverslips, and nail polish.
8. Optic fiber (200 μm in diameter, *see* **Note 1**).
9. Olympus IX81 confocal laser scanning microscope with a 20× objective.
10. Leica DM6000B epifluorescence microscope with a 20× objective.
11. 980 nm laser resource (MDL-980, Changchun New Industries CNI Laser) under control of a function generator.
12. Ultra-fine sheathed thermocouple sensor (K-type, Aeropak Nano, Okazaki, outer diameter: 150 μm, sheath length: 50 mm).
13. K-type thermocouple thermometer (AD-5601A, A&D).
14. Miniature thermocouple connector (SMPW-KJ-M-ROHS, Omega).

3 Methods

3.1 Stereotactic Injection

1. Anaesthetize the mouse using 500 mg/kg avertin.
2. Shave the fur on the skull, clean the skin with 70% ethanol and place the animal in the stereotaxic apparatus.
3. Measure the position of the x and y coordinates of bregma under a dissecting microscope, and calculate the coordinates of the target injection area (for VTA, AP: −3.5 mm, ML: +0.4 mm, DV: −4.2 mm).

4. Perform craniotomy above the target area (*see* **Note 2**). Keep both the skull and exposed dura moist with PBS.
5. Fill the 10 μL Hamilton microsyringe with a beveled 33 gauge needle (*see* **Note 3**) with AAVs or UCNPs, and assemble the microsyringe to the stereotaxic frame.
6. Move the needle to the injection coordinates, slowly lower it to the target depth (*see* **Note 4**), and remain it in place for 10 min before the beginning of the injection.
7. Use the microsyringe pump and its controller to control the speed of the injection. AAVs were typically injected at a speed of 100 nL/min, while UCNPs at 30 nL/min (*see* **Note 4**).
8. Remove the needle 5 min after infusion was complete (*see* **Note 4**).
9. Clean the injection site with moist cotton swabs, suture the skin, and apply antibiotic ointment to the wound.
10. Inject sterile PBS (30 mL/kg body weight) to avoid dehydration of the animal. Keep the animal warm until it fully recovers.

3.2 Fiber Photometry for Confirming In Vivo Upconversion Emission

1. Anesthetize the mouse and inject 900 nL of 200 mg/mL $NaYF_4$:Yb/Tm@SiO_2 UCNPs into the VTA using the protocol in Subheading 3.1.
2. Clean the injection site and the surrounding area on the skull with moist cotton swabs, and maintain the anesthesia with 0.5–2% isoflurane inhalation.
3. Insert an optic fiber, connected to the fiber photometric system, to the VTA for the collection of upconversion emission (Fig. 1b) (*see* **Note 5**).
4. Place the tip of a NIR-transmitting optic fiber 2 mm above the skull at an angle to target the NIR laser from the laser generator to the UCNP injection site in the VTA (Fig. 1b) (*see* **Note 6**).
5. Generate pulsed NIR laser (980 nm) by the laser source under the control of a function generator (*see* **Note** 7).
6. Use the fiber photometric system to detect upconversion emission from the UCNPs. Record the data by a LabVIEW program at a sampling frequency of 1 kHz.
7. Use a blue LED irradiating 0.13 mW/mm^2 light to calibrate the detected blue light intensity.
8. Sacrifice the animal after the experiment (*see* **Note 8**).

3.3 In Vivo NIR Stimulation of VTA Dopamine Neurons

1. Inject 300 nL of AAV.EF1a.DIO.hChR2(H134R).EYFP (3×10^8 vg) bilaterally into the VTA of the TH-Cre mouse using the protocol in Subheading 3.1 (Fig. 2a).
2. Four weeks later inject 900 nL of 200 mg/mL $NaYF_4$:Yb/Tm@SiO_2 UCNPs bilaterally into the VTA (Fig. 2a).

3. Clean the injection site and the surrounding area on the skull with moist cotton swabs, and maintain the anesthesia with 0.5–2% isoflurane inhalation.
4. Place the tip of a NIR-transmitting optic fiber 2 mm above the skull at an angle to target the NIR laser from the laser generator to the UCNP injection site in the VTA (Fig. 2a).
5. Deliver 15 ms pulses of transcranial NIR irradiation at 20 Hz and 3.0 W (peak power) for 3 s every 3 min over the course of 30 min (average power 15 mW).
6. Sacrifice the mouse 90 min later. Transcardially perfuse the mouse and postfix the brain with 4% PFA in 0.1 M PBS. Follow the protocol in Subheading 3.5 for c-Fos mapping of NIR-activated neurons (Fig. 2b, c).

3.4 In Vivo Fast-Scan Cyclic Voltammetry

1. Inject 300 nL of AAV.EF1a.DIO.hChR2(H134R).EYFP (3×10^8 vg) bilaterally into the VTA of the TH-Cre mouse using the protocol in Subheading 3.1.
2. Four weeks later inject 900 nL of 200 mg/mL $NaYF_4$:Yb/Tm@SiO_2 UCNPs bilaterally into the VTA (Fig. 3a).
3. Clean the injection site and the surrounding area on the skull with moist cotton swabs, and maintain the anesthesia with 0.5–2% isoflurane inhalation.
4. Perform craniotomy above the ventral striatum (AP: +1.3 mm, ML: +0.8 mm, DV: −4.2 mm).
5. Lower a carbon fiber electrode into the ventral striatum.
6. Insert one Ag wire (auxiliary electrode) and one Ag/AgCl wire (reference electrode) into the cortex of the contralateral hemisphere.
7. Administer 10 mg/kg nomifensine to the animals via intraperitoneal injection (*see* **Note 9**).
8. Place the tip of a NIR-transmitting optic fiber 2 mm above the skull at an angle to target the NIR laser from the laser generator to the UCNP injection site in the VTA.
9. Record fast-scan cyclic voltammetry with the potentiostat. Each recording session is 20 s, with NIR stimulation (15 ms pulses at 20 Hz over 2 s, 700 mW peak power) applied at second five. Set the amplifier gain as 500 nA/V and the low-pass filter time-constant as 0.2 ms. Apply voltage sweeps from −0.4 V to 1.3 V and back vs Ag/AgCl reference to the carbon fiber electrode. Repeat this triangle-positive waveform at 10 Hz (Fig. 3c).
10. Voltammetric data is first output in the software as background-subtracted currents relative to the mean of 10 waveform applications 1 s before stimulation. The background-subtracted currents

are analyzed against calibration templates of known values of dopamine, pH, and adenosine using principal component regression to isolate the dopamine signal from other electroactive factors and convert it to concentration values (Fig. 3d).

11. Sacrifice the animal after the experiment (*see* **Note 8**).

3.5 Immunohistochemistry

1. Transcardially perfuse the mouse with 4% PFA in 0.1 M PBS.
2. Postfix the brain in 4% PFA and prepare 50 μm-thick vibratome sections.
3. After 3 × 10 min PBS rinses, block the sections in PBST with 3% normal donkey serum for 1.5 h, and incubate in primary antibody (Table 1) at 4 °C for 20 h.
4. After 3 × 10 min washes in PBST, incubate the sections with secondary antibody (Table 1) for 2 h at room temperature followed by 3 × 10 min washes in PBST.
5. Mount the sections onto microscope slides and allow to air dry. Next, pipet the mounting medium for fluorescence with DAPI, place glass coverslip on top, and seal edges with nail polish.
6. Acquire confocal fluorescence images on the Olympus microscope with a 20× objective. Acquire luminescent images of UCNPs on the Leica microscope with a 20× objective under 25 W/mm^2 980 nm NIR excitation (*see* **Note 10**). Merge the luminescent image with confocal fluorescence images of same sections (Fig. 2b).
7. To analyze the overlap between fluorescent markers (e.g., anti-cFos and DAPI, Iba1 and DAPI etc.), use a *z*-stack function to montage 5 optical stacks (1 μm each, step size 4 μm). Digitally combine filtered fluorescent images to produce composite images. Count the number of cells with overlapping signals in a region of interest (ROI) of 200 μm with ImageJ. Select three ROIs on each section and then average (Fig. 2c).

3.6 Cytotoxicity Assay

1. Inject 900 nL of 200 mg/mL UCNPs into mouse VTA using the protocol in Subheading 3.1.
2. After various periods (e.g., 1 day, 2 weeks, and 4 weeks) following the surgery, transcardially perfuse the mouse with 4% PFA in 0.1 M PBS.
3. Perform immunohistochemistry with Iba1 and GFAP antibodies (Table 1), using the protocol in Subheading 3.5. Acquire and merge confocal fluorescence images and luminescent images of UCNPs.
4. Count the numbers of Iba1+ microglia and GFAP+ astrocytes in a 200 μm vicinity of the UCNP injection site and calculate their percentages in cell population indicated by DAPI.

3.7 Evaluation of NIR Heating Effect

1. Anaesthetize the mouse using 500 mg/kg avertin and place it in the stereotaxic frame. Maintain the anesthesia with 0.5–2% isoflurane inhalation.
2. Place the tip of a NIR-transmitting optic fiber 2 mm above the skull at an angle to target the NIR laser from the laser generator to the VTA (Fig. 4a).
3. Insert the ultrafine sheathed thermocouple sensor, which is connected to a K-type thermocouple thermometer via a miniature thermocouple connector, into the target brain region or at the brain surface (*see* **Note 11**).
4. Measure the local temperature in the brain while delivering transcranial NIR pulses. For measuring brain temperature change as a function of depth, elevate the tip of the thermocouple sensor stepwise in a series of measurements (Fig. 4b).
5. Sacrifice the animal after the experiment (*see* **Note 8**).

4 Notes

1. In order to achieve high light transmission, the tips of optic fibers have to be polished and the transmission rate should be examined.
2. Thin the skull over the target area using a handheld drill. Stop when the bone is very thin. Take a small needle (27 G) and carefully perforate the edges of the craniotomy; next, with the tip of the needle, flip up a piece of the thinned bone and then carefully remove it with fine forceps.
3. The UCNP solution used in this protocol has a high concentration. Therefore, once left in air for minutes, dried UCNPs would clog up the needle. For this reason, needles with smaller diameters than 33 gauge are not recommended.
4. Before lowering the needle or after raising it out of the brain, confirm whether the virus or UCNPs can flow out under the microscope while controlling the microsyringe pump.
5. The optic fiber for emission detection is suggested to be inserted from an angle rather than from a perpendicular direction (Fig. 1b). This requires a careful design of the craniotomy. The insertion should be sufficiently slow to avoid bleeding.
6. Maintain a thin water layer above the skull under the tip of the NIR-transmitting optic fiber. This is very important to avoid overheating of the bone, because dry bone accumulates heat easily.
7. After insertion of the optic fiber, wait for 10 min to stabilize the implantation before NIR delivery and emission measurement.

8. If necessary, transcardially perfuse the mouse and postfix the brain with 4% PFA in 0.1 M PBS. Prepare vibratome sections to examine the track of the implantation or the injection site of UCNPs.
9. This step may be skipped. The application of the dopamine transporter blocker is used to enhance the detectable dopamine signal.
10. Ideally the microscope can be customized to accommodate a NIR excitation channel. An alternative way is to place a NIR-transmitting fiber nearly perpendicular above the sample yet out of the microscope light path for irradiating NIR light to the sample, so that emission could be acquired by the microscope.
11. The thermocouple sensor for temperature detection is suggested to be inserted from an angle rather than from a perpendicular direction (Fig. 4a). This requires a careful design of the craniotomy. The insertion should be sufficiently slow to avoid bleeding.

Acknowledgments

This work was supported by Human Frontier Science Program Postdoctoral Fellowship (LT 000579/201) (S.C.), JSPS (Japan Society for the Promotion of Science) Postdoctoral Fellowship (16F16386) (S.C.), RIKEN Special Postdoctoral Researchers Program (S.C.), Grant-in-Aid for Young Scientists from MEXT (the Ministry of Education, Culture, Sports, Science and Technology of Japan) (16K18373, 18K14857) (S.C.), RIKEN Incentive Research Project Grant for Individual Germinating Research (S.C.), Narishige Neuroscience Research Foundation Grant (S.C.), and Nakatani Foundation Grant Program (S.C.).

References

1. Fenno L, Yizhar O, Deisseroth K (2011) The development and application of optogenetics. Annu Rev Neurosci 34:389–412
2. Lin JY, Knutsen PM, Muller A, Kleinfeld D, Tsien RY (2013) ReaChR: a red-shifted variant of channelrhodopsin enables deep transcranial optogenetic excitation. Nat Neurosci 16 (10):1499–1508
3. Hamblin MR (2016) Shining light on the head: photobiomodulation for brain disorders. BBA Clin 6:113–124
4. Chuong AS et al (2014) Noninvasive optical inhibition with a red-shifted microbial rhodopsin. Nat Neurosci 17(8):1123–1129
5. Klapoetke NC et al (2014) Independent optical excitation of distinct neural populations. Nat Methods 11(3):338–346
6. Rajasethupathy P et al (2015) Projections from neocortex mediate top-down control of memory retrieval. Nature 526(7575):653–659
7. Yizhar O, Fenno LE, Davidson TJ, Mogri M, Deisseroth K (2011) Optogenetics in neural systems. Neuron 71(1):9–34
8. Zhang F et al (2008) Red-shifted optogenetic excitation: a tool for fast neural control derived from Volvox carteri. Nat Neurosci 11 (6):631–633

9. Chen S et al (2018) Near-infrared deep brain stimulation via upconversion nanoparticle-mediated optogenetics. Science 359 (6376):679–684
10. Chen G, Qiu H, Prasad PN, Chen X (2014) Upconversion nanoparticles: design, nanochemistry, and applications in theranostics. Chem Rev 114(10):5161–5214
11. Zhou B, Shi B, Jin D, Liu X (2015) Controlling upconversion nanocrystals for emerging applications. Nat Nanotechnol 10 (11):924–936
12. Wanat MJ, Willuhn I, Clark JJ, Phillips PE (2009) Phasic dopamine release in appetitive behaviors and drug addiction. Curr Drug Abuse Rev 2(2):195–213
13. Boyden ES, Zhang F, Bamberg E, Nagel G, Deisseroth K (2005) Millisecond-timescale, genetically targeted optical control of neural activity. Nat Neurosci 8(9):1263–1268
14. Gunaydin LA et al (2014) Natural neural projection dynamics underlying social behavior. Cell 157(7):1535–1551

Chapter 5

Lab-Scale Production of Recombinant Adeno-Associated Viruses (AAV) for Expression of Optogenetic Elements

Janina Haar, Chiara Krämer, and Dirk Grimm

Abstract

Optogenetics, that is, the use of photoswitchable/−activatable moieties to precisely control or monitor the activity of cells and genes at unprecedented spatiotemporal resolution, holds tremendous promise for a wide array of applications in fundamental and clinical research. To fully realize and harness this potential, the availability of gene transfer vehicles ("vectors") that are easily produced and that allow to deliver the essential components to desired target cells in an efficient manner is key. For in vivo applications, it is, moreover, important that these vectors exhibit a high degree of cell specificity in order to reduce the risk of adverse side effects in off-targets and to minimize manufacturing costs. Here, we describe a set of basic protocols for the cloning, production, purification, and quality control of a particular vector that can fulfill all these requirements, that is, recombinant adeno-associated viruses (AAV). The latter are very attractive owing to their apathogenicity, their compatibility with the lowest biosafety level 1 conditions, their occurrence in multiple natural variants with distinct properties, and their exceptional amenability to engineering of the viral capsid and genome. The specific procedures reported here complement alternative protocols for AAV production described by others and us before, and, together, should enable any laboratory to generate these vectors on a small-to-medium scale for ex vivo or in vivo expression of optogenetic elements.

Key words AAV, Adeno-associated virus, Cesium chloride, CNS, CsCl, Gene expression, Optogenetics, Transduction, Ultracentrifugation

1 Introduction

Over the last two decades, the field of optogenetic research has experienced significant and rapid progress that was essentially driven by, and concurrently resulted in, an ever-increasing experimental tool kit [1, 2]. A seminal player is photoswitchable or photoactivatable proteins, whose advent has opened up entirely novel options to investigate cellular circuits and to study the role of single cellular subtypes and genes in a complex organ or in a whole organism. In parallel to the classical microbial opsins, non-microbial counterparts are increasingly being exploited as powerful and flexible tools for spatiotemporal, noninvasive, and reversible

Dominik Niopek (ed.), *Photoswitching Proteins: Methods and Protocols*, Methods in Molecular Biology, vol. 2173,
https://doi.org/10.1007/978-1-0716-0755-8_5,

control of gene expression or of protein localization or activity ex vivo or in vivo, complementing the existing arsenal of biological, chemical, or physical modalities. For instance, as recently jointly reported by the Niopek laboratory and us, photoswitchable protein domains such as LOV2 (Light-Oxygen-Voltage domain 2) can be harnessed to gain exogenous control over the activity of anti-CRISPR (Acr) proteins [3]. This, in turn, allows for the blue light-mediated regulation of CRISPR/Cas9 activity in living cells, thus offering a flurry of innovative new concepts to control basic or therapeutic (epi)genetic editing in space and time. Of note, this application merely exemplifies the vast potential of this transformative technology, and the reader will find numerous other exciting uses in the literature (see also below) and in the various chapters in this issue. Collectively, it is clear that the repertoire of photoswitching proteins has revolutionized our options to unravel fundamental biology or disease progression and continues to accelerate the development of new therapeutic modalities.

However, to fully realize this enormous potential and promise, it will be pivotal to combine these elements with gene transfer vectors permitting their in vivo delivery to, and expression in, a desired target cell, ideally in an efficient and specific manner that limits vector dissemination to unwanted off-target tissues and thus also restricts adverse side effects. To this end, vectors based on recombinant adeno-associated viruses (AAV) have emerged as a most promising and most versatile modality for in vivo delivery of foreign DNA including optogenetic elements. AAV are small, non-pathogenic viruses that are widely used as vectors in human gene therapy owing to their inherent safety and their amenability to genetic engineering of the viral genome and the tropism-determining capsid [4–8]. Further boosting their potential and adding to their appeal as vectors is the availability of a large variety of natural AAV isolates that differ in cell specificity and reactivity with preexisting anti-AAV antibodies in the human population. As a whole, the broad pool of natural AAV variants, combined with the portfolio of diverse technologies for capsid modification and repurposing, nowadays offers manifold opportunities to isolate designer vectors that are tailored for a given application, including targeted delivery of optogenetic elements and related expression cassettes.

In this respect, a particularly well-suited target for AAV vector-mediated gene transfer involving photoswitchable proteins is the eye, due to its rather immunoprivileged nature and its ease of access, in turn providing multiple delivery options such as intravitreal or subretinal vector injection. Best exemplifying the enormous potential of using AAV for gene therapy in the eye is the FDA (Food and Drug Administration) approval and commercialization of Luxturna, that is, a recombinant AAV serotype 2 (AAV2) vector for the treatment of Leber's congenital amaurosis and other RPE65-dependent eye disorders [9]. It is thus not surprising that a wealth of reports have already provided proofs-of-concept for the power of combining AAV vectors and optogenetic elements in various small or large animal models and in different cell types in the eye [10–17].

Besides the eye, many other cell and tissue types are also highly amenable to AAV-mediated gene transfer and thus represent good targets for expression of photoswitchable proteins, as already demonstrated, for instance, in neurons or other cells in the central nervous system (CNS) [18–29]. Additionally, others have exemplified the usefulness of optogenetic AAV expression vectors in further tissues including the heart, the ear, the respiratory network or the liver [19, 30–36]. Finally, we again refer the reader to the other chapters in this issue for a comprehensive overview of additional targets and applications of optogenetic elements, including in an AAV vector context.

Here, our aim is to provide a basic protocol for the generation and production of AAV vectors for ex vivo or in vivo expression of photoswitchable proteins, comprising guidelines for the purification of these vectors by cesium chloride (CsCl) density gradient centrifugation and for quality control by silver staining of protein gels. We note that the protocols described in the following are useful for the production and application of AAV vectors on the scale of a regular laboratory but are neither intended nor suitable for clinical-grade AAV manufacturing. In addition, we encourage readers to thoroughly study the available literature related to their preferred experimental setting, such as specific target cell or tissue types or routes of administration, and then judiciously select the best-suited natural or synthetic AAV capsid variant. Importantly, the protocols provided below were purposely selected to be compatible with all AAV capsid variants, as neither CsCl purification nor silver staining is specific for any particular AAV sequence, unlike for example affinity chromatography purification or detection of capsid proteins by Western blotting.

Last but not least, it is also important to point out that our protocols are not intended to be exclusive, and that the literature including this book series provides numerous permutations that readers may also want to consider and either use to replace some of the steps suggested here or combine with our protocols. As two examples, we have previously reported a procedure for iodixanol-based purification of AAV vectors that can substitute for the present CsCl protocol [37], while the Vandenberghe laboratory has recently published a detailed protocol for the use of droplet digital (dd)PCR as a means for AAV vector titration [38], which perfectly complements the silver staining procedure described below.

2 Materials

2.1 Cloning of AAV Vector Plasmids for Subsequent Packaging

1. AAV transgene expression vector with AAV inverted terminal repeats (ITRs; a typical example from our laboratory with a modular design is shown in Fig. 1, plasmid B) (*see* **Note 1**).
2. Template containing transgene DNA and/or promoter sequence (1 ng/μL dilution).
3. Phusion Hot Start II High-Fidelity DNA polymerase kit (Thermo Fisher Scientific).
4. 10 mM dNTPs.
5. Eppendorf MasterCycler or equivalent.
6. Restriction enzymes (for the example used here: EcoRI-HF, NotI-HF, ClaI; all from New England Biolabs [NEB]).
7. CutSmart® buffer (10×, NEB).
8. Thermomixer at 37 °C and at 23 °C (or 16 °C).
9. Antarctic phosphatase (NEB).
10. QIAquick PCR purification kit (Qiagen).
11. QIAquick gel extraction kit (Qiagen).
12. T4 DNA ligase (NEB).
13. T4 DNA ligase reaction buffer (10×, NEB).
14. Electrocompetent *E. coli* DH5α (e.g., ElectroMAX DH5α-E; Thermo Fisher Scientific).

2.2 Transient Transfection

AAV vector production is routinely accomplished through triple-transfection of cultured human embryonic kidney cells (HEK293 or HEK293T) with three plasmids encoding (1) all AAV genes (Fig. 1, construct A, note that the *aap* open reading frame is not shown as it overlaps with *cap*), (2) the transgene cassette including promoter and polyadenylation (pA) site (Fig. 1, construct B), or (3) all necessary "helper" genes for AAV production from Adenovirus (wild-type AAV depends on coinfection with Adenovirus or other viruses for its own propagation) (Fig. 1, construct C). For a detailed description of the reagents used for transfection of these three plasmids and for the subsequent cell harvest, we refer the reader to our recent publication by Fakhiri et al. in this book series [37].

2.3 CsCl Gradient Centrifugation

To simplify buffer preparation, stock solutions of reagents 4 to 8 are used for some of the buffers listed below.

1. Bandelin Sonorex Sonication bath or equivalent.
2. Water bath at 37 °C.
3. Ice box.

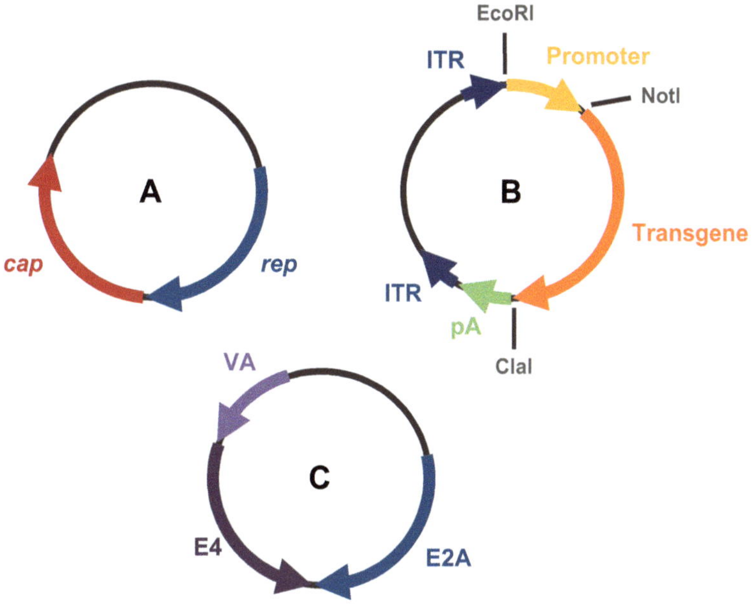

Fig. 1 Plasmids for AAV vector production. Depicted are the three basic plasmids needed to produce AAV vectors through triple-transfection of HEK293(T) cells: (**a**) an AAV helper plasmid encoding replication (*rep* gene), capsid (*cap* gene), and AAP (*aap* open reading frame, overlapping with *cap*; not shown) proteins, (**b**) an AAV vector plasmid encoding a transgene expression cassette (or, more generally, a foreign DNA) flanked by AAV ITRs and containing various restriction sites to simplify cloning, and (**c**) an adenoviral helper plasmid encoding several RNAs (VA) and proteins (E2A, E4) that are essential for AAV vector production. *ITR* inverted terminal repeat, *pA* polyadenylation site

4. $MgCl_2$ (1 M): To make 1 L of a 1 M $MgCl_2$ solution, dissolve 203.3 g of $MgCl_2{\cdot}6H_2O$ in 1 L H_2O and sterile-filter through a bottle-top filter (0.22 μm pore size).
5. NaCl (5 M): To make 1 L of a 5 M NaCl solution, dissolve 292.2 g of NaCl in 1 L H_2O and sterile-filter through a bottle top filter (0.22 μm pore size).
6. $CaCl_2$ (1 M): To make 1 L of a 1 M $CaCl_2$ solution, dissolve 147.02 g of $CaCl_2{\cdot}2H_2O$ in 1 L H_2O and sterile-filter through a bottle-top filter (0.22 μm pore size).
7. EDTA·2Na (0.5 M), pH 8.0 (VWR).
8. UltraPure™ 1 M Tris–HCI, pH 8.0 (Invitrogen).
9. Benzonase buffer (50 mM Tris–HCl, 2 mM $MgCl_2$, 150 mM NaCl, pH 8.5): Dissolve 8.8 g NaCl in 900 mL double-distilled (dd)H_2O, then add 2 mL 1 M $MgCl_2$ and 50 mL 1 M Tris–HCl. Adjust pH to 8.5 with 1 M NaOH and fill up to 1 L with ddH_2O.
10. Benzonase nuclease (Merck, 250 U/μL).

11. Beckman Avanti® centrifuge or equivalent.
12. 40% PEG8000/1.915 M NaCl: Dissolve 400 g PEG8000 in 617 mL ddH_2O and 383 mL 5 M NaCl (*see* **Note 2**).
13. Na-HEPES resuspension buffer (50 mM HEPES, 0.15 M NaCl, 25 mM EDTA): Dissolve 11.92 g HEPES in 920 mL ddH_2O, then add 30 mL 5 M NaCl and 50 mL 0.5 M EDTA.
14. CsCl.
15. CsCl topping solution: 0.55 g/mL CsCl in Na-HEPES resuspension buffer (refractive index [RI] = 1.3710 at room temperature, determined with refractometer).
16. Exacta Optech RMI Refractometer or equivalent.
17. Ultracentrifuge plus Beckman 70Ti rotor or equivalent.
18. 21-gauge (G) needles.
19. OptiSeal™ polypropylene centrifugation tubes (Beckman Coulter, nominal capacity 29.9 mL).
20. 10× PBS: Dissolve 160 g NaCl, 4 g KCl, 36.1 g $Na_2HPO_4(\times 2H_2O)$ and 4.8 g KH_2PO_4 in 1.8 L ddH_2O. Control that pH is 7.4 (this should not need further adjustment) and fill up to 2 L with ddH_2O, then autoclave.
21. 15 mL and 50 mL Falcon tubes (BD Falcon).

2.4 AAV Dialysis and Concentration

1. Slide-A-Lyzer™ Dialysis Cassette (20,000 molecular weight cutoff [MWCO], 15 mL).
2. 1 L beaker
3. Magnetic stirrer.
4. Beckman Avanti® centrifuge or equivalent.
5. Amicon® Ultra Centrifugal Filters (10,000 MWCO, 15 mL).

2.5 AAV Quality Control by Silver Staining

1. MiniProtean TGX precast 4–15% gradient gel (Bio-Rad) (*see* **Note 3**).
2. Tris-glycine SDS (TGS) buffer (10×) (Bio-Rad).
3. Bio-Rad PAGE equipment.
4. PageRuler Plus Prestained Protein Ladder (Thermo Fisher Scientific).
5. Laemmli buffer (4×) (Bio-Rad) supplemented with 100 μL 2-mercaptoethanol per 900 μL buffer.
6. SilverQuest™ Silver Staining Kit (Invitrogen).
7. Glacial acetic acid (*see* **Note 4**).
8. Ethanol absolute.
9. ddH_2O.

3 Methods

3.1 Cloning of Transgenes into AAV Expression Vectors

As noted before (Subheading 2.2), a standard protocol for AAV vector production starts with a transient triple transfection of cultured cells with three plasmids, one of which is a scAAV or ssAAV (*see* **Note 1**) vector plasmid. In the latter, the transgene (e.g., a photoswitching protein or domain) is typically expressed from a ubiquitously active or a cell–/tissue-specific promoter. The reader will find many examples in the literature, and corresponding cloning templates can be obtained from plasmid repositories such as Addgene or from the authors.

Here, we will use one of our own ssAAV vector plasmids as a representative example of a highly modular AAV cloning template permitting rapid and simple swapping of the two most critical elements, that is, promoter and transgene, facilitated by flanking unique restriction sites. The following steps specifically refer to this plasmid and thus need to be adapted to alternative AAV vector templates.

1. To clone a transgene cDNA into the AAV vector plasmid, design primers for its amplification containing NotI (5′) or ClaI (3′) restriction sites within the primer overhangs (for details on primer design, *see* **Note 5**).
2. To PCR-amplify the desired open reading frame, set up a mix containing 10 μL of 5× Phusion HF buffer, 1 μL 10 mM dNTPs, 2.5 μL of forward primer, and 2.5 μL of reverse primer (both 10 μM), 0.5 μL Phusion Hot Start II DNA polymerase (2 U/μL), 32.5 μL nuclease-free H_2O, and 1 μL of diluted template DNA (1 ng/μL), resulting in a final volume of 50 μL. Run the PCR as specified in the Phusion Hot Start II High-Fidelity DNA polymerase kit manual (*see* **Note 6**).
3. Purify the PCR product using the QIAquick PCR purification kit according to the manufacturer's protocol.
4. Digest 1–5 μg of the AAV vector plasmid backbone and 0.5–1 μg of the PCR product by setting up two reactions with 1 μL NotI, 1 μL ClaI, and 5 μL 10× CutSmart buffer in a 50 μL total reaction volume. Incubate at 37 °C for 1 h (*see* **Note 7**).
5. The digested plasmid can be dephosphorylated to reduce the risk of religation during subsequent ligation. For this purpose, it is sufficient to add 1 μL of Antarctic phosphatase (5 U/μL) directly to the plasmid restriction mix and continue incubating for another hour at 37 °C.
6. Purify the vector backbone (4257 bp in the shown example) and the PCR product using the QIAquick gel extraction kit according to the manufacturer's protocol.

7. Ligate 100 ng of the vector backbone with the PCR product at a 1:3 molar ratio in a total volume of 20 μL containing 2 μL 10× T4 DNA ligase buffer and 1 μL T4 DNA ligase.
8. Incubate at 23 °C for 1 h or, to improve ligation efficiency, at 16 °C overnight. Additionally, set up a ligation with vector backbone only, to estimate the residual religation background (*see again* **Note 7**).
9. Electroporate 2 μL of the ligation mixture in electrocompetent *E. coli* DH5α according to the manufacturer's protocol (in the case of commercial cells) and grow colonies on LB-agar plates supplemented with ampicillin (or with the appropriate antibiotics, if another plasmid is used).

The principle described above can also be applied to exchange a promoter within an existing AAV expression vector. In the shown example, EcoRI and NotI restriction sites are used for promoter cloning and included in the primers (EcoRI in the forward and NotI in the reverse primer). For further information on the plasmids used for AAV vector production, *see* **Notes 8–10**.

3.2 Transient Transfection

As noted in Subheading 2.2, a detailed description of a standard protocol for triple-transfection and ensuing cell harvest is found in our recent publication in this book series [37].

3.3 CsCl Gradient Centrifugation

The following protocol provides instructions for purification of recombinant AAV vectors using CsCl density gradient ultracentrifugation that are suitable for in vivo applications (*see* **Note 11**). Typically, AAV lysates from 30 to 40 15 cm dishes of transfected cells are purified over one CsCl gradient. An overview of the main steps of this protocol including time lines is shown in Fig. 2.

1. Resuspend HEK293(T) cell pellet in 5 mL benzonase buffer per 10 transfected 15 cm dishes.

   Use 50 mL Falcon tubes for subsequent steps (*see* **Note 12**).
2. Freeze the cell suspension in liquid nitrogen for 5 min. Alternatively, use a mix of dry ice and ethanol.
3. Thaw the solution at 37 °C in a water bath for 10 min.
4. Repeat **steps 2** and **3** until five freeze–thaw cycles are completed (*see* **Note 13**).
5. Sonicate sample for 80 s in a sonication bath.
6. Add 75 U/mL of benzonase to remove residual plasmid DNA and contaminating genomic DNA from the cell lysate (*see* **Note 14**).
7. Incubate in a water bath at 37 °C for 1 h. Invert every 15 min to ensure homogeneous digestion. Do not vortex.

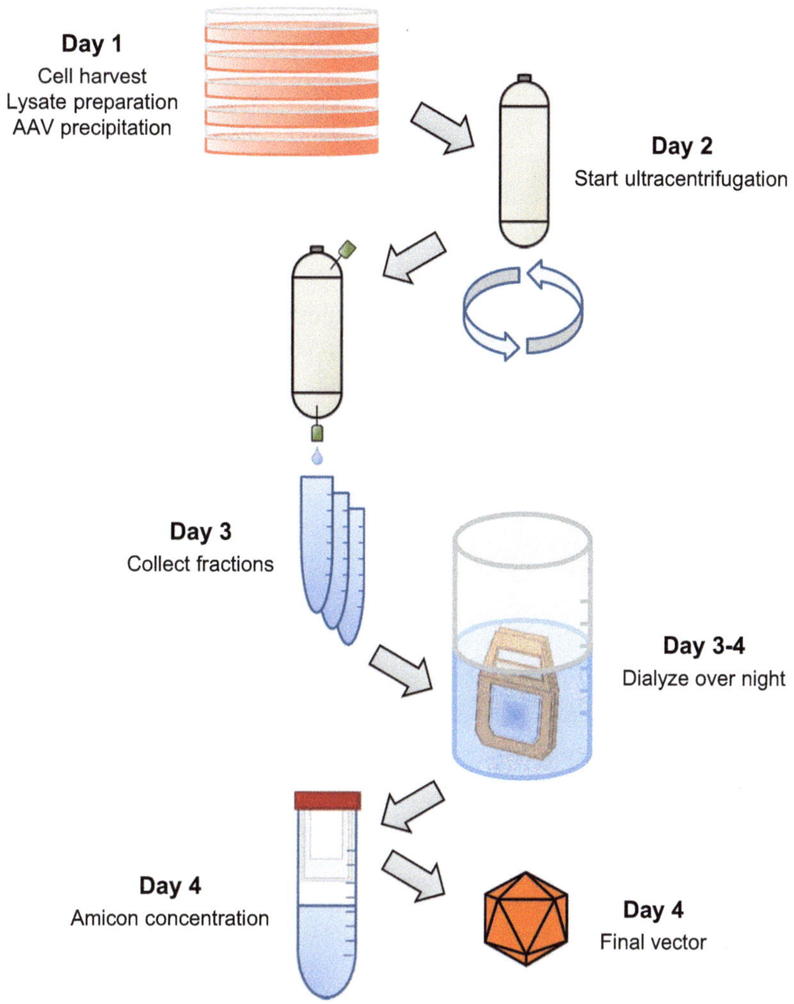

Fig. 2 Workflow for AAV purification by CsCl density gradient centrifugation. Depicted are the main steps including a minimal timeline

8. Centrifuge at 4000 rcf for 15 min. Transfer AAV-containing supernatant to a new falcon tube.
9. Precipitate proteins with 1/39 volume of 1 M $CaCl_2$ (*see* **Note 15**).
10. Incubate on ice for 1 h.
11. Centrifuge at 10,000 rcf for 15 min at 4 °C. Transfer AAV-containing supernatant to a new Falcon tube.
12. Precipitate AAV particles by adding 1/4 volume 40% PEG8000/1.915 M NaCl.
13. Incubate on ice for 3 h. Alternatively, precipitate overnight.
14. Centrifuge at 2500 rcf for 30 min at 4 °C and discard supernatant.

15. Resuspend AAV-containing pellet in 10 mL Na-HEPES resuspension buffer (*see* **Note 16**).
16. Centrifuge at 2500 rcf for 30 min at 4 °C. Transfer AAV-containing supernatant to a new falcon tube.
17. Adjust the total volume to 24 mL with Na-HEPES resuspension buffer.
18. Add 0.55 g/mL CsCl and invert. This corresponds to 13.2 g CsCl for 24 mL (*see* **Note 17**).
19. Adjust the RI of the solution to 1.3710. If the initial RI is too high, add Na-HEPES resuspension buffer in ~100 μL steps. If it is too low, add CsCl.
20. Transfer the solution to an OptiSeal™ centrifugation tube. Try to avoid bubbles and remove remaining bubbles with a sterile pipette tip.
21. Fill the tube to its neck with topping solution. Seal the tube with a cap.
22. Balance multiple tubes to a weight difference not more than 0.01 g. Prepare balance tube containing topping solution when running a single gradient (*see* **Note 18**).
23. Load tubes into ultracentrifuge plus Beckman 70Ti rotor or equivalent.
24. Centrifuge at 200,000 × *g* for 21–23 h at 21 °C, with the centrifuge set to high acceleration and low deceleration (*see* **Note 19**).
25. Collect fractions in 15 mL falcon tubes after puncturing the gradient tube with a 21G needle from the bottom. Recommended fraction volumes and representative RI values per fraction are shown in Fig. 3.
26. Mix all fractions containing full AAV capsids in one 15 mL Falcon tube. Top up to a total volume of 9 mL with 1× PBS (*see* **Note 20**).

3.4 AAV Dialysis and Concentration

Cool down sufficient amounts of 1× PBS at 4 °C in advance for dialysis and buffer exchange.

1. Transfer collected CsCl fractions containing full AAV capsids into a Slide-A-Lyzer™ Dialysis Cassette (20,000 MWCO) with a sterile pipette as described in the manufacturer's manual.
2. Submerge the dialysis cassette in a 1 L beaker filled with 700 mL 1× PBS. Leave standing at room temperature for 30 min. Do not stir at this step, as diffusing CsCl sinks to the bottom, thereby creating a concentration gradient.
3. Change buffer to fresh 1× PBS and slowly stir on a magnetic stirrer at 4 °C for 1 h.

Fraction	Volume	RI (example)
1.	3 ml	
2.	3 ml	
3.	0.5 ml	1.3785
4.	0.5 ml	1.3777
5.	0.5 ml	1.3768
6.	5 ml	1.3740
7.	0.5 ml	1.3718
8.	0.5 ml	1.3717
9.	0.5 ml	1.3715
10.	3 ml	1.3696
	rest	1.3663 (empty capsids)
Pool RI 1.3766-1.3711 (usually fractions 5-9)		

Fig. 3 Fractionation of CsCl density gradients. Shown are recommended fraction volumes (the numbering starts from the bottom of the gradient) with examples of typical refractive indexes (RI) as determined using a refractometer. The fractions/RIs shaded in grey usually contain the full, that is, vector DNA-containing AAV particles. Empty particles are found in the low-density fractions toward the top of the gradient and should be discarded

4. Repeat buffer exchange twice and in between slowly stir at 4 °C for 2 h.
5. Exchange buffer again and slowly stir at 4 °C overnight.
6. Exchange buffer once more on the following day and slowly stir at 4 °C for 2 h.
7. Equilibrate Amicon® Ultra Centrifugal Filters (10,000 MWCO) twice by centrifugation with 15 mL 1× PBS at 1000 rcf for 2 min. Discard flow-through.
8. Place dialyzed AAV sample on top of the filter and reduce volume to 1–1.5 mL AAV stock solution by repeatedly centrifuging at 500 rcf (*see* **Note 21**).
9. Before collecting the sample from the filter, pipet up and down to remove attached AAV particles. Transfer to a 2 mL tube and store at −80 °C. Freeze a separate aliquot for titration and silver stain analysis.

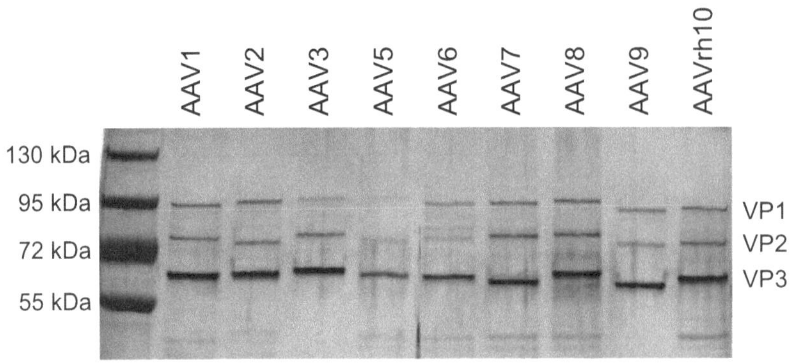

Fig. 4 Representative silver staining of various purified AAV serotypes. Shown here is a typical result that illustrates the high degree of purity that can be obtained using the protocols described here. In each lane, three predominant bands are visible that correspond to the three AAV capsid proteins VP1, VP2, and VP3 (usually found in the shown 1:1:10 ratio). Due to minor differences in protein size and posttranslational modifications, the sizes of the corresponding capsid proteins vary slightly between the different AAV serotypes

3.5 AAV Quality Control by Silver Staining

A variety of methods are available for the quantification and quality assessment of purified AAV particles. The most widely used quantification methods are quantitative PCR-based or, more recently, droplet digital (dd)PCR-based titration. For respective detailed protocols, we refer the reader to recent other articles in this series [37, 38]. Moreover, to distinguish empty from full capsids, ELISA protocols are available [39, 40]. Another highly sensitive method to control for AAV integrity and purity is the visualization of AAV capsid proteins by silver staining, which we describe below. Additionally, Fig. 4 shows an exemplary silver staining of different AAV serotypes purified using the protocol described above (Subheadings 3.3 and 3.4).

1. For sample preparation, mix 1×10^{10} viral genomes with 5 μL of 4× Laemmli buffer and H_2O in a total volume of 20 μL and denature at 95 °C for 5 min. Briefly cool down on ice.
2. Load 2 μL PageRuler Plus Prestained Protein Ladder (diluted 1:10 in 20 μL H_2O) in one lane as a size marker.
3. Resolve AAV proteins on a MiniProtean TGX precast 4–15% gradient gel in 1× TGS Buffer at 90 V until the running front reaches the gel bottom.
4. While the gel is running, prepare fresh solutions of all reagents needed for the silver staining (100 mL each) (*see* **Note 22**).
5. Stain the gel using the SilverQuest™ Silver Staining Kit (Invitrogen) as described in the manufacturer's protocol (*see* **Note 23**).
6. After development, the gels can be wrapped in foil and are stable for several weeks at 4 °C (*see* **Note 24**).

4 Notes

1. Wild-type AAV contains a single-stranded (ss)DNA genome with a length of ~4.7 kilobases (kb), comprising three genes (*rep*, *cap*, *aap*) flanked by the inverted terminal repeats (ITRs), that is, hairpin-like structures that are critical for genome replication, expression and packaging. In conventional ssAAV vectors, all viral genes are replaced by a foreign DNA, and only the ITRs are maintained in the recombinant vector as *cis*-regulatory elements. These vectors can accommodate up to ~4.7 kb of exogenous cargo but, because of their ssDNA nature, typically express their transgene relatively slowly (the exact kinetics depend on dose, capsid, and target cell or tissue), which can limit specific applications. Therefore, an alternative option is the use of the so-called self-complementary (sc)AAV genomes in which one of the ITRs is truncated, leading to encapsidation of a partially replicated genome containing three ITRs and two inverted copies of the same transgene [41]. Upon transduction, these two copies immediately fold back onto each other and thus create an expression-competent double-stranded (ds)DNA, explaining the much faster kinetics of scAAV versus ssAAV vectors. However, a downside of this vector design is the lower cargo capacity of only ~2.2 kb, caused by the packaging of two transgene copies into the same capsid. Thus, users need to consider the total size of their gene expression cassette and then choose the most suitable AAV vector design. Wherever possible, we recommend the use of scAAV vectors due to their faster and often also more robust transgene expression, but ssAAV is the only option for cassettes greater than 2.2 kb. In fact, there are even possibilities to deliver expression cassettes that exceed the cargo capacity of a single AAV capsid (as noted, ~4.7 kb), such as split and self-reconstituting AAV vector designs [42], albeit their expression levels may be compromised as compared to classical sc or ssAAV vectors. Importantly, the reader will readily find cloning templates for sc or ssAAV vectors other than the specific example mentioned here (as noted, a ssAAV vector) for example on Addgene or in the literature.
2. For faster dissolving, heat to a maximum of 80 °C. Slow pipetting is recommended due to the high viscosity of PEG.
3. Alternatively, homemade 8% polyacrylamide gels can be used when prepared with deionized H_2O (e.g., nuclease-free water, Qiagen).
4. When handling concentrated acids, always take appropriate safety precautions, that is, wear safety goggles, a lab coat, and gloves, and work in a chemical hood.

5. The primers recommended here usually consist of three elements. The restriction site itself is flanked at the 5′ end by 3–6 nucleotides (nt) of a random leader sequence, which is crucial to ensure efficient cleavage and which is followed by 18–20 nt of a template-complementary sequence required for specific primer–template hybridization. Accordingly, in forward primer TAGATC-GCGGCCGC-$(N)_{18-20}$, TAGATC is the leader, the underlined sequence is the NotI recognition site, and $(N)_{18-20}$ is the template-specific sequence. Similarly, in reverse primer TAGATC-ATCGAT -$(N)_{18-20}$, TAGATC is the leader, the underlined sequence is the ClaI recognition site, and $(N)_{18-20}$ is the template-specific sequence (in reverse-complemented orientation).
6. The initial annealing temperature chosen to facilitate hybridization should be determined only based on the primer part corresponding to the template DNA. During the PCR run, it can be increased after 5–10 cycles to the final annealing temperature of the complete primer sequence including the restriction sites (which have by then become part of the template).
7. If high degrees of vector self-ligation or low cloning efficiencies are observed, extend the incubation time with the restriction enzyme (e.g., digest overnight). Also, if using a plasmid backbone other than the one recommended here, ensure that the restriction enzyme target sites are not *dam*/*dcm*-methylated. For instance, ClaI sites are *dam*-methylated when they are flanked by a G (5′) or a C (3′), as this will create a 5′-GATC-3′ consensus site for the Dam methylase. In these cases, switch to another enzyme, if possible, or regrow the acceptor plasmid in *dam*/*dcm*-negative bacteria.
8. Recombination often occurs when cloning with restriction sites that bind and cut close to both ITRs, due to the repetitive nature of the latter. Therefore, a subsequent control of ITR integrity is necessary, for instance by restriction digest of selected clones with XmaI that cuts twice in each ITR. In contrast, direct sequencing of the ITRs is not recommended since their strong secondary structure often produces ambiguous results.
9. A large variety of AAV transgene expression cassettes is already available publicly (e.g., from authors or through repositories such as Addgene) or commercially. Likewise, there are numerous sources for relevant cDNAs; for example, opsin constructs can be requested from the Optogenetics Resource Center (Deisseroth Lab) at the University of Stanford (CA, USA).
10. Similarly, there are many sources for AAV helper plasmids encoding different AAV capsid variants. The final choice depends on the application and on the experimental conditions

including administration route, which all dictate the selection of the optimal capsid. For instance, numerous naturally occurring serotypes including AAV2 or AAV8 work well in the eye, whereas synthetic capsids such as AAV-PHP.B can provide interesting features not found in nature (e.g., the ability to effectively cross the blood–brain barrier coupled with a highly efficient transduction of cells in the CNS) [43].

11. Here, we suggest the use of CsCl gradients for AAV vector purification because these yield high purities, by effectively removing cellular contaminants and by allowing to separate full (i.e., vector DNA-containing) from empty AAV particles (the latter usually make up the majority of an unpurified AAV vector stock). Nonetheless, we note that there are many alternative methods, including iodixanol gradients as described by us in an earlier article in the same book series [37]. In direct comparison, iodixanol gradients are much faster than CsCl gradients but provide a lower degree of purification; hence, users should pick the methodology that best matches their capacities and desires.
12. Falcon tubes from different brands can sometimes break during freeze/thaw cycles. Therefore, test material stability during freezing/thawing with cell culture medium or buffer prior to use with actual AAV lysates.
13. The lysate can be stored at −80 °C at any of the freeze cycles.
14. For 30 plates, this corresponds to 15 mL benzonase buffer and 4.5 μL benzonase (250 U/μL).
15. For 30 plates, this corresponds to 15 mL benzonase buffer and 0.385 mL 1 M $CaCl_2$, resulting in a final 25 mM $CaCl_2$ concentration.
16. The precipitate is typically very sticky. Start to resuspend with 1–2 mL buffer and a 1000 μL pipette tip, and then add the remaining volume after the pellet has detached from the Falcon tube and a homogeneous suspension is obtained.
17. The solution will get cold while the CsCl is dissolving. Wait until the solution returns to room temperature before proceeding. Do not incubate at 37 °C.
18. Always weigh gradients together with rotor tube caps.
19. Run the centrifuge at room temperature, since CsCl precipitates at low temperatures.
20. Minimize AAV exposure to CsCl as the infectivity of many AAV serotypes can become impaired upon prolonged exposure.
21. The exact conditions will vary with each sample; thus, we recommend to begin with a short centrifugation to obtain a feeling for the optimal centrifugation time for the specific

sample. Nonetheless, multiple subsequent centrifugation steps are typically required to approximate the final volume.

22. The following solutions need to be prepared before starting the staining protocol. (1) Fixative: Mix 50 mL ddH_2O, 40 mL ethanol, and 10 mL glacial acetic acid. (2) Sensitizing solution: Mix 60 mL ddH_2O, 30 mL ethanol, and 10 mL Sensitizer (from kit). (3) Staining solution: Mix 99 mL ddH_2O with 1 mL Stainer (from kit). (4) Developing solution: Mix 90 mL ddH_2O, 10 mL Developer, and one drop of Enhancer (from kit). Additionally, prepare 200 mL of 30% ethanol for washing steps. Instead of the 100 mL recommended by the manufacturer, the use of 50 mL volumes per gel at each step of the protocol works well in our hands.
23. We recommend the protocol version without using a microwave. Although it is more time-consuming, the resulting staining is very sensitive and typically produces a low background.
24. Silver-containing solutions need to be disposed of separately. Accordingly, consult with a person in charge of waste disposal at your institute when performing silver staining.

Acknowledgments

We kindly acknowledge support of this work by the German Research Foundation (DFG, EXC81 [Cluster of Excellence CellNetworks]; SFB1129 [Collaborative Research Center 1129] to D.G. [TP2/16, Projektnummer 240245660]; and TRR179 [Transregional Collaborative Research Center 179] to D.G. [TP18, Projektnummer 272983813]). D.G. is moreover grateful for funding and other support from the MYOCURE project and the ERA-NET Neuron project KARTLE. MYOCURE has received funding from the European Union's Horizon 2020 research and innovation programme under grant agreement No 667751. D.G. is also thankful for support by the German Center for Infection Research (DZIF, BMBF [TTU-HIV 04.803 and TTU-HIV 04.815]). Finally, we apologize to all authors whose work on optogenetic elements and/or AAV vector technologies we unfortunately had to omit due to space constraint.

References

1. Deisseroth K (2015) Optogenetics: 10 years of microbial opsins in neuroscience. Nat Neurosci 18:1213–1225. https://doi.org/10.1038/nn.4091
2. Grosenick L, Marshel JH, Deisseroth K (2015) Closed-loop and activity-guided optogenetic control. Neuron 86:106–139. https://doi.org/10.1016/j.neuron.2015.03.034
3. Bubeck F, Hoffmann MD, Harteveld Z et al (2018) Engineered anti-CRISPR proteins for optogenetic control of CRISPR-Cas9. Nat Methods 15:924–927. https://doi.org/10.1038/s41592-018-0178-9

4. Grimm D, Zolotukhin S (2015) E pluribus Unum: 50 years of research, millions of viruses, and one goal--tailored acceleration of AAV, evolution. Mol Ther 23:1819–1831. https://doi.org/10.1038/mt.2015.173
5. Kotterman MA, Schaffer DV (2014) Engineering adeno-associated viruses for clinical gene therapy. Nat Rev Genet 15:445–451. https://doi.org/10.1038/nrg3742
6. Buchholz CJ, Friedel T, Buning H (2015) Surface-engineered viral vectors for selective and cell type-specific gene delivery. Trends Biotechnol 33:777–790. https://doi.org/10.1016/j.tibtech.2015.09.008
7. Grimm D, Buning H (2017) Small but increasingly mighty: latest advances in AAV vector research, design, and evolution. Hum Gene Ther 28:1075–1086. https://doi.org/10.1089/hum.2017.172
8. Lee EJ, Guenther CM, Suh J (2018) Adeno-associated virus (AAV) vectors: rational design strategies for capsid engineering. Curr Opin Biomed Eng 7:58–63. https://doi.org/10.1016/j.cobme.2018.09.004
9. Russell S, Bennett J, Wellman JA et al (2017) Efficacy and safety of voretigene neparvovec (AAV2-hRPE65v2) in patients with RPE65-mediated inherited retinal dystrophy: a randomised, controlled, open-label, phase 3 trial. Lancet 390:849–860. https://doi.org/10.1016/S0140-6736(17)31868-8
10. Ganjawala TH, Lu Q, Fenner MD, Abrams GW, Pan ZH (2019) Improved CoChR variants restore visual acuity and contrast sensitivity in a mouse model of blindness under ambient light conditions. Mol Ther 27:1195–1205. https://doi.org/10.1016/j.ymthe.2019.04.002
11. Khabou H, Garita-Hernandez M, Chaffiol A et al (2018) Noninvasive gene delivery to foveal cones for vision restoration. JCI Insight 3:96029. https://doi.org/10.1172/jci.insight.96029
12. Chaffiol A, Caplette R, Jaillard C et al (2017) A new promoter allows Optogenetic vision restoration with enhanced sensitivity in macaque retina. Mol Ther 25:2546–2560. https://doi.org/10.1016/j.ymthe.2017.07.011
13. Sengupta A, Chaffiol A, Mace E et al (2016) Red-shifted channelrhodopsin stimulation restores light responses in blind mice, macaque retina, and human retina. EMBO Mol Med 8:1248–1264. https://doi.org/10.15252/emmm.201505699
14. Pan ZH, Lu Q, Bi A, Dizhoor AM, Abrams GW (2015) Optogenetic approaches to restoring vision. Annu Rev Vis Sci 1:185–210. https://doi.org/10.1146/annurev-vision-082114-035532
15. Mace E, Caplette R, Marre O et al (2015) Targeting channelrhodopsin-2 to ON-bipolar cells with vitreally administered AAV restores ON and OFF visual responses in blind mice. Mol Ther 23:7–16. https://doi.org/10.1038/mt.2014.154
16. Cronin T, Vandenberghe LH, Hantz P et al (2014) Efficient transduction and optogenetic stimulation of retinal bipolar cells by a synthetic adeno-associated virus capsid and promoter. EMBO Mol Med 6:1175–1190. https://doi.org/10.15252/emmm.201404077
17. Byrne LC, Khalid F, Lee T et al (2013) AAV-mediated, optogenetic ablation of Muller glia leads to structural and functional changes in the mouse retina. PLoS One 8:e76075. https://doi.org/10.1371/journal.pone.0076075
18. Fomicheva A, Zhou C, Sun QQ, Gomelsky M (2019) Engineering adenylate cyclase activated by near-infrared window light for mammalian Optogenetic applications. ACS Synth Biol 8:1314–1324. https://doi.org/10.1021/acssynbio.8b00528
19. Fortuna MG, Kugler S, Hulsmann S (2019) Probing the function of glycinergic neurons in the mouse respiratory network using optogenetics. Respir Physiol Neurobiol 265:141–152. https://doi.org/10.1016/j.resp.2018.10.008
20. Keaveney MK, Tseng HA, Ta TL, Gritton HJ, Man HY, Han X (2018) A MicroRNA-based gene-targeting tool for virally Labeling interneurons in the rodent cortex. Cell Rep 24:294–303. https://doi.org/10.1016/j.celrep.2018.06.049
21. Mondello SE, Sunshine MD, Fischedick AE et al (2018) Optogenetic surface stimulation of the rat cervical spinal cord. J Neurophysiol 120:795–811. https://doi.org/10.1152/jn.00461.2017
22. Redchuk TA, Karasev MM, Omelina ES, Verkhusha VV (2018) Near-infrared light-controlled gene expression and protein targeting in neurons and non-neuronal cells. Chembiochem 19:1334–1340. https://doi.org/10.1002/cbic.201700642
23. Watanabe M, Narita M, Hamada Y et al (2018) Activation of ventral tegmental area dopaminergic neurons reverses pathological allodynia resulting from nerve injury or bone cancer. Mol Pain 14:1744806918756406. https://doi.org/10.1177/1744806918756406
24. Yazdan-Shahmorad A, Tian N, Kharazia V et al (2018) Widespread optogenetic expression in macaque cortex obtained with MR-guided,

convection enhanced delivery (CED) of AAV vector to the thalamus. J Neurosci Methods 293:347–358. https://doi.org/10.1016/j.jneumeth.2017.10.009
25. El-Shamayleh Y, Kojima Y, Soetedjo R, Horwitz GD (2017) Selective optogenetic control of Purkinje cells in monkey cerebellum. Neuron 95:51–62 e54. https://doi.org/10.1016/j.neuron.2017.06.002
26. Mendoza SD, El-Shamayleh Y, Horwitz GD (2017) AAV-mediated delivery of optogenetic constructs to the macaque brain triggers humoral immune responses. J Neurophysiol 117:2004–2013. https://doi.org/10.1152/jn.00780.2016
27. Jin L, Lange W, Kempmann A et al (2016) High-efficiency transduction and specific expression of ChR2opt for optogenetic manipulation of primary cortical neurons mediated by recombinant adeno-associated viruses. J Biotechnol 233:171–180. https://doi.org/10.1016/j.jbiotec.2016.07.001
28. Gompf HS, Budygin EA, Fuller PM, Bass CE (2015) Targeted genetic manipulations of neuronal subtypes using promoter-specific combinatorial AAVs in wild-type animals. Front Behav Neurosci 9:152. https://doi.org/10.3389/fnbeh.2015.00152
29. Osawa S, Iwasaki M, Hosaka R et al (2013) Optogenetically induced seizure and the longitudinal hippocampal network dynamics. PLoS One 8:e60928. https://doi.org/10.1371/journal.pone.0060928
30. Ambrosi CM, Sadananda G, Han JL, Entcheva E (2019) Adeno-associated virus mediated gene delivery: implications for scalable in vitro and in vivo cardiac optogenetic models. Front Physiol 10:168. https://doi.org/10.3389/fphys.2019.00168
31. Keppeler D, Merino RM, de la Morena DL et al (2018) Ultrafast optogenetic stimulation of the auditory pathway by targeting-optimized Chronos. EMBO J 37:e99649. https://doi.org/10.15252/embj.201899649
32. Duarte MJ, Kanumuri VV, Landegger LD et al (2018) Mol Ther 26:1931–1939. https://doi.org/10.1016/j.ymthe.2018.05.023
33. Vajanthri KY, Yadav P, Poddar S, Mahto SK (2018) Development of optically sensitive liver cells. Tissue Cell 52:129–134. https://doi.org/10.1016/j.tice.2018.05.004
34. Yu L, Zhou L, Cao G et al (2017) Optogenetic modulation of cardiac sympathetic nerve activity to prevent ventricular arrhythmias. J Am Coll Cardiol 70:2778–2790. https://doi.org/10.1016/j.jacc.2017.09.1107
35. Richter C, Bruegmann T (2019) No light without the dark: perspectives and hindrances for translation of cardiac optogenetics. Prog Biophys Mol Biol. https://doi.org/10.1016/j.pbiomolbio.2019.08.013
36. Vogt CC, Bruegmann T, Malan D, Ottersbach A, Roell W, Fleischmann BK, Sasse P (2015) Systemic gene transfer enables optogenetic pacing of mouse hearts. Cardiovasc Res 106:338–343. https://doi.org/10.1093/cvr/cvv004
37. Fakhiri J, Nickl M, Grimm D (2019) Rapid and simple screening of CRISPR guide RNAs (gRNAs) in cultured cells using Adeno-associated viral (AAV) vectors. Methods Mol Biol 1961:111–126. https://doi.org/10.1007/978-1-4939-9170-9_8
38. Sanmiguel J, Gao G, Vandenberghe LH (2019) Quantitative and digital droplet-based AAV genome titration. Methods Mol Biol 1950:51–83. https://doi.org/10.1007/978-1-4939-9139-6_4
39. Grimm D, Kern A, Pawlita M, Ferrari F, Samulski R, Kleinschmidt J (1999) Titration of AAV-2 particles via a novel capsid ELISA: packaging of genomes can limit production of recombinant AAV-2. Gene Ther 6:1322–1330. https://doi.org/10.1038/sj.gt.3300946
40. Kuck D, Kern A, Kleinschmidt JA (2007) Development of AAV serotype-specific ELISAs using novel monoclonal antibodies. J Virol Methods 140:17–24. https://doi.org/10.1016/j.jviromet.2006.10.005
41. McCarty DM, Monahan PE, Samulski RJ (2001) Self-complementary recombinant adeno-associated virus (scAAV) vectors promote efficient transduction independently of DNA synthesis. Gene Ther 8:1248–1254. https://doi.org/10.1038/sj.gt.3301514
42. Carvalho LS, Turunen HT, Wassmer SJ, Luna-Velez MV, Xiao R, Bennett J, Vandenberghe LH (2017) Evaluating efficiencies of dual AAV approaches for retinal targeting. Front Neurosci 11:503. https://doi.org/10.3389/fnins.2017.00503
43. Deverman BE, Pravdo PL, Simpson BP et al (2016) Cre-dependent selection yields AAV variants for widespread gene transfer to the adult brain. Nat Biotechnol 34:204–209. https://doi.org/10.1038/nbt.3440

Chapter 6

AAV-Mediated Gene Delivery to Foveal Cones

Stéphane Bertin, Elena Brazhnikova, Céline Jaillard, José-Alain Sahel, and Deniz Dalkara

Abstract

Adeno-associated virus (AAV) has emerged as the vector of choice for delivering genes to the mammalian retina. From the first gene therapy to receive FDA approval for the inherited retinal disease (Luxturna™) to more recent clinical trials using microbial opsins to regain light sensitivity, therapeutic transgenes rely on AAV vectors for safe and efficient gene delivery to retinal cells. Such vectors are administered to the retina via subretinal (SR) injection or intravitreal (IVT) injection routes depending on the targeted retinal cell type. An attractive target for gene therapy is the fovea, bearing the highest concentration of cone cells responsible for our high acuity daylight vision. However, previous clinical trials and large animal studies reported that SR administration of vector under the cone-exclusive fovea disrupts its fine structure and might impair visual acuity. Due to its technical difficulty and potential risks, alternatives to vector injection under this delicate region have been investigated by using novel AAV capsid variants identified via rational design or directed evolution. We recently established new vector–promoter combinations to overcome the limitations associated with AAV-mediated cone transduction in the fovea. Our methods provide efficient foveal cone transduction without detaching this delicate region and rely on the use of engineered AAVs and optimal promoters compatible with optogenetic vision restoration. Here we describe in detail our AAV vectors, methods for intravitreal and subretinal injections as well as pre- and postoperative procedures as performed in cynomolgus macaques.

Key words Subretinal injection, Fovea, AAV, Gene delivery, Optogenetics, Cone Photoreceptors, Novel surgical technique, Noninvasive retina imaging

1 Introduction

The fovea—located at the center of the macula—is a specialized region of the retina that provides high acuity color vision in primates [1]. The highest density of cones is found at the center of the fovea (<0.3 mm from the foveal center), and this region is devoid

Electronic Supplementary Material: The online version of this chapter (https://doi.org/10.1007/978-1-0716-0755-8_6) contains supplementary material, which is available to authorized users.

Stéphane Bertin, Elena Brazhnikova and Céline Jaillard contributed equally to this work.

Dominik Niopek (ed.), *Photoswitching Proteins: Methods and Protocols*, Methods in Molecular Biology, vol. 2173,
https://doi.org/10.1007/978-1-0716-0755-8_6, © Springer Science+Business Media, LLC, part of Springer Nature 2020

of any other neurons including rod photoreceptors [2]. Cone density decreases dramatically as one moves away from the center of the fovea. Naturally, signals originating from the cone cells in the fovea dominate our visual perception and are the primary targets of gene therapies aiming to treat inherited retinal diseases like mid-stage retinitis pigmentosa and achromatopsia [3–5]. Currently, the most common approach to transduce cones is to inject AAV vectors subretinally in the region of interest. However, when the region of interest involves the fovea, there are increased risks of retinal detachment damaging the cone ultrastructure [6] particularly in eyes with retinal degeneration [7]. The early clinical trials using subretinal injections of adeno-associated virus (AAV) to deliver a healthy retinal pigment epithelium-specific 65 kilodalton protein (RPE65) gene in Leber's congenital Amaurosis-2 patients led to some improvements in vision, but the treatment was, in certain cases, complicated by macular holes and increased macular thinning in the case of subfoveal injections [8–11]. Furthermore, contrary to the surrounding regions, there were no treatment benefits in the fovea [11]. Similarly one latter clinical trial for achromatopsia reported visual acuity loss in the treated eye compared with his untreated eye for one patient [12]. With more gene therapies reaching clinical stages of application, there is a growing need to find new methods for delivering gene therapy to the fovea without detaching this delicate region. Since the only mammals having a fovea are primates, we developed and tested viral vector improvements to achieve foveal gene delivery in this species [13].

We propose a combination of novel AAV capsids and promoters that can provide gene delivery to foveal cones after injection into the vitreous cavity or by SR injections in the periphery. We identified PR1.7 [14] as a cone specific promoter capable of driving high-level transgene expression in these cells without leading to off-target expression in neighboring cells regardless of the delivery route. We combined this promoter with 2 potent AAV capsid variants: AAV2-7m8 that provides gene delivery to cone photoreceptors when injected intravitreally or AAV9-7m8 that performs this task via distal SR injections outside the fovea.

We validated the utility of these gene delivery strategies in the context of cone-directed optogenetic therapy [15] in vivo, and ex vivo imaging techniques in mouse, macaque, and human retinal tissue. Here we describe in detail our protocols for surgical delivery and in vivo imaging along with our AAV compositions used in the primate retina (Video S1). We hope these protocols will be useful to investigators aiming to transduce the primate foveal cones efficiently and specifically.

2 Materials

2.1 Animal Preparation and Postoperative Care

Examples of commercial products are in parentheses:

1. For sedation and anesthesia: ketamine, 10 mg/kg (Imalgene 1000, *Merial*), xylazine 0.5 mg/kg (Rompun 2%, *Bayer*); propofol, 1 mL/kg/h (PropoVet Multidose 10 mg/mL, *Zoetis*).
2. Scale.
3. Needles with syringes for IM injections.
4. Ophthalmic solution for pupil dilation: 0.5% Tropicamide (Mydriaticum 0.5%, *Thea*).
5. IV Catheter and syringes.
6. Syringe pump (Perfusor® Space).
7. Endotracheal tubing.
8. Topical anesthetic spray (Xylocaine 5%, Nebulizer, *AstraZeneca*).
9. Body temperature maintenance system for animals.
10. Monitoring system (PM-9000 Vet, Veterinary Portable Multi-Parameter Patient Monitor, *Shenzhen Mindray, Biomedical Electronics Co, LTD*).
11. Antiseptic solution: 5% povidone–iodine (Vetadin 10%, *Vé toquenol*).
12. Physiological solutions: sodium chloride (0.9%) or Balanced Salt Solution (BSS), (*Alcon*).
13. Ophthalmic gel: Lubrithal (*Centravet Plancoët)* or other with irrigating cannula.
14. Sterile surgical and ophthalmic drapes.
15. Speculum.
16. Anesthetic ophthalmic solution (oxybuprocaine hydrochloride1.6 mg/0.4 mL, *THEA*).

2.2 Intravitreal Injection

1. Ophthalmic surgical microscope (preferably with digital video) (Zeiss Opmi Lumera 700).
2. 26 or 30G needle with 1 mL syringes.
3. Surgical forceps.
4. Contact flat lens (MV LV40, *Phakos*) with sterile viscose ophthalmic solution (like gel Z-CELCOAT (*Carl Zaiss Meditec AG*) or indirect ophthalmoscope with 20D lens, or noncontact vitrectomy viewing system.
5. Ophthalmic ointment containing an antibiotic and a corticosteroid (Fradexam: Framycétine 3105UI, Dexamethasone 0.76 mg, *TVM lab*).

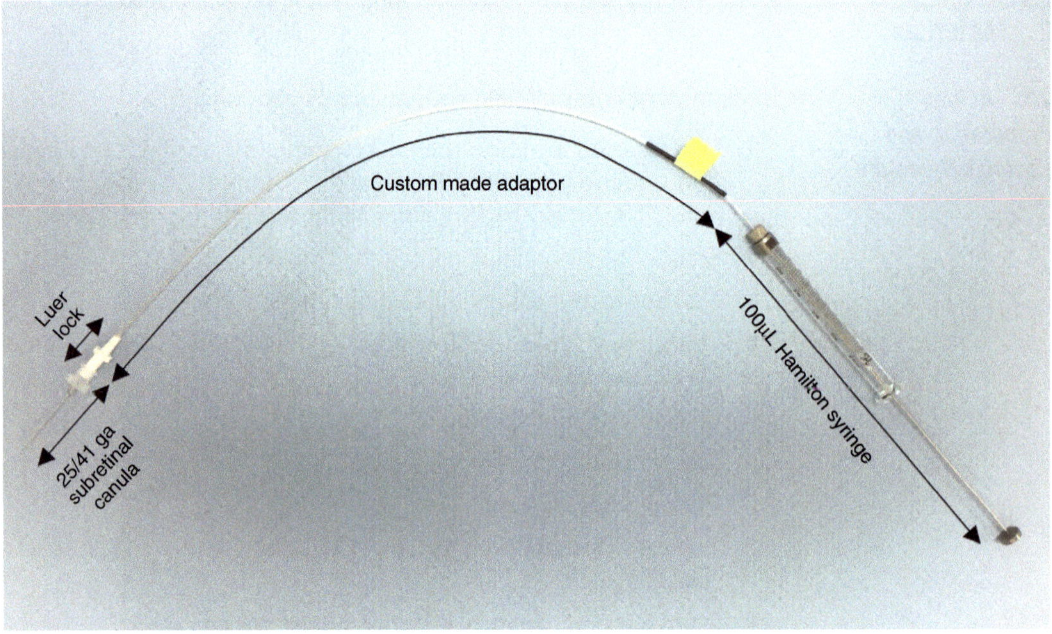

Fig. 1 Custom made adaptor connected to subretinal cannula and Hamilton syringe

2.3 Subretinal Injection

1. Ophthalmic surgical microscope (with digital video is preferred) (Zeiss Opmi Lumera 700).
2. Vitrectomy system (Constellation Vitrectomy System, *Alcon*).
3. 25G Valved Trocars (*Alcon*).
4. 25G intraocular illumination probe—25G Endoilluminator (*Alcon*).
5. BSS Solution, 15 mL (*Alcon*).
6. Vitrectomy flat contact lens (MV LV40, *Phakos)* or noncontact vitrectomy viewing system.
7. Ophthalmic gel Z-CELCOAT (*Carl Zaiss Meditec AG*).
8. 25/41G subretinal injection cannula (*Baush&Lomb*).
9. 100 μL Hamilton syringe.
10. Custom-made extension adaptor (Fig. 1): 300 μm diameter tubing (Phymep), connected to a male/female Luer lock. The length of the tubing is 30 cm and from polyethylene PE20.
11. 8-0 Vicryl sutures
12. Ophthalmic ointment containing an antibiotic and a corticosteroid (Fradexam: Framycétine 3105UI, Dexamethasone 0.76 mg, *TVM lab*).

2.4 AAV Preparations

1. For IVT injections a sterile vial containing 110 μL of AAV2-7m8-PR1.7-GFP at a concentration of 1×10^{12} vg/mL (void volume 10 μL).

2. For SR injections a sterile vial containing 110 uL of AAV9-7m8-PR1.7-GFP at a concentration of 1×10^{11} vg/mL (void volume 60 μL).

2.5 Retinal Imaging

1. Spectralis HRA+OCT system (Heidelberg Engineering).

3 Methods

3.1 Animal Preparation (Common Procedure Before the Injections and Imaging)

1. Seronegative animals are selected (*see* **Note 1**) and fasted the day prior to procedures.
2. On the day of surgery perform baseline of fluorescence fundus imaging and optical coherence tomography (OCT) (*see* **Note 2**).
3. The intraocular injection has to be performed in the surgery room under sterile conditions.
4. Anesthesia is induced by IM injection of ketamine–xylazine mixture 10 mg/kg and 0.5 mg/kg.
   (a) Anesthesia is maintained by IV perfusion of propofol 1 mL/kg/h (*see* **Note 3**).
   (b) The animal is intubated as a precautionary measure during surgery for the subretinal injection.
   (c) During surgery, animals are supervised and kept warm until they wake up.
5. Dilate the pupil by applying drops of Ophthalmic Solution.
6. Protect contralateral eye by ophthalmic gel application or by closing the eyelid.
7. Decontaminate the surrounding area prior to surgical intervention: use antiseptic (**5%** Vetadin povidone–iodine dilution, prepared extempore **10%**).
8. Decontaminate the conjunctival cavity by applying the same antiseptic solution for 2 min, then rinse the conjunctival cavity with the normal saline solution or BSS.
9. Use a sterile ophthalmological drape to cover surfaces or operative field, which provide a barrier against infectious agents.
10. Insert the lid speculum to keep the lid open during the procedure.
11. Apply the anesthetic eye drops for the local anesthesia.
12. Perform the injection under direct observation through the operating microscope.

3.2 Intravitreal Injection

1. Use the 26-30G needle with the syringe filled with the 100 μL of AAV solution to inject (the void volume has to be taken into account). Insert needle ~2 mm posterior to the limbus with a 45° angle to the sclera and in parallel position to the limbus.

Once the tip of the needle is inserted into the sclera, turn the needle directly toward the center of the ocular globe.

2. Advance the needle to the center of the vitreous body taking care not to touch the lens. Slowly inject the solution into the vitreous.
3. After 4 or 5 s remove slowly the needle and immediately close the site of injection with forceps to prevent reflux of the intraocular liquid (*see* **Note 4**). If the volume of intravitreal injection exceeds 100 μL, *see* **Note 5**.
4. Observe the central retinal vein through the contact lens, noncontact vitrectomy viewing system, or indirect ophthalmoscope.

   If the central retinal vein is closed with absence of vein pulsation, proceed with anterior chamber parenthesis procedure: puncture the cornea with open insulin syringe equipped with a 26-30-gauge needle at 9 o'clock for the right eye and 3 o'clock for the left one with following passive aqueous humor aspiration (great care should be taken so as not to touch the lens).

   If the pulsation of the central vein is present or the vein is open, apply the corticosteroid and antibiotic in the form of ophthalmic ointment.

3.3 Subretinal Injection Technique

1. Insert 25-gauge trocars from the vitrectomy set ~2 mm posteriorly to the limbus at superior-temporal and superior-nasal areas. While inserting into the sclera, the position of the needle of the trocar should be at about 45° to the sclera and parallel to the limbus. After the tip of the trocar needle has entered the sclera, but not further, turn the trocar directly toward the center of the ocular globe and push it.
2. Use the contact lens with ophthalmic gel or noncontact vitrectomy viewing system to visualize the eye fundus.
3. Injecting only 100 μL, we do not perform the vitrectomy before the viral solution delivery (*see* **Note 6**).
4. Fill Hamilton syringe with solution and connect it to homemade adaptor and to 41 G subretinal cannula (Fig. 1) (*see* **Note 7**).
5. Test the injection system under the microscope by pushing the Hamilton syringe plunger and taking care not to contaminate anything by the virus. A small drop of the viral solution has to appear on the tip of the subretinal cannula.
6. Use one port for the endo-illumination probe and the other one for inserting the subretinal cannula.
7. Advance the tip of the subretinal cannula and guide it to the retina in the region selected for injection.

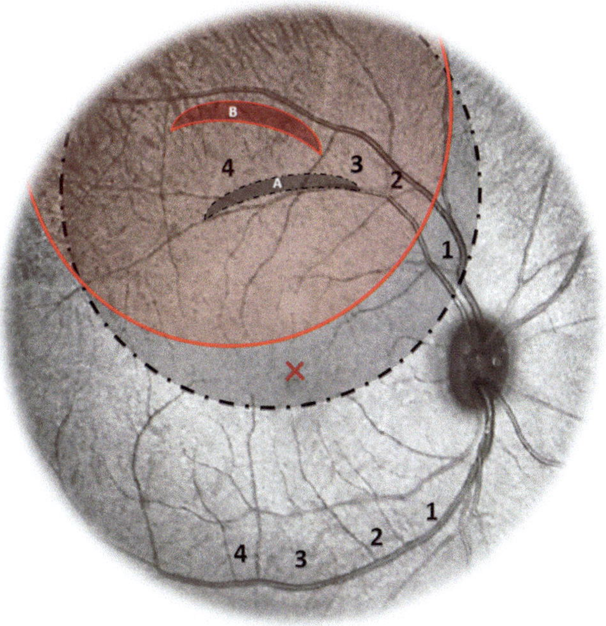

Fig. 2 Localization of injection sites to deliver the AAV solution under the retina, excluding or including the fovea. This chart is representative for the 100 μL injection volume. 1, 2, 3, 4—the number of shapes named "Bertin's square." (**a**) Zone of the injection site to deliver the viral solution under the retina including the fovea; (**b**) Zone to deliver the viral solution under the retina close to the fovea but without foveal detachment

8. Depending on the researcher's needs the bleb might be created in superior nasal, superior temporal, inferior temporal, and inferior nasal areas of the eye fundus. The choice of the insertion point of the cannula (injection site) is very important especially for the creation of retinal detachment with fovea off or involving the fovea. In general, it is preferable to make the retinal puncture in the region of the vascular arcades, but avoiding the blood vessels. A schematic presentation of injection sites in the posterior pole of the eye is detailed in Fig. 2 (*see* **Note 8**). To create the peripheral retinal detachment, the injection site has to be localized just outside of peripheral side of 4th "Bertin's quadrant" (*see* Fig. 2).
9. The insertion angle of the tip of the cannula into the retina is about 45°. When a mild pressure is applied to the cannula in contact with the retina, the whitish tint appears first behind the tip of the cannula and then in front of it with a tiny fold on the retina. This is the moment to start the injection (Fig. 3).

   Take care to penetrate only the tip of the subretinal cannula, but not more, under the retina. Ask an experienced assistant to push the plunger of the Hamilton syringe. The

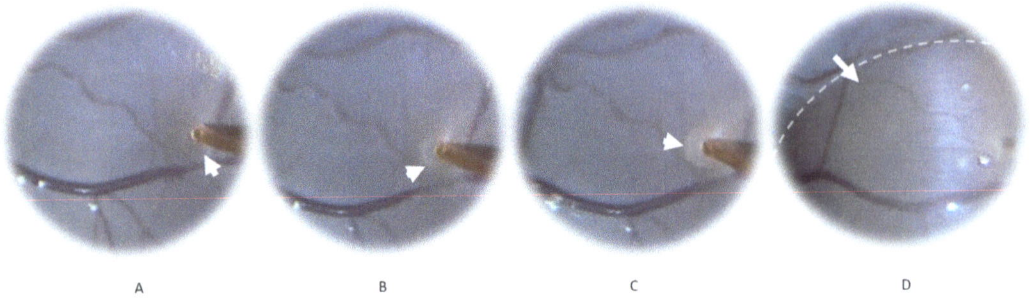

Fig. 3 Flowchart depicting the subretinal injection technique. Indicated by arrows: (**a**) Whitish area behind the cannula tip after an application of the mild pressure to the tip of the canula; (**b**) Whitish area around the cannula tip after continuation of the application of the mild pressure to the tip of the cannula; (**c**) Tiny retinal fold in front of cannula tip (moment when the injection begins); (**d**) Bleb due to retinal detachment

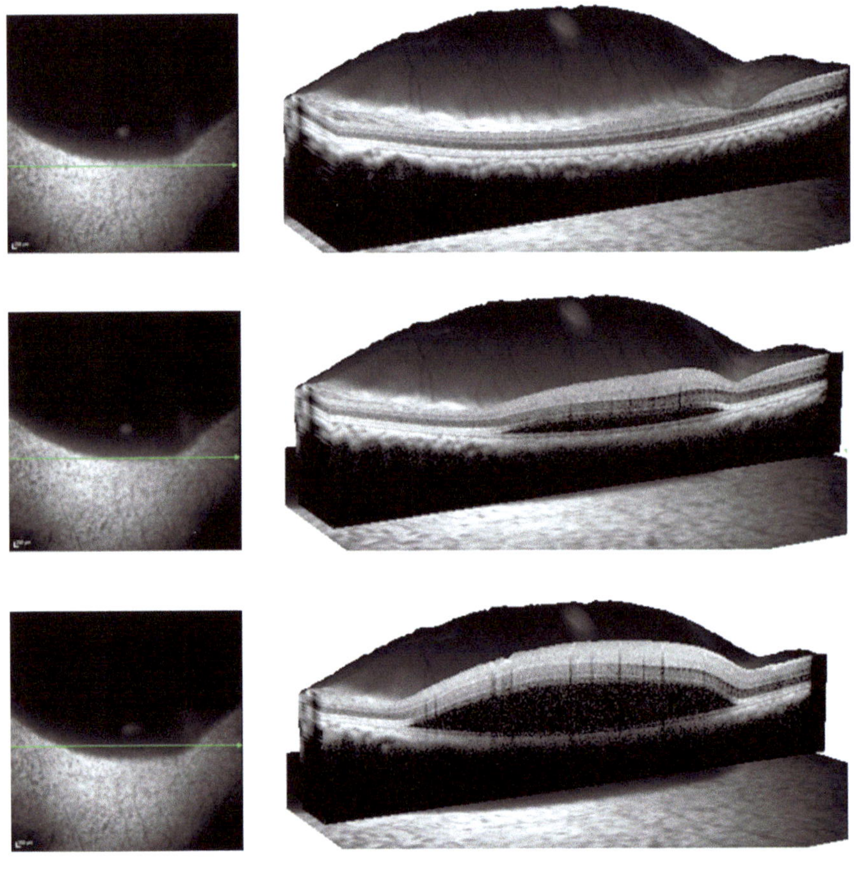

Fig. 4 Selected samples of segmented scans from a single 3D view of a subretinal bleb

initial pressure to detach the retina has to be greater than the injection flow rate (*see* **Note 9**). The resulting bleb of retinal detachment after the injection of 100 μL of viral solution measures approximately 8–10 mm in diameter (Fig. 4)

10. Wait 4–5 s for the subretinal pressure to stabilize before withdrawing the cannula from the retina and then from the eye. Remove the trocars from the eye. Check if the sclerotomies are hermetically sealed. If there is a flux of the intraocular liquid, close the sclerotomies with the conjunctiva by 8.0 vicryl ophthalmic sutures.
11. Apply corticosteroid and antibiotic in the form of ophthalmic ointment.
12. Remove catheter and tubing from the animal. Keep the animal under warm cover until the first signs of the awakening.

3.4 Imaging

After pupil dilation, Spectralis HRA+OCT system (Heidelberg Engineering) was used to image fundus eye and OCT. This Spectralis is a combination of a confocal laser scanning system and spectral domain optical coherence tomography that allows high-speed, high-resolution cross-sectional imaging of the retina.

During imaging, the cornea is continuously hydrated manually with a BSS solution.

3.4.1 Enter New Animal Data

Open program, click the New Patient icon. Enter, at the minimum, the animal name, date of birth, and name of the project into the appropriate fields.

3.4.2 Align eye of Animal with Camera

Turn on the camera and adjust head of animal and the joystick of the apparatus so the pupil is aligned with the center of the live window.

3.4.3 Acquire an IR Image (Wavelength 820 nm)

1. Push the base of the camera forward toward the animal's pupil in keeping image centered.
2. Play with gain control to adjust image brightness. Animal's fundus becomes visible on the computer screen.
3. Align the camera to the eye so that no black corners reside on the computer screen. Adjust the focus using the knob on the back of the camera until the image is brightest.
4. Once the image is positioned and focused properly, turn on the automatic real-time tracking (ART) by pressing the black gain control button on the touch panel. Allow the ART bar to fill and press Acquire when the image looks optimal (Fig. 5a).

3.4.4 Acquire an OCT Image

1. Press IR+OCT button on the touch panel. Proceed as for acquisition of IR image. When fundus image is illuminated, slide the base of the camera forward until the OCT scan is within the blue markers in the top one-third of the acquisition screen.
2. Once the saturation and placement of the scan is optimal, activate the ART and press acquire (Fig. 5b, c).

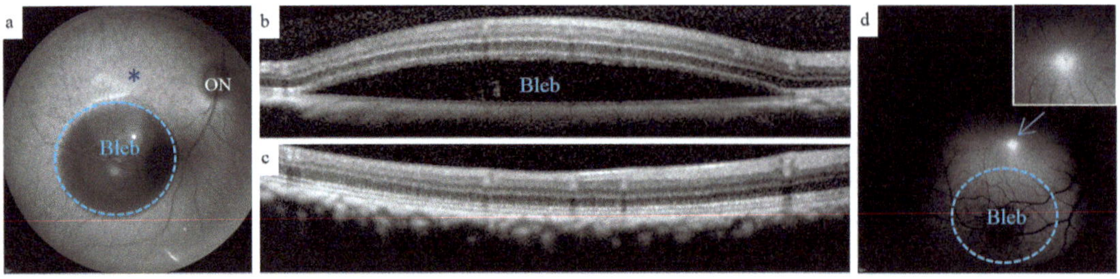

Fig. 5 Transduction of the fovea via distal subretinal administration of AAV9-7m8-PR1.7-Jaws-GFP. After pupil dilation, infrared, fluorescent images of GFP and OCT images were recorded using a Spectralis HRA+OCT system (Heidelberg Engineering). (**a**) Eye fundus infrared image and (**b**) optical coherence tomography (OCT) image immediately after subretinal delivery of AAV9-7m8-Jaws-GFP in peripheral retina. (**c**) OCT image 1 day after subretinal injection showing reattachment of photoreceptors to RPE. (**d**) Eye fundus fluorescence image 1 month after injection shows strong GFP expression in subretinal bleb and away from the injection site, including fovea. PR1.7, promoter (1.7 kb) based on the human red opsin gene enhancer and promoter sequences. *IR* infrared, *RPE* retinal pigment epithelium, *ON* optic nerve

3.4.5 Acquire an Autofluorescence Image (Wavelength 488 nm)

1. Immediately after acquisition of IR image, press FA on the touch panel.
2. Adjust the brightness so the vasculature is visible and turn on ART. Acquire when the image looks optimal (Fig. 5d).

4 Notes

1. Test for the presence of neutralizing antibodies as described in Desrosiers and Dalkara, Methods in Molecular Biology [15].
2. Baseline fluorescence and optical coherence tomography (OCT) imaging can be performed by using a Scanning Laser Ophthalmoscope, such as Heidelberg Spectralis®.
3. Ventilation with isoflurane (1.5%) can also be used to maintain anesthesia [16].
4. Injection angle of 45° with changing of direction of the needle toward the center of the ocular globe is a part of the reflux preservation of the intraocular liquid at the time of needle withdrawal.
5. Prior to injection of a considerable volume of viral solution (more than 100 μL) it is advisable to use preoperative antihypertensive drug's induced intraocular pressure decline or mechanical intraocular pressure decline (permanent pressure or finger massage). If it is planned to inject more than 300 μL, anterior chamber parenthesis are applied before the injection.
6. To perform the subretinal injection, in our opinion, the vitrectomy is not necessary in the case of macaque experiments for the following reasons:

(a) Intervention without vitrectomy is less invasive.

(b) Complications related to the vitrectomy (e.g., vessel bleeding) are avoided [16].

(c) The retina is protected by vitreous from the excessive retinal stretching during subretinal injection (retinal stretch leads to mechanical disruption of nerve fibers and neuronal synapses and a secondary retinal hole formation) [17].

7. The subretinal injection device (homemade connecting adaptor) facilitates the manipulation of the cannula by the surgeon.

8. With rare exceptions, the vessels of retinal arcades have geometric forms such as quadrangles and triangles. Shapes 3 and 4 (Fig. 2) which start from the optic disk usually take shapes close to a square. We named those shapes the "Bertin's quadrant," after the description of Dr. Stéphane Bertin. Usually the sites of subretinal injection are localized within the Bertin's quadrants 3 or 4 which are situated at the aligning with fovea in a distance 2 to 2.5 optic disk diameter or a little bit further.

   The zone A on the Fig. 2, which is as close as possible to the proximal vessel of the Bertin's quadrant but avoiding the retinal vessels, marks the injection site for delivery of viral solution under the retina including the fovea.

   The zone B (the central area close to the distal vessel of the Bertin's quadrant) marks the injection site for delivery of the viral solution under the retina close to the fovea but without foveal detachment.

   If the vessel square's shaping is not evident to the experimenter, injection at 2.5–3 optic disk diameter and further away from the fovea normally leads to delivery of the viral solution without the foveal detachment. An injection less than 2 optic disk diameter from the fovea leads to delivery of the virus under the retina including the fovea.

9. Another option to create retinal detachment is to use the injection system using viscous fluid control (VFC) port with the incorporation of foot-pedal control proposed by [17].

   A manual assistance by an experienced person is also suitable, as mentioned above. The initial pressure to detach the retina has to be greater (about 30–60 psi) than injection flow rate (12–20 psi). One must pay attention to slowly inject the viral solution, especially at the moment of creation of the foveal detachment.

   The infusion pressure rate can be determined by using the viscous fluid control (VFC) system connected directly to the

Subretinal 25G cannula with a 41G tip for the creation of retinal detachment by BSS (without the virus) subretinal delivery.

References

1. Sinha R, Hoon M, Baudin J et al (2017) Cellular and circuit mechanisms shaping the perceptual properties of the primate fovea. Cell 168 (3):413–426. e412. https://doi.org/10.1016/j.cell.2017.01.005
2. Kolb H, Zhang L, Dekorver L et al (2002) A new look at calretinin-immunoreactive amacrine cell types in the monkey retina. J Comp Neurol 453(2):168–184. https://doi.org/10.1002/cne.10450
3. Komaromy AM, Alexander JJ, Rowlan JS et al (2010) Gene therapy rescues cone function in congenital achromatopsia. Hum Mol Genet 19 (13):2581–2593. https://doi.org/10.1093/hmg/ddq136
4. Busskamp V, Duebel J, Balya D et al (2010) Genetic reactivation of cone photoreceptors restores visual responses in retinitis Pigmentosa. Science 329(5990):413–417. https://doi.org/10.1126/science.1190897
5. Byrne LC, Dalkara D, Luna G et al (2015) Viral-mediated RdCVF and RdCVFL expression protects cone and rod photoreceptors in retinal degeneration. J Clin Invest 125 (1):105–116. https://doi.org/10.1172/JCI65654
6. Fisher SK, Lewis GP, Linberg KA et al (2005) Cellular remodeling in mammalian retina: results from studies of experimental retinal detachment. Prog Retin Eye Res 24 (3):395–431. https://doi.org/10.1016/j.preteyeres.2004.10.004
7. Kolstad KD, Dalkara D, Guerin K et al (2010) Changes in Adeno-associated virus-mediated gene delivery in retinal degeneration. Hum Gene Ther 21(5):571–578. https://doi.org/10.1089/hum.2009.194
8. Bainbridge JWB, Smith AJ, Barker SS et al (2008) Effect of gene therapy on visual function in Leber's congenital amaurosis. New Engl J Med 358(21):2231–2239. https://doi.org/10.1056/NEJMoa0802268
9. Cideciyan AV, Aleman TS, Boye SL et al (2008) Human gene therapy for RPE65 isomerase deficiency activates the retinoid cycle of vision but with slow rod kinetics. Proc Natl Acad Sci U S A 105(39):15112–15117. https://doi.org/10.1073/pnas.0807027105
10. Maguire AM, Simonelli F, Pierce EA et al (2008) Safety and efficacy of gene transfer for Leber's congenital amaurosis. New Engl J Med 358(21):2240–2248. https://doi.org/10.1056/NEJMoa0802315
11. Jacobson SG, Cideciyan AV, Ratnakaram R et al (2012) Gene therapy for Leber congenital Amaurosis caused by RPE65 mutations safety and efficacy in 15 children and adults followed up to 3 years. Arch Ophthalmol 130(1):9–24. https://doi.org/10.1001/archophthalmol.2011.298
12. MacLaren RE, Groppe M, Barnard AR et al (2014) Retinal gene therapy in patients with choroideremia: initial findings from a phase 1/2 clinical trial. Lancet 383 (9923):1129–1137. https://doi.org/10.1016/S0140-6736(13)62117-0
13. Khabou H, Garita-Hernandez M, Chaffiol A et al (2018) Noninvasive gene delivery to foveal cones for vision restoration. JCI Insight 3(2):e96029. https://doi.org/10.1172/jci.insight.96029
14. Ye GJ, Budzynski E, Sonnentag P et al (2016) Safety and biodistribution evaluation in cynomolgus macaques of rAAV2tYF-PR1.7-hCNGB3, a recombinant AAV vector for treatment of Achromatopsia. Hum Gene Ther Clin Dev 27(1):37–48. https://doi.org/10.1089/humc.2015.164
15. Desrosiers M, Dalkara D (2018) Neutralizing antibodies against Adeno-associated virus (AAV): measurement and influence on retinal gene delivery. Methods Mol Biol 1715:225–238. https://doi.org/10.1007/978-1-4939-7522-8_16
16. Boye SE, Alexander JJ, Boye SL et al (2012) The human rhodopsin kinase promoter in an AAV5 vector confers rod- and cone-specific expression in the primate retina. Hum Gene Ther 23(10):1101–1115. https://doi.org/10.1089/hum.2012.125
17. Xue K, Groppe M, Salvetti AP et al (2017) Technique of retinal gene therapy: delivery of viral vector into the subretinal space. Eye 31 (9):1308–1316. https://doi.org/10.1038/eye.2017.158

Chapter 7

Engineering Optogenetic Protein Analogs

Bei Liu, Daniel J. Marston, and Klaus M. Hahn

Abstract

This chapter provides an overview of the technologies we have developed to control proteins with light. First, we focus on the LOV domain, a versatile building block with reversible photo-response, kinetics tunable through mutagenesis, and ready expression in a broad range of cells and animals. Incorporation of LOV into proteins produced a variety of approaches: simple steric block of the active site released when irradiation lengthened a linker (PA-GTPases), reversible release from sequestration at mitochondria (LOV-TRAP), and Z-lock, a method in which a light-cleavable bridge is placed where it occludes the active site. The latter two methods make use of Zdk, small engineered proteins that bind selectively to the dark state of LOV. In order to control endogenous proteins, inhibitory peptides are embedded in the LOV domain where they are exposed only upon irradiation (PKA and MLCK inhibition). Similarly, controlled exposure of a nuclear localization sequence and nuclear export sequence is used to reversibly send proteins into the nucleus. Another avenue of engineering makes use of the heterodimerization of FKBP and FRB proteins, induced by the small molecule rapamycin. We control rapamycin with light or simply add it to target cells. Incorporation of fused FKBP-FRB into kinases, guanine exchange factors, or GTPases leads to rapamycin-induced protein activation. Kinases are engineered so that they can interact with only a specific substrate upon activation. Recombination of split proteins using rapamycin-induced conformational changes minimizes spontaneous reassembly. Finally, we explore the insertion of LOV or rapamycin-responsive domains into proteins such that light-induced conformational changes exert allosteric control of the active site. We hope these design ideas will inspire new applications and broaden our reach towards dynamic biological processes that unfold when studied in vivo.

Key words Optogenetics, Chemogenetics, LOV, RapR, LOVTRAP, Z-lock, Zdk, Engineered extrinsic disorder

1 Introduction

In this chapter we will provide a brief overview of the approaches we have developed to control proteins with light, in single cells and in animals. Most of the methods we will describe make use of the light-sensing domain that plants employ for phototropism, the light oxygen voltage domain (LOV2, from *Avena sativa*) [1, 2]. As we will show, it is a versatile building block that can be the basis of designs with complementary strengths, providing access to diverse protein structures. When irradiated between

Dominik Niopek (ed.), *Photoswitching Proteins: Methods and Protocols*, Methods in Molecular Biology, vol. 2173,
https://doi.org/10.1007/978-1-0716-0755-8_7,

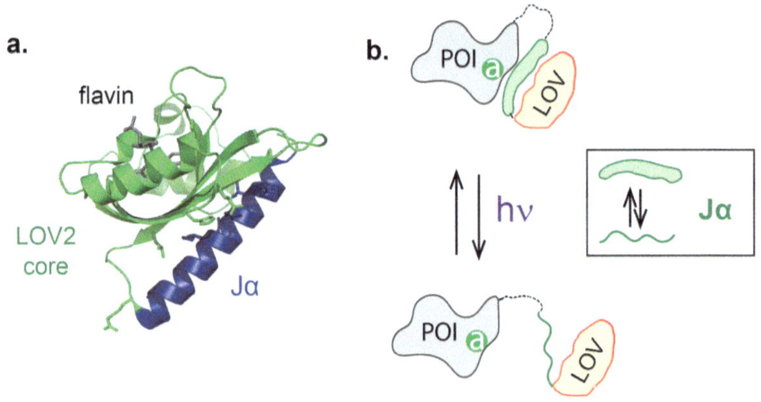

Fig. 1 (**a**) The LOV2 domain. The green globular portion contains a flavin that absorbs light between 400 and 500 nm. Light absorption initiates changes in structure that ultimately lead to unwinding of the Jα helix (blue). This helix is at the C terminus of LOV2 where it can readily be appended to target proteins. (**b**) Protein photoactivation using the LOV domain as a steric block. The Jα helix is fused to the N terminus of the protein of interest (POI). When the Jα helix is coiled, LOV covers the active site of the POI. Upon irradiation the Jα helix unwinds, placing the LOV globular domain at the end of a longer linker and freeing the POI active site

400 and 500 nm, the LOV domain undergoes a dramatic conformational change that affects several portions of its structure [2–4]. Our methods harness the light-induced unwinding of the Jα helix (Fig. 1a). LOV is well suited for incorporation into different target proteins: the Jα helix is positioned at the C terminus where it can be fused to targets. Because the N and C termini are close together, LOV can be spliced into surface loops of target proteins, and the Jα helix can withstand substantial structure manipulation while maintaining light-controlled unfolding. The light-absorbing flavin cofactor of LOV is incorporated during protein synthesis, making LOV a good choice for optogenetic tools to be used in animals.

Although LOV undergoes its light-induced conformational change in less than 2 ms [5], it returns to its dark state with a half-life of 27 s [6]. This limits the precision with which the kinetics of protein activity can be controlled, and also limits spatial control because substantial diffusion can occur while optogenetic analogs are turning off. Fortunately, point mutations can be used to adjust the rate of return to the dark state, producing half-lives ranging from 1.7 to 496 s [7, 8]. For precise kinetics the most rapid possible return is useful, but slow return kinetics can also be valuable. In animal studies, for example, they enable the activated state to be maintained simply using brief light pulses every ten minutes. The LOV conformational change has been induced using scanning confocal microscopes, simple mercury arc lamp illumination, turning on room lights, or more sophisticated equipment for precise control, including feedback from biological readouts to maintain fixed levels of activation [9]. Scanning instruments activate more slowly, as each pixel receives illumination only periodically during the scan.

LOV is always in equilibrium between the lit and dark states, with a K_d that is affected by light (i.e., irradiation shifts the K_d to favor the lit conformation). Because of this, most optogenetic protein analogs show a small amount of activity even under conditions meant to keep them in the off state. This "leakiness" is an important issue whose impact on biological studies depends on the effectiveness of inactivation, the increase in activity upon irradiation, and the biological characteristics of the target protein. In some cases, expression must be carefully controlled. A major advantage of optogenetics is the ability to suddenly trigger a change in protein activity, without giving the cell enough time to compensate, and expression of a leaky protein can lead to compensation. It is often valuable to stably express optogenetic proteins under the control of an inducible promoter, to prevent compensation, and for more convenient tissue culture. When the protein is expressed, cells can sometimes be cultured in TC dishes covered with foil, but it usually best to handle them solely under red light for 24 h or more before the experiment.

The LOV domain is a versatile building block that can be incorporated in proteins to produce very different photomodulation approaches, providing routes to widely varying protein structures. It can be used for photoactivation or photoinhibition, to send proteins to particular regions of the cell, to produce protein analogs that when activated act at multiple sites, or to control protein fragments that modulate endogenous targets. In the sections below we illustrate LOV's versatility by contrasting different designs and highlighting considerations that determine when each can best be used.

2 Photoactivatable GTPases

In our first foray into the field, we sought to photoactivate a protein that was not an ion channel. Optogenetics to that point had used light-responsive ion channels to control nerve conduction, but we needed to manipulate GTPases, our primary research focus. We devised a strategy to reversibly position the LOV domain over the active site of the GTPase Rac1. We reasoned that a steric block, not dependent on specific binding interactions, could be applied to multiple closely related GTPases. The Jα helix was conveniently placed at the end of the LOV sequence so that we could attach it directly to the N terminus of Rac. The coiled helix would hold LOV over the active site, but unwinding the helix with light would produce a long linker that enabled Rac to interact with its downstream effectors. Optimizing the residues connecting the Jα helix to Rac1 produced the optogenetic protein analog we sought (photoactivatable Rac1, PA-Rac1) [10, 11] (Fig. 1b).

Using point mutations we knocked out upstream regulatory sites so that the protein would be controlled only by irradiation. Rac1 and other Rho family GTPases act at the plasma membrane but are sequestered in the cytoplasm when inactive. To focus our light where it was needed, and to eliminate the influence of mechanisms that translocate Rac1 into the membrane upon activation, we used a constitutively membrane-anchored version of Rac1. We did not yet know about LOV mutations that sped the return to the dark state (*see* above), so PA-Rac1 diffused from the point of activation over tens of seconds while returning to the dark state, limiting our spatial resolution to about 10 μm. Irradiating off the cell edge could narrow the region of the cell experiencing activation.

PA-Rac1 proved effective not only at controlling the morphology and movements of single mammalian cells [10, 11] but could also guide the movements of cells within Drosophila [11, 12] and zebrafish [13]. In the brains of mice, PA-Rac1 was used to control specific memories [14] or to modulate cocaine addiction [15]. Notably, Rac activation alone could in some cases guide cell movement, but sometimes Rac1 could produce protrusion only in specific subcellular regions (e.g., the front of polarized fibroblasts), likely because Rac1 signaling was superimposed on inhibitory programs or over-riding signals controlling polarization.

Because PA-Rac1 relied on a steric block of the active site, we initially thought the strategy would be applicable to many GTPases, but we were surprised to find that neither RhoA nor Cdc42 could be controlled. Although LOV's block of the Rac1 active site did depend on the state of the Jα helix, molecular modeling and mutagenesis showed that it also relied on a surprising binding interaction between LOV and the Rac1 surface, even though these proteins are not thought to interact in nature. By mutating the surface of Cdc42 we could produce a PA-Cdc42 [10], but only at the expense of perturbing Cdc42 interactions that may determine target specificity. Only recently we have understood the LOV-GTPase interface sufficiently to predict which GTPases can be controlled by attaching LOV, and thereby produced new PA-GTPases (unpublished data).

3 Applications of Zdk, a Building Block that Binds LOV only in the Dark

Given that the PA-Rac1 approach appeared initially to be suitable only for Rac1, we searched for a route to access other Rho GTPases. This family of molecules is regulated by controlling their access to

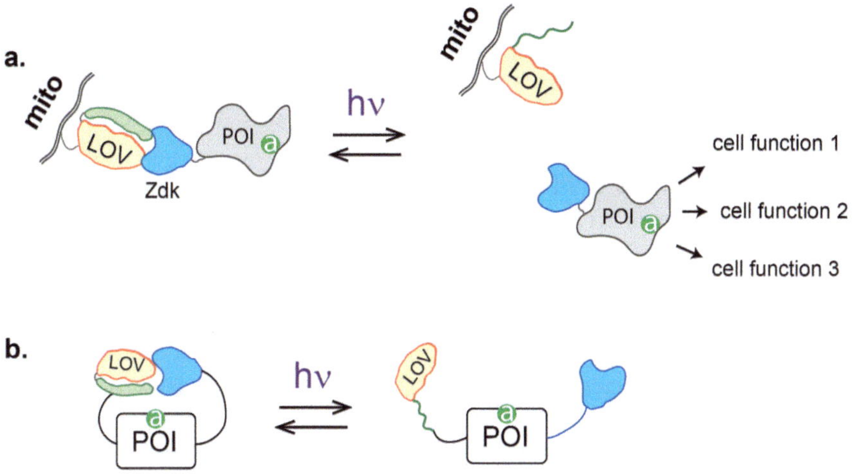

Fig. 2 (**a**) LOVTRAP. LOV is anchored at a site in the cell away from the POI site of action. The POI is fused to Zdk, a small protein that binds LOV only in the dark. Upon irradiation, Zdk releases LOV, freeing the POI to move to its site of action. Rather than controlling the localization of active protein, LOVTRAP releases an active protein that is free to act at multiple sites. (**b**) Z-lock. LOV and Zdk are attached to the N and C termini of the target protein, positioned so they block the active site when they are bound to one another. Irradiation breaks the LOV-Zdk linkage, freeing the active site. The papers describing LOVTRAP and Z-lock provide a library of Zdk molecules with varying affinity

the plasma membrane. When in the membrane they interact with effectors to exert activity but otherwise are held inactive in the cytosol, where the lipid anchors they use for membrane interaction are buried in guanine dissociation inhibitors. We took advantage of this regulatory mechanism by using light to control the GTPases' subcellular distribution, and in the process generated a broadly applicable method much easier to apply to many targets. This method was named LOVTRAP [7, 8], because it was based on trapping the protein of interest (POI) at a subcellular site where it could not reach its targets (Fig. 2a). Trapping was effective in the dark, but the protein was reversibly released whenever the cell was irradiated. LOVTRAP depended on a new reagent named Zdk that bound selectively to the dark conformation of LOV. This was created by screening a large library based on the Z domain of protein A. A number of Zdk reagents, with differing affinity and binding sites on LOV, proved useful in different optogenetic methods [8, 16].

In the LOVTRAP method, Zdk was appended to the POI, and the LOV domain was anchored at mitochondria. In the dark, the POI was sequestered at mitochondria, where it could not access its targets (e.g., GEFs and GTPases that act at the plasma membrane). Irradiation broke the linkage between the POI-Zdk fusion and LOV, releasing the POI. Proteins could also be anchored at other subcellular sites. LOVTRAP was especially useful because it could be applied to a wide range of protein structures; Zdk simply had to be attached to either the N or C terminus. Importantly LOVTRAP could overcome the "leakiness" that plagued other methods (*see* Subheading 1); LOV at the mitochondria could be expressed in excess over the POI so that there was always excess dark state LOV to trap the POI.

LOVTRAP had clear advantages and disadvantages. It worked well to precisely control the kinetics of protein activation, as illustrated by our studies of oscillations governing Rho signaling pathways [8]. However, because the POI had to diffuse from mitochondria to the target site, spatial accuracy was compromised. For large nerve cells, localized irradiation of mitochondria provided sufficient spatial control [17]. Some users reported incomplete release from mitochondria, preventing them from reaching effective concentrations of free POI. By attaching LOV to the POI, and Zdk to the sequestration site (reversing the positions described above), more complete release could be achieved, but at the expense of additional leakiness [7]. In this configuration, LOV had to be attached to the POI's N terminus because all Zdk variants bind near LOV's C terminus.

The Zdk reagent was also used in an approach called Z-lock (Fig. 2b) [16]. As in the PA-GTPases described above, the target's active site was occluded by a reversible steric block. Zdk was attached to one terminus of the protein and LOV to the other. With appropriate linker length the Z-lock and LOV bound each other in the dark, forming a bridge over the active site that was broken upon irradiation. Unlike the steric blocking used in the PA-GTPases, Z-lock promises to be a versatile tool because it does not require specific interaction with any part of the target. Because modification of both termini is required, it will likely not be so useful in controlling intact protein analogs, but rather can be most readily applied to protein fragments with a single active site and a single activity (e.g., enzymes). To date, Z-lock has been applied to control the activity of the actin severing protein cofilin and αTat, a protein that acetylates microtubules. In some cases circular permutation may be useful to position the termini. Because the LOV-Zdk interaction in Z-lock is intramolecular, release in the light was not very effective. We produced a tool kit of Zdk reagents with reduced affinity that can be used to build other Z-lock proteins.

4 Loopology

We found that we could insert either the LOV domain or an engineered drug-responsive domain into proteins to allosterically control the active site (Fig. 3a). Insertion could be far from important binding sites, and in many cases the conformations induced by the inserted domain mimicked the native on and off conformations

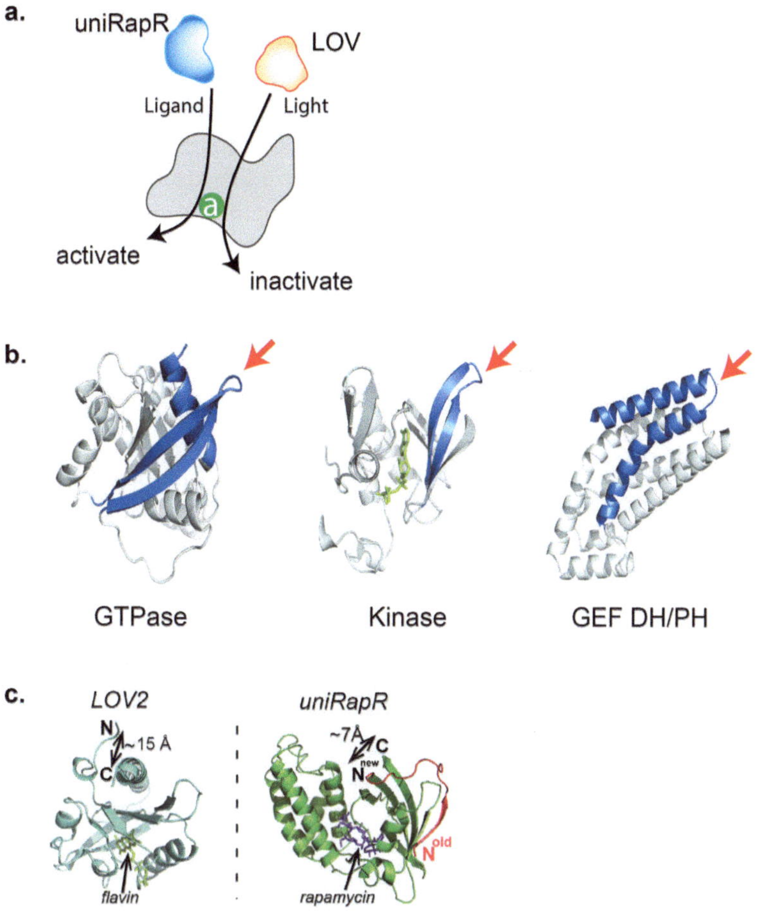

Fig. 3 (**a**) Loopology. For this approach, one first identifies surface loops that are allosterically connected to the active site, ideally far from any important binding sites. Domains can then be inserted into the loops to confer sensitivity to small molecules (uniRapR, small molecules induce protein activation), or to light (LOV, irradiation leads to protein inhibition). By using light to inhibit autoinhibitory domains, light-induced activation is possible. (**b**) Examples of surface loops that are part of secondary structure motifs. Antiparallel helices and other configurations span the interior of the protein and reach from surface loops to the active site. In some cases, perturbation of the surface loop travels through more than one such motif. (**c**) The spacing between LOV and uniRapR's termini matches the length of surface loops

of the target protein. To confer response to either light or small molecules, engineered domains were inserted into small "tight loops" on the protein surface (Fig. 3b). These loops are an essential part of many secondary structure motifs, acting as short connecting strands between antiparallel helices, segments of a β-pleated sheet, and so on. These motifs are largely buried within proteins, but the tight loops often project out at the surface. A scan of structures in the protein database (PDB) revealed that tight loops are restricted to a narrow range of lengths. The insertable domains were selected or engineered so that the spacing between their N and C termini matched the length of typical tight loops (approximately 10 Å), enabling us to insert them into diverse protein structures. To date we have used two different insertable domains: For response to light we used the LOV domain, whose N and C termini are less than 15 Å apart (Fig. 3c, left). To cause proteins to respond to small molecules we used the 12-kDa FK506 binding protein (FKBP12), which binds to the membrane-permeable drug rapamycin. The rapamycin interaction mediates binding to a second protein, FRB, which for some targets was important to the activation mechanism (*see* below). The FKBP N terminus was truncated to bring the N and C termini within 7 Å, to make insertable FKBP (iFKBP) [18] (Fig. 3c, right). The rapamycin-responsive analogs were controlled by light using a "caged" analog of rapamycin (rapamycin bearing a light-cleavable protecting group) [19] (Fig. 4a).

When a tight loop was replaced with the LOV domain, dark state LOV could effectively substitute for the loop. Upon irradiation, the LOV Jα helix unwound, increasing the spacing across the loop and the disorder of the associated protein structure. Through proper selection of the insertion site, this disorder could be communicated to the active site, effectively inhibiting the target protein with light [20]. Molecular modeling, NMR, and crystallography studies indicated that target proteins were being driven between two energy minima, their naturally occurring active and inactive conformations [20].

Unlike LOV, insertion of the rapamycin-responsive iFKBP domain *inhibited* protein activity; Addition of rapamycin restored the target protein to normal (Fig. 4b). Molecular modeling of rapamycin-regulated kinases (RapR kinases) indicated that a mechanism similar to that of LOV insertion was at play. In the absence of rapamycin, iFKBP was loosely folded and had substantial molecular motion; this was transferred to a critical ATP binding site in the kinase, killing activity [18]. Rapamycin addition caused the iFKBP to fold tightly and bind coexpressed FRB protein, reducing molecular motion and thereby restoring the normal mobility of the ATP binding site. Activation was essentially irreversible, but the Karginov laboratory showed that a functionally silent mutation developed by Shokat et al. could be used to make RapR kinases susceptible to an inhibitor [21].

Our initial success with kinases led us to develop a straightforward way to identify insertion sites for diverse protein structures [22]. Briefly put, surface loops are identified through inspection or by calculating solvent exposure, and highly conserved loops are dropped from contention. The remaining loops are screened for allosteric interactions with the active site using either molecular dynamics simulations or a simpler approach based on the distance between internal residues. Optimization of the linkers connecting

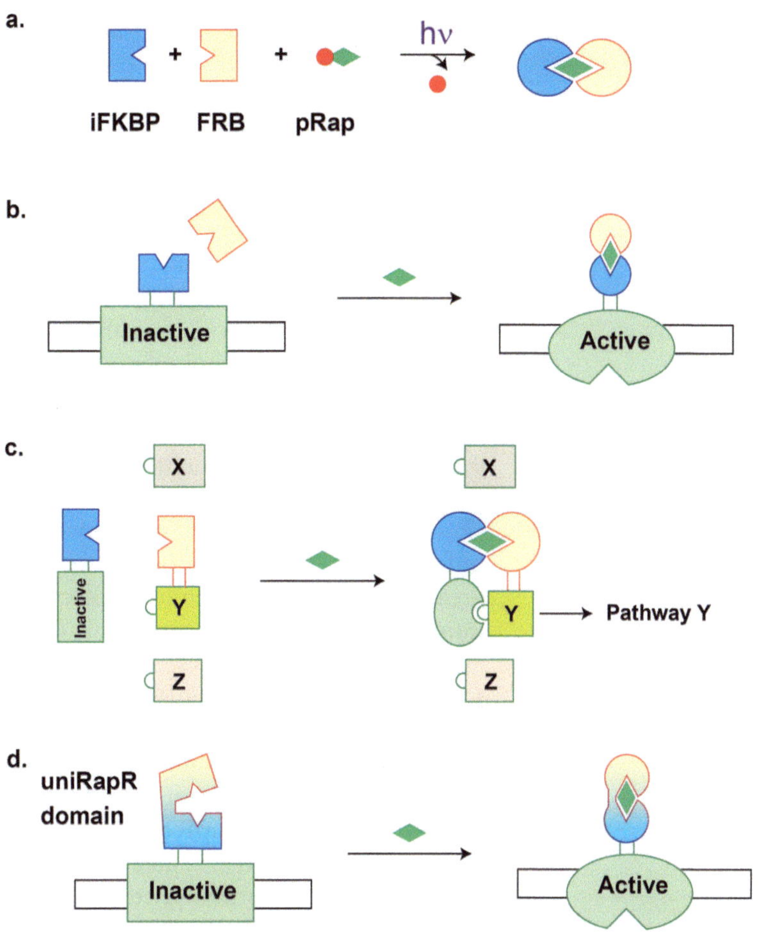

Fig. 4 Variations on a theme—applications of iFKBP. (**a**) iFKBP binds FRB in the presence of the membrane-permeable small molecule rapamycin. "Caged" rapamycin, bearing a photocleavable protecting group, does not interact with the proteins until it is irradiated. (**b**) When iFKBP is introduced in an appropriate surface loop, it knocks out the activity of the modified protein. In the presence of rapamycin, this iFKBP binds coexpressed FRB, causing protein activation. (**c**) When the activated protein has multiple downstream effectors, FRB can be attached to one so that activation of the protein leads to interaction with only one specific downstream target. (**d**) The iFKBP and FRB groups can be fused to generate the uniRapR domain. Using this leads to less heterogeneous responses than when FRB and an iFKBP-POI fusion are coexpressed

the inserted domain to the target is important, but initial screens can identify promising loops before optimization. This approach has been used for photoinhibition of GEFs, GTPases, and kinases, and for rapamycin activation of kinases and GEFs. By inhibiting the autoinhibitory domain, it was also possible to photoactivate Vav2.

There are now several variants of the rapamycin-mediated activation, each with unique abilities. By attaching FRB to a specific downstream target of an activatable kinase, we could direct the kinase to interact with only one specific target upon activation (Fig. 4c). This "rapamycin-regulated targeted activation of pathways" (RapRTAP) approach was used to show the different roles of p130Cas activation and focal adhesion kinase activation downstream of Src [23]. iFKBP was also fused to FRB to produce "uniRapR" kinases that could be used without coexpressing FRB (Fig. 4d) [24]. This eliminated heterogenous responses produced by variations in the relative expression of the two proteins. Finally, we used the iFKBP group to enhance the control of split proteins [25]. In a variety of techniques proteins are split into two parts that can be induced to reassemble and restore activity. Spontaneous reassembly, in the absence of stimulation, has been a persistent problem with these techniques. iFKBP and FRB were placed on the two protein halves to control reassembly with rapamycin (Fig. 5). Before rapamycin was added, the iFKBP domain partially unfolded the attached protein, so spontaneous reassembly was greatly reduced. This technique, known as SPELL, was used for a number of proteins, and specifically for light activation of Vav2.

One advantage of these techniques is their essentially absolute specificity. Genetic manipulations lead to compensation, and drugs can usually not distinguish between very similar isoforms or structures. With the protein analogs described here only the modified protein is affected by light. We demonstrated the power of this specificity by making RapR analogs of very closely related Src family kinases (e.g., Fyn and Src, greater than 70% homology). Activation of each produced a strikingly different phenotype and translocated to different subcellular locations upon activation [26].

Photoinhibition, as opposed to photoactivation, presents unique challenges. With photoactivation, if the protein can be expressed at levels where "leakiness" has little effect, the cell can be considered unaffected prior to irradiation. In contrast, photoinhibitable protein analogs are active until they are turned off, so simple expression essentially produces reversible overexpression. In some cases (especially with GEFs) we have found that the cell compensates for fluorescent analog expression by downregulating the endogenous protein. In other cases, knockdown/rescue or

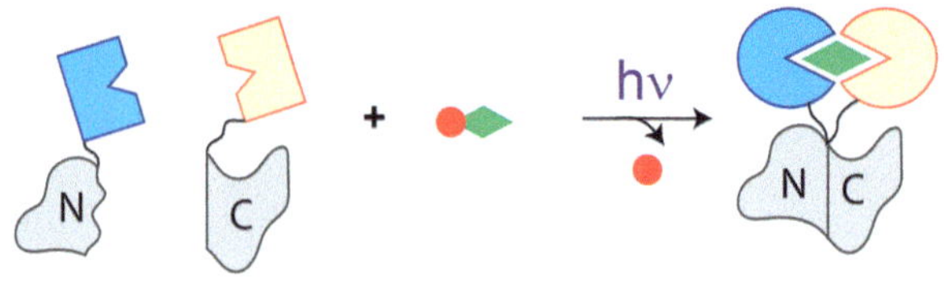

Fig. 5 SPELL. When iFKBP and FRB are placed on the two halves of a split protein, activity can be restored by adding rapamycin to reunite the halves. When not bound to rapamycin, the iFKBP causes partial unfolding of the half to which it is attached, reducing undesirable spontaneous reassembly

replacement of endogenous genes with optogenetic analogs could be used. For photoactivated proteins, one can knock out upstream regulation through point mutation or truncation so that protein activity is solely in control of the user. For photoinhibitable proteins, it may instead be wise to leave regulation intact, as the protein analog will substitute for native protein.

5 Embedded Peptides

The Jα helix of the LOV domain can withstand extensive modifications without perturbing critical interactions that maintain helical structure in the dark state [27, 28]. This has been used for light-induced dimerization, by incorporating a peptide in the Jα helix where it is unable to interact with its binding partners while the helix is folded [27–31]. When light causes the helix to unwind, the peptide is freed to interact with another protein or peptide that is anchored at a specific site in the cell (e.g., the plasma membrane). This light-induced dimerization has been used to direct GEFs to the plasma membrane, where they interact with targets.

In another application, a nuclear import peptide was embedded in the helix such that it could interact with the nuclear import machinery only when Jα was unwound [32–36]. This construct was attached to proteins for light-controlled nuclear import. Leakiness occurred both because of LOV unfolding in the dark, and because of simple diffusion into the nucleus. This was controlled by including a weak nuclear export signal, always exposed, that could be overcome by the nuclear localization signal. The Zdk group was used to further reduce leakiness by anchoring it at the mitochondria [37]. In the dark, Zdk bound to the NLS–LOV combination; upon irradiation, the NLS was exposed and LOV was released by Zdk, too (Fig. 6a).

Hiding peptides in the LOV domain presents an exciting path for future research. Biologists often seek the ability to inhibit specific proteins, but the use of photoinhibitable protein analogs is complicated by the presence of endogenous protein (*see* above). We have explored the photoactivation of inhibitory peptides as an alternative to photoinhibitable protein analogs [38] (Fig. 6b). Peptides that inhibit either myosin light chain kinase or protein kinase A were embedded in LOV where they could only interact with their targets upon irradiation. This enabled photoinhibition of endogenous kinases.

An attempt to inhibit a GTPase by using Z-lock to control a fragment of a downstream protein was less successful, and held a lesson for this approach. Although biochemical assays showed light-controlled binding to a GTPase region required for downstream interactions, no effect was seen in living cells. This was perhaps because fragments of downstream targets have to compete

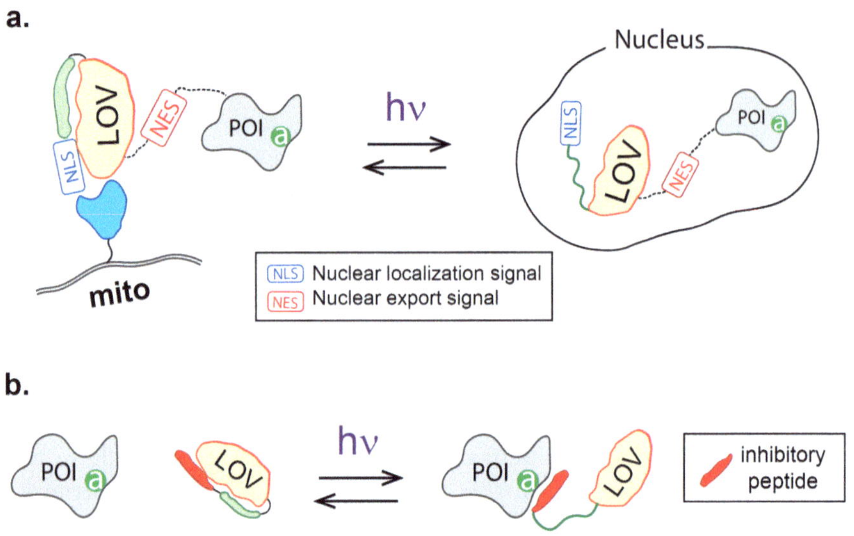

Fig. 6 Controlling embedded peptides. (**a**) For photocontrol of nuclear import a nuclear localization sequence is embedded in the Ja helix where it is exposed only upon irradiation. To hold the construct out of the nucleus in the dark, a weaker nuclear export sequence is always exposed, and Zdk anchored at mitochondria binds LOV only in the dark. (**b**) Other peptides can be embedded in the helix as well, including inhibitors of endogenous proteins

with endogenous molecules. In the future, it will be important to find inhibitors that can be expressed at effective levels while not producing effects due to leakiness.

6 Conclusion

This chapter was certainly not intended as a review but rather as a discussion of the things we have learned by developing and comparing designs. Hopefully it will spark new ideas, and the caveats and considerations we have discussed will save time for future protein engineers. These designs were made possible through the extensive help and collaboration of multiple friends and colleagues, to whom we are very grateful.

Acknowledgments

We are grateful to the National Institutes of Health for funding (GM-R35GM122596 to K.M.H.) and thank Andrei Karginov, Onur Dagliyan, and Brian Kuhlman for helpful suggestions.

References

1. Christie JM, Salomon M, Nozue K, Wada M, Briggs WR (1999) LOV (light, oxygen, or voltage) domains of the blue-light photoreceptor phototropin (nph1): binding sites for the chromophore flavin mononucleotide. Proc Natl Acad Sci U S A 96:8779–8783
2. Harper SM, Neil LC, Gardner KH (2003) Structural basis of a phototropin light switch. Science 301:1541–1544
3. Swartz TE, Corchnoy SB, Christie JM, Lewis JW, Szundi I, Briggs WR et al (2001) The photocycle of a flavin-binding domain of the blue light photoreceptor phototropin. J Biol Chem 276:36493–36500
4. Zayner JP, Antoniou C, Sosnick TR (2012) The amino-terminal helix modulates light-activated conformational changes in AsLOV2. J Mol Biol 419:61–74
5. Eitoku T, Nakasone Y, Matsuoka D, Tokutomi S, Terazima M (2005) Conformational dynamics of phototropin 2 LOV2 domain with the linker upon photoexcitation. J Am Chem Soc 127:13238–13244
6. Salomon M, Christie JM, Knieb E, Lempert U, Briggs WR (2000) Photochemical and mutational analysis of the FMN-binding domains of the plant blue light receptor, phototropin. Biochemistry 39:9401–9410
7. Wang H, Hahn KM (2016) LOVTRAP: a versatile method to control protein function with light. Curr Protoc Cell Biol 73:21.10.1–21.10.14
8. Wang H, Vilela M, Winkler A, Tarnawski M, Schlichting I, Yumerefendi H et al (2016) LOVTRAP: an optogenetic system for photoinduced protein dissociation. Nat Methods 13:755–758
9. Toettcher JE, Gong D, Lim WA, Weiner OD (2011) Light-based feedback for controlling intracellular signaling dynamics. Nat Methods 8:837–839
10. Wu YI, Frey D, Lungu OI, Jaehrig A, Schlichting I, Kuhlman B et al (2009) A genetically encoded photoactivatable Rac controls the motility of living cells. Nature 461:104–108
11. Wu YI, Wang X, He L, Montell D, Hahn KM (2011) Spatiotemporal control of small GTPases with light using the LOV domain. Methods Enzymol 497:393–407
12. Wang X, He L, Wu YI, Hahn KM, Montell DJ (2010) Light-mediated activation reveals a key role for Rac in collective guidance of cell movement in vivo. Nat Cell Biol 12:591–597
13. Yoo SK, Deng Q, Cavnar PJ, Wu YI, Hahn KM, Huttenlocher A (2010) Differential regulation of protrusion and polarity by PI3K during neutrophil motility in live zebrafish. Dev Cell 18:226–236
14. Hayashi-Takagi A, Yagishita S, Nakamura M, Shirai F, Wu YI, Loshbaugh AL et al (2015) Labelling and optical erasure of synaptic memory traces in the motor cortex. Nature 525:333–338
15. Dietz DM, Sun H, Lobo MK, Cahill ME, Chadwick B, Gao V et al (2012) Rac1 is essential in cocaine-induced structural plasticity of nucleus accumbens neurons. Nat Neurosci 15:891–896
16. Stone OJ, Pankow N, Liu B, Sharma VP, Eddy RJ, Wang H et al (2019) Optogenetic control of cofilin and TAT in living cells using Z-lock. Nat Chem Bio 15(12):1183–1190
17. Takano T, Wu M, Nakamuta S, Naoki H, Ishizawa N, Namba T et al (2017) Discovery of long-range inhibitory signaling to ensure single axon formation. Nat Commun 8:33
18. Karginov AV, Ding F, Kota P, Dokholyan NV, Hahn KM (2010) Engineered allosteric activation of kinases in living cells. Nat Biotechnol 28:743–747
19. Karginov AV, Zou Y, Shirvanyants D, Kota P, Dokholyan NV, Young DD et al (2011) Light regulation of protein dimerization and kinase activity in living cells using photocaged rapamycin and engineered FKBP. J Am Chem Soc 133:420–423
20. Dagliyan O, Tarnawski M, Chu PH, Shirvanyants D, Schlichting I, Dokholyan NV et al (2016) Engineering extrinsic disorder to control protein activity in living cells. Science 354:1441–1444
21. Klomp JE, Huyot V, Ray AM, Collins KB, Malik AB, Karginov AV (2016) Mimicking transient activation of protein kinases in living cells. Proc Natl Acad Sci U S A 113:14976–14981
22. Dagliyan O, Dokholyan NV, Hahn KM (2019) Engineering proteins for allosteric control by light or ligands. Nat Protoc 14:1863–1883
23. Karginov AV, Tsygankov D, Berginski M, Chu PH, Trudeau ED, Yi JJ et al (2014) Dissecting motility signaling through activation of specific Src-effector complexes. Nat Chem Biol 10:286–290
24. Dagliyan O, Shirvanyants D, Karginov AV, Ding F, Fee L, Chandrasekaran SN et al (2013) Rational design of a ligand-controlled

protein conformational switch. Proc Natl Acad Sci 110:6800–6804
25. Dagliyan O, Krokhotin A, Ozkan-Dagliyan I, Deiters A, Der CJ, Hahn KM et al (2018) Computational design of chemogenetic and optogenetic split proteins. Nat Commun 9:4042
26. Chu PH, Tsygankov D, Berginski ME, Dagliyan O, Gomez SM, Elston TC et al (2014) Engineered kinase activation reveals unique morphodynamic phenotypes and associated trafficking for Src family isoforms. Proc Natl Acad Sci U S A 111:12420–12425
27. Lungu Oana I, Hallett Ryan A, Choi Eun J, Aiken Mary J, Hahn Klaus M, Kuhlman B (2012) Designing photoswitchable peptides using the AsLOV2 domain. Chem Biol 19:507–517
28. Zimmerman SP, Kuhlman B, Yumerefendi H (2016) Engineering and application of LOV2-based photoswitches. Methods Enzymol 580:169–190
29. Guntas G, Hallett RA, Zimmerman SP, Williams T, Yumerefendi H, Bear JE et al (2015) Engineering an improved light-induced dimer (iLID) for controlling the localization and activity of signaling proteins. Proc Natl Acad Sci USA 112:112–117
30. Hallett RA, Zimmerman SP, Yumerefendi H, Bear JE, Kuhlman B (2016) Correlating in vitro and in vivo activities of light-inducible dimers: a cellular optogenetics guide. ACS Synth Biol 5:53–64
31. Strickland D, Lin Y, Wagner E, Hope CM, Zayner J, Antoniou C et al (2012) TULIPs: tunable, light-controlled interacting protein tags for cell biology. Nat Methods 9:379–384
32. Di Ventura B, Kuhlman B (2016) Go in! Go out! Inducible control of nuclear localization. Curr Opin Chem Biol 34:62–71
33. Niopek D, Wehler P, Roensch J, Eils R, Di Ventura B (2016) Optogenetic control of nuclear protein export. Nat Commun 7:10624
34. Niopek D, Benzinger D, Roensch J, Draebing T, Wehler P, Eils R et al (2014) Engineering light-inducible nuclear localization signals for precise spatiotemporal control of protein dynamics in living cells. Nat Commun 5:4404
35. Lerner AM, Yumerefendi H, Goudy OJ, Strahl BD, Kuhlman B (2018) Engineering improved photoswitches for the control of nucleocytoplasmic distribution. ACS Synth Biol 7:2898–2907
36. Yumerefendi H, Dickinson DJ, Wang H, Zimmerman SP, Bear JE, Goldstein B et al (2015) Control of protein activity and cell fate specification via light-mediated nuclear translocation. PLoS One 10:e0128443
37. Yumerefendi H, Wang H, Dickinson DJ, Lerner AM, Malkus P, Goldstein B et al (2018) Light-dependent cytoplasmic recruitment enhances the dynamic range of a nuclear import photoswitch. Chembiochem 19:2898–2907
38. Yi JJ, Wang H, Vilela M, Danuser G, Hahn KM (2014) Manipulation of endogenous kinase activity in living cells using photoswitchable inhibitory peptides. ACS Synth Biol 3:788–795

Chapter 8

Optogenetic Control of Nucleocytoplasmic Protein Transport

Daniel Weis and Barbara Di Ventura

Abstract

The transport of proteins between the nucleus and the cytosol is a vital process regulating cellular activity. The ability to spatiotemporally control the nucleocytoplasmic transport of a protein of interest allows for elucidating its function taking into account the dynamic and heterogeneous nature of biological processes contrary to conventional knockin, knockout, and chemically induced overexpression strategies. We recently developed two optogenetic tools, called LINuS and LEXY, for reversibly controlling with blue light the nuclear import and export of proteins of interest, respectively. Here we describe how to use them to control the localization of a protein of interest in cultured mammalian cells using a fluorescence microscope.

Key words Optogenetics, Nuclear protein import, Nuclear protein export, Synthetic biology, AsLOV2, Protein engineering

1 Introduction

Eukaryotic cells are highly compartmentalized. Membrane-bound organelles as well as various proteinaceous compartments help keep cellular processes well organized, efficient, and isolated from the surrounding cellular activity. A prerequisite for cellular compartments to be effective is that the entrance, permanence and exit of proteins and other molecules from their interior be under tight control. For proteins, this is typically achieved by localization tags, posttranslational modifications, or association with other proteins [1–3]. The nucleus is arguably the most important cellular compartment as it houses the most precious molecule of all: the DNA. Trafficking in and out of the nucleus is finely regulated to make sure that only the needed molecules are present in this compartment at any time point. The nuclear envelope is the barrier that cytosolic/nuclear proteins need to pass in order to gain access to the nucleus/cytosol [4]. This double membrane contains structures, called nuclear pore complexes (NPCs), through which molecules can come in and go out of the nucleus [5]. While proteins smaller than ~60 kDa can passively diffuse through the NPCs, larger

Dominik Niopek (ed.), *Photoswitching Proteins: Methods and Protocols*, Methods in Molecular Biology, vol. 2173, https://doi.org/10.1007/978-1-0716-0755-8_8,

proteins require most often the engagement of the specialized, energy-consuming import–export machinery [6]. In this latter case, the protein to be imported/exported, called the cargo, must contain a distinctive signal: the nuclear localization signal (NLS) or nuclear export signal (NES). NLSs and NESs are recognized by specialized nuclear transport receptors that escort the cargo through the NPCs toward the destination compartment [4]. If the NLS/NES is present on the cargo but concealed, for instance due to a certain structural conformation, the receptor cannot bind and the cargo is not imported/exported. Many natural processes such as gene expression are regulated this way because NLSs/NESs present on regulatory proteins, for instance transcription factors, become exposed only in response to the appropriate stimuli [1]. Typically, this process is reversible, allowing for the relocalization of the regulatory proteins to the compartments they resided in before arrival of the signal.

We have recently developed two optogenetic tools based on this very concept: we have exploited a small protein domain that naturally responds to blue light with a drastic conformational change to reversibly control the exposure of a NLS and NES, respectively. The tools, called LINuS (for blue light-inducible nuclear import) [7] and LEXY (for blue light-inducible nuclear export) [8], are based on the second light-oxygen-voltage (LOV) domain of *Avena sativa* phototrophin 1 (*As*LOV2). Specifically, the *As*LOV2 domain is modified to bear within its C-terminal Jα helix a NLS/NES (Fig. 1). In the dark, the Jα helix is folded and bound to the core *As*LOV2 domain [9–11], effectively concealing the NLS/NES from the endogenous import–export machinery [7, 8]. Upon blue light exposure, the Jα helix unfolds and undocks from the core *As*LOV2 domain [9–11], allowing for the recognition and binding of the NLS/NES by the endogenous import–export receptor [7, 8]. Notably, once back in the dark, the *As*LOV2 domain reacquires within less than a minute its closed conformation [12], thus masking the NLS/NES once again. This reversibility at the level of the *As*LOV2 domain is most often reflected by the reacquisition by the protein of interest fused to LINuS/LEXY of its localization prior to light exposure.

LINuS and LEXY are small, single-component optogenetic tools that allow for dictating with high spatiotemporal precision if a protein of interest should be nuclear or cytoplasmic. Interestingly, the group of Brian Kuhlmann has independently developed LANS and LINX, two optogenetic tools identical in concept to LINuS and LEXY, albeit slightly different in the realization [13, 14]. The advantage of these *As*LOV2-based tools is that the chromophore, flavin mononucleotide (FMN), is made by all cells and does not need external supply. The disadvantage is, however, that blue light can be toxic to cells if not carefully administered and poorly penetrates into biological tissues.

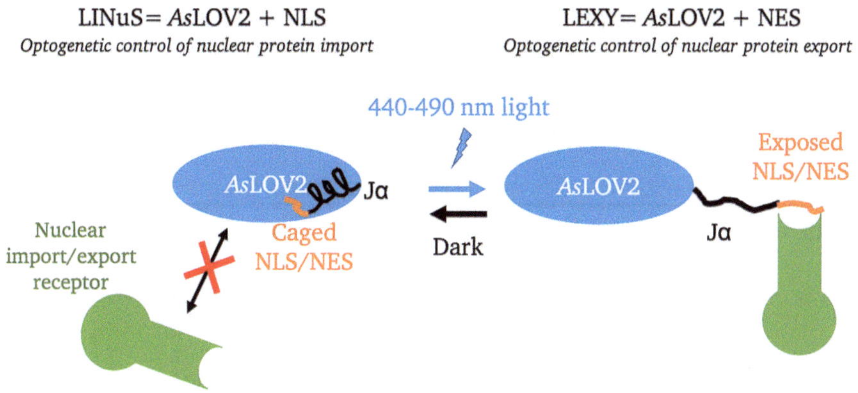

Fig. 1 LINuS and LEXY are based on the *As*LOV2 domain. Schematic drawing showing the conformational change that the modified *As*LOV2 domain within LINuS and LEXY undergoes upon blue light absorption leading to the exposure of the NLS/NES

In this chapter, we describe an easy protocol to assess whether a protein of interest fused to LINuS/LEXY responds as expected to light by accumulating into the nucleus/cytoplasm and to darkness by relocating back into the cytosol/nucleus. This implies that a fluorescent protein (*see* **Note 1**) be included in the final construct to monitor the localization of the fusion protein with live fluorescence microscopy.

2 Materials

1. Construct expressing a protein fused to –mCherry-LINuS or –mCherry-LEXY.
2. Shuttling control vectors NES-mCherry-LINuS (e.g., Addgene #61346) or NLS-mCherry-LEXY (Addgene #72655).
3. Stuffer DNA (e.g., pBluescript).
4. HEK293T cells (*see* **Note 2**).
5. JetPrime (Polyplus Transfection) (*see* **Note 3**).
6. Cell culture medium (e.g., DMEM) supplemented with 10% fetal bovine serum (*see* **Note 4**).
7. 35 mm glass-bottom microscope dishes (Greiner One) (*see* **Note 5**)
8. 1.5 mL tubes.
9. Microcentrifuge.
10. Vortex.
11. Tissue culture incubator.

12. Epifluorescence microscope (*see* **Note 6**) equipped with GFP (*see* **Note 7**) and mCherry filter sets.
13. Computer with ImageJ (or other appropriate image analysis software).

3 Methods

While the following protocol has been optimized for HEK293T cells, it can be easily adapted to other cell lines such as HeLa, HepG2, or H1299 (*see* **Note 2**).

1. Seed ~350,000 HEK293T cells into a 35 mm microscopy dish using 2 mL culture medium.
2. The following day, add 50–100 ng (*see* **Note 8**) of LINuS/LEXY vector or shuttling control vector (*see* **Note 9**) and 400–450 ng stuffer DNA to 200 μL JetPrime Buffer in a 1.5 mL tube. Vortex sample for 10 s and briefly spin down.
3. Add 2 μL JetPrime reagent to the DNA mix. Vortex sample for 10 s and briefly spin down. Incubate transfection mix at RT for 10 min.
4. Add transfection mix dropwise to the cells (*see* **Note 10**).
5. Incubate cells in a tissue culture incubator at 37 °C and 5% CO_2 for about 16 h (*see* **Note 11**). Shield samples from light (also white light) during incubation.
6. Switch the microscope on and prewarm the incubation chamber (*see* **Note 12**) to 37 °C for at least 20 min prior to use. Adjust CO_2 levels to 5%.
7. Take one sample from the incubator and transport it to the microscope wrapped in aluminum foil or within a light-protected box. Place sample in sample holder without exposing it to light (*see* **Note 13**).
8. Use red-filtered transmission light or the mCherry channel to focus the cells (*see* **Note 14**). Find a field of view with several mCherry-expressing cells.
9. Take a snapshot or a short time-lapse movie of the cells with the mCherry filter set. This image/short movie contains the information about the initial localization of the fusion protein.
10. Set microscope software to acquire a time-lapse movie in the GFP channel (*see* **Note 15**) to activate LINuS/LEXY. If interested in the kinetics of import/export of the POI, take also images in the mCherry channel during this activation phase.
11. If no mCherry images were acquired during the activation phase, acquire an image in the mCherry channel right at the end of the activation phase (*see* **Note 16**).

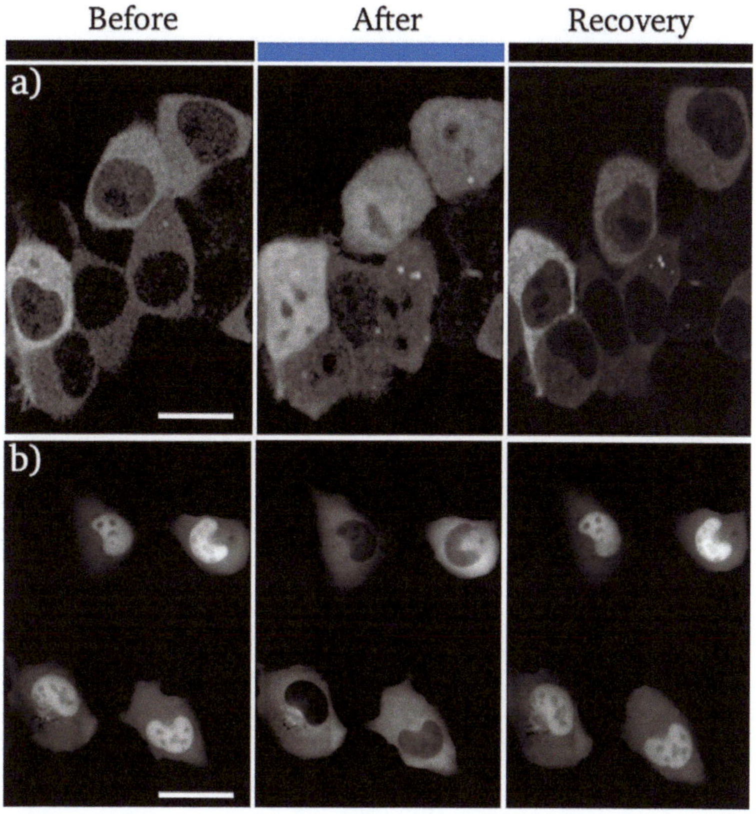

Fig. 2 LINuS and LEXY allow for controlling the nuclear/cytoplasmic accumulation of a protein of interest with blue light. (**a**) Representative images of HEK293T cells transfected with NES-mCherry-LINuS before receiving light (Before), after a 20-min activation phase with blue light (After) and at the end of a 20-min recovery phase in the dark (Recovery). NES, IDELAKLLPDLNLD; LINuS = biLINuS22. Scale bar, 20 μm. (**b**) Representative images of H1299 cells transfected with NLS-mCherry-LEXY before receiving light (Before), after a 5-min activation phase with blue light (After) and at the end of a 15-min recovery phase in the dark (Recovery). NLS = $cMyc^{P1A}$. Scale bar, 50 μm

12. Set microscope software to acquire a time-lapse movie in the mCherry channel (*see* **Note 17**) to follow the localization of the POI-mCherry-LINuS/LEXY fusion during dark recovery.
13. Before proceeding with the quantification, it should be already possible for you to visually assess if translocation occurred and was reversible (Fig. 2).
14. Open the acquired images/movies with ImageJ.
15. From the "Analyze" menu, select "Set measurements" and make sure that "mean grey value" is checked.
16. Select a region of the image without cells. Draw here a polygon or an irregular shape (ROI), then press "m" (*Analyze > Measure*) to calculate the mean grey value inside the ROI. Do this a

couple of times selecting different regions in the image deprived of cells.

17. Transfer the measurements to a spreadsheet and calculate the average background intensity.
18. Select a cell. Measure its cytosolic signal by drawing a polygon or an irregular shape (ROI) inside the cytoplasm, then pressing "m" (*Analyze* > *Measure*) to calculate the mean grey value inside the ROI. Make sure not to encompass with the ROI any part of the nucleus or of the extracellular space. Measure the nuclear signal of the cell by repeating the same procedure, this time drawing an ROI inside the nucleus taking care not to encompass any part of the cytoplasm. Do this for several cells (at least 20). Copy the measurements of the mean grey values to the spreadsheet.
19. Subtract the mean fluorescence intensity of the background from the mean nuclear and cytosolic fluorescence intensities.
20. In case of LINuS, calculate the relative nuclear localization for each cell by dividing the background-subtracted mean nuclear fluorescence intensity by the background-subtracted mean cytosolic fluorescence intensity. In case of LEXY, calculate the relative cytosolic localization for each cell by dividing the background-subtracted mean cytosolic fluorescence intensity by the background-subtracted mean nuclear fluorescence intensity.
21. Calculate the fold change in nuclear/cytosolic intensity by dividing the background-subtracted nuclear/cytosolic fluorescence intensity at a specific time point by the background-subtracted nuclear/cytosolic fluorescence intensity of time point 0.

4 Notes

1. The fluorescent protein should have an excitation wavelength > 510 nm to allow for monitoring the localization over time of the fusion protein within cells without activating the *As*LOV2 domain within LINuS/LEXY. We routinely use the monomeric red fluorescent protein mCherry, because it has satisfactory brightness, good photostability and relatively fast maturation time [15].
2. When using cell lines other than HEK293T, alternative culture media, transfection reagents (e.g., Lipofectamine 2000/3000) and seeding/transfection procedures might be required.
3. Transfection reagent may be unnecessary if the LINuS/LEXY-bearing construct is stably integrated in the genome. However, considering that some optimization of the construct may be

needed, it is most common to first test a LINuS/LEXY fusion protein expressed from a plasmid transiently transfected into cells.

4. A special medium and a special supplement for use in optogenetic experiments have been recently developed [16]. While for the protocol described here we do not expect any substantial advantage in using this medium and this supplement instead of the commonly used ones, they may be critical for long-term experiments, especially with primary cells.
5. Use glass-bottom instead of plastic-bottom microscopy dishes, since plastic materials show significant autofluorescence when excited by near-UV or visible radiation.
6. It is also possible to use the confocal microscope to test if nuclear/cytoplasmic translocation occurs. In case of laser-scanning confocal microscopy, it is important to scan with the laser large enough a region of interest (ROI) to activate sufficient molecules to detect a macroscopic translocation. This is especially important when the POI-mCherry-LINuS/LEXY fusion does not freely diffuse in the cytoplasm/nucleus due to binding to relatively static molecules such as the DNA. We recommend using a confocal microscope only if single-cell activation is required.
7. The CFP filter set can also be used to activate LINuS/LEXY. However, since in this case more energetic wavelengths are used, phototoxicity will increase.
8. The amount of DNA to transfect has to be calculated in a way to achieve a mild overexpression of the construct. It therefore depends on the strength of the promoter driving expression of the construct, as well as other factors regulating, for instance, translation efficiency of the mRNA. Most often, it has to be empirically determined. Strong overexpression of the LINuS/LEXY fusion protein has a negative effect on translocation due to saturation of the import–export machinery on the one hand and increase in spontaneous binding events between the NLS/NES and the import–export receptors in the dark on the other. Stuffer DNA can be used to reach the necessary DNA amount recommend in the transfection protocol.
9. Always include NES-mCherry-LINuS (e.g., Addgene #61346) or NLS-mCherry-LEXY (Addgene #72655) as control of the procedure alongside the fusion between your protein of interest (POI) and mCherry-LINuS/LEXY (POI-mCherry-LINuS/LEXY) which you wish to test. This should reveal if the POI interferes with the translocation. In the case of LINuS, we advise to test more than one variant selecting from those based on monopartite and bipartite NLSs [7]. In the case of LEXY, this is typically not needed as the variant simply called

LEXY in the original publication [8] has proven quite robust with different POIs [17, 18]. Additionally, mutagenesis of endogenous NES(s)/NLS(s) or both might be necessary to allow for effective translocation of the POI-mCherry-LINuS/LEXY by light. Note that LINuS/LEXY can be fused also N-terminally to the POI.

10. The transfection can be performed under normal light conditions.
11. The duration of this incubation step depends on your cell line and the transfection method used. Shorter incubation times are advisable for more sensitive cell lines.

    If cells are going to be imaged for a long period (>5 h), the cell medium with the transfection reagent should be exchanged with fresh cell medium before starting the imaging.
12. Should your microscope not be equipped with a dark incubation chamber, cover this latter in aluminum foil or keep the entire room in the dark and work under red LED safelight.
13. In case of unwanted light contamination, it is possible to incubate the sample in the dark for some time (e.g., 1 h) to restore cytoplasmic/nuclear localization of the POI-mCherry-LINuS/LEXY construct. However, while the translocation of mCherry-LINuS/LEXY is known to be reversible, it is not possible to know a priori if the fusion between the POI and mCherry-LINuS/LEXY maintains this feature. It can be that the POI interacts with other proteins that might act as nuclear/cytoplasmic anchors thus not consenting to the fusion protein to relocate to the compartment it resided in prior to illumination. When performing the initial assessment of the POI, it is therefore advisable to simply repeat the experiment if light contamination occurred at any stage.
14. Take care not to accidentally use any light between 350–510 nm (DAPI, CFP, GFP or YFP filter sets) since these excitation wavelengths will activate LINuS/LEXY. If the sample has been exposed to light, storing it for 1 h in the dark might reset the localization prior to activation. However, as the reversibility of the POI-mCherry-LINuS/LEXY construct may not be known at this stage (*see* **Note 13**), the experiment may need to be repeated.
15. Exposure times for GFP and mCherry depend upon the available filter sets, the light source and the camera. In our experiments performed on a ZEISS Axio Observer inverted microscope equipped with a cooled CCD-Camera (Axiocam 506 mono) and the Colibri LED illumination system, we typically take an image in the GFP channel setting the exposure time to 100 ms, and the light intensity to 2–4% (corresponding to ~300 W/m^2 of light) every 30 s. It is important to keep the

interval between the activation steps (=GFP images) shorter than half the life-time of the photocycle of the *As*LOV2 domain.

16. Do not let the cells in the dark for more than a couple of seconds prior to acquiring the mCherry image, as relocalization of the fusion protein may quickly occur, obfuscating the real extent of nuclear/cytoplasmic translocation that took place after illumination.
17. Adjust the intensity of the light source and the exposure times to obtain a good comprise between photobleaching and image quality. Exposure times depend upon the filter set used, the light source and the camera. We typically set the exposure time to 500 ms, and the light intensity to 50%. For first experiments, we suggest acquiring images every 3–5 min for 1 h.

Acknowledgments

This work was supported by the German Research Foundation (DFG) (Excellence Strategy BIOSS-EXC-294 and grant VE 776/ 3-1 within the SPP1926).

References

1. Ramakrishnan G, Davaakhuu G, Chung WC et al (2015) AKT and 14-3-3 regulate Notch4 nuclear localization. Sci Rep 5:8782. https://doi.org/10.1038/srep08782
2. Sahin U, de The H, Lallemand-Breitenbach V (2014) PML nuclear bodies: assembly and oxidative stress-sensitive sumoylation. Nucleus 5 (6):499–507. https://doi.org/10.4161/19491034.2014.970104
3. Whiteside ST, Goodbourn S (1993) Signal transduction and nuclear targeting: regulation of transcription factor activity by subcellular localisation. J Cell Sci 104(Pt 4):949–955
4. Gorlich D, Mattaj IW (1996) Nucleocytoplasmic transport. Science 271(5255):1513–1518
5. Beck M, Hurt E (2017) The nuclear pore complex: understanding its function through structural insight. Nat Rev Mol Cell Biol 18 (2):73–89. https://doi.org/10.1038/nrm.2016.147
6. Timney BL, Raveh B, Mironska R et al (2016) Simple rules for passive diffusion through the nuclear pore complex. J Cell Biol 215 (1):57–76. https://doi.org/10.1083/jcb.201601004
7. Niopek D, Benzinger D, Roensch J et al (2014) Engineering light-inducible nuclear localization signals for precise spatiotemporal control of protein dynamics in living cells. Nat Commun 5:4404. https://doi.org/10.1038/ncomms5404
8. Niopek D, Wehler P, Roensch J et al (2016) Optogenetic control of nuclear protein export. Nat Commun 7:10624. https://doi.org/10.1038/ncomms10624
9. Harper SM, Neil LC, Gardner KH (2003) Structural basis of a phototropin light switch. Science 301(5639):1541–1544. https://doi.org/10.1126/science.1086810
10. Peter E, Dick B, Baeurle SA (2010) Mechanism of signal transduction of the LOV2-Jalpha photosensor from Avena sativa. Nat Commun 1:122. https://doi.org/10.1038/ncomms1121
11. Zayner JP, Antoniou C, Sosnick TR (2012) The amino-terminal helix modulates light-activated conformational changes in AsLOV2. J Mol Biol 419(1–2):61–74. https://doi.org/10.1016/j.jmb.2012.02.037
12. Swartz TE, Corchnoy SB, Christie JM et al (2001) The photocycle of a flavin-binding domain of the blue light photoreceptor phototropin. J Biol Chem 276(39):36493–36500. https://doi.org/10.1074/jbc.M103114200

13. Yumerefendi H, Dickinson DJ, Wang H et al (2015) Control of protein activity and cell fate specification via light-mediated nuclear translocation. PLoS One 10(6):e0128443. https://doi.org/10.1371/journal.pone.0128443
14. Yumerefendi H, Lerner AM, Zimmerman SP et al (2016) Light-induced nuclear export reveals rapid dynamics of epigenetic modifications. Nat Chem Biol 12(6):399–401. https://doi.org/10.1038/nchembio.2068
15. Shaner NC, Campbell RE, Steinbach PA et al (2004) Improved monomeric red, orange and yellow fluorescent proteins derived from Discosoma sp. red fluorescent protein. Nat Biotechnol 22(12):1567–1572. https://doi.org/10.1038/nbt1037
16. Stockley JH, Evans K, Matthey M et al (2017) Surpassing light-induced cell damage in vitro with novel cell culture media. Sci Rep 7 (1):849. https://doi.org/10.1038/s41598-017-00829-x
17. Baarlink C, Plessner M, Sherrard A et al (2017) A transient pool of nuclear F-actin at mitotic exit controls chromatin organization. Nat Cell Biol 19(12):1389–1399. https://doi.org/10.1038/ncb3641
18. Hooikaas PJ, Martin M, Muhlethaler T et al (2019) MAP7 family proteins regulate kinesin-1 recruitment and activation. J Cell Biol 218 (4):1298–1318. https://doi.org/10.1083/jcb.201808065

Chapter 9

Light-Inducible CRISPR Labeling

Mareike D. Hoffmann, Felix Bubeck, and Dominik Niopek

Abstract

CRISPR labeling is a powerful technique to study the chromatin architecture in live cells. In CRISPR labeling, a catalytically dead CRISPR-Cas9 mutant is employed as programmable DNA-binding domain to recruit fluorescent proteins to selected genomic loci. The fluorescently labeled loci can then be identified as fluorescent spots and tracked over time by microscopy. A limitation of this approach is the lack of temporal control of the labeling process itself: Cas9 binds to the g(uide)RNA-complementary target loci as soon as it is expressed. The decoration of the genome with Cas9 molecules will, however, interfere with gene regulation and—possibly—affect the genome architecture itself. The ability to switch on and off Cas9 DNA binding in CRISPR labeling experiments would thus be important to enable more precise interrogations of the chromatin spatial organization and dynamics and could further be used to study Cas9 DNA binding kinetics directly in living human cells.

Here, we describe a detailed protocol for light-inducible CRISPR labeling. Our method employs CASANOVA, an engineered, optogenetic anti-CRISPR protein, which efficiently traps the *Streptococcus pyogenes* (*Spy*)Cas9 in the dark, but permits Cas9 DNA targeting upon illumination with blue light. Using telomeres as exemplary target loci, we detail the experimental steps required for inducible CRISPR labeling with CASANOVA. We also provide instructions on how to analyze the resulting microscopy data in a fully automated fashion.

Key words CRISPR labeling, Telomere, LOV2 domain, Photoreceptor, Anti-CRISPR, CASANOVA, Automated image analysis, KNIME

1 Introduction

The development of CRISPR (clustered regularly interspaced short palindromic repeats)-Cas9 technologies revolutionized our ability to target and modify the genome in living mammalian cells [1–3]. CRISPR-Cas9 tools typically comprise two components: The Cas9 endonuclease and a g(uide)RNA, which directs the Cas9 enzyme to a selected genomic site [1]. Cas9 can be programmed to target in principle any genomic locus simply by altering the target site complementary part within the gRNA sequence. The sole sequence constraint is the required presence of a so-called protospacer adjacent motif (PAM; NGG in case of *Spy*Cas9) at the

Dominik Niopek (ed.), *Photoswitching Proteins: Methods and Protocols*, Methods in Molecular Biology, vol. 2173,
https://doi.org/10.1007/978-1-0716-0755-8_9,

target locus, that is, next to the ~20 nucleotides gRNA-complementary region.

While the default mode of Cas9 action is the induction of DNA double-strand breaks, catalytically inactive variants of the enzyme have been created by mutating its RuvC and HNH catalytic domains. These so-called d(ead)Cas9s can be employed as programmable DNA binding domains to recruit fused effector domains including transcriptional activators and repressors [4–7], epigenetic modifiers [8, 9] or DNA base-modifying enzymes [10, 11] to selected genomic loci. An additional, particularly useful application of dCas9 is the fluorescent labeling and subsequent microscopy-based tracking of endogenous genomic loci in living cells, a technique known as CRISPR labeling [12]. In CRISPR labeling, dCas9 is fused to a fluorescent protein such as green fluorescent protein (GFP) or mCherry and coexpressed with gRNAs that target the dCas9–fluorophore fusion to desired loci. Using fluorescence microscopy, the labeled target locus can then be identified as fluorescent spot within the cell nucleus and its localization can be followed in time. Provided the genomic target locus comprises repetitive sequences, as is the case for telomeres or centromeres, a single gRNA is sufficient to recruit multiple Cas9 molecules to the target site(s) and thereby achieve a high signal-to-noise ratio. However, if nonrepetitive loci are to be labeled, multiple gRNAs [12] in combination with fluorescence signal amplification strategies [13] might be required to achieve an acceptable signal-to-noise ratio.

In typical CRISPR labeling experiments, a constitutively active dCas9 is employed, that is, the dCas9–fluorophore fusion will bind to its corresponding target locus right upon expression. However, decorating a genomic locus with dCas9 molecules will almost inevitably perturb gene expression [14] and—possibly—the genome architecture to be studied. Thus, in ideal CRISPR labeling experiments, dCas9 would only be recruited to the genomic DNA right before starting imaging.

We recently developed CASANOVA (CRISPR-Cas9 activity switching via an optogenetic variant of AcrIIA4, [15]), an engineered, light-dependent inhibitor of the SpyCas9 most commonly employed for CRISPR labeling. CASANOVA consists of an anti-CRISPR protein (AcrIIA4) [16], a bacteriophage-derived *Spy*Cas9 inhibitor, fused to the LOV2 domain from *Avena sativa* phototropin-1. In the dark, the AcrIIA4 part in CASANOVA efficiently traps *Spy*Cas9-gRNA complexes, thereby preventing them from binding DNA. Upon illumination with blue light, however, the LOV2 part in CASANOVA induces a conformational change on the fused AcrIIA4, thereby releasing Cas9. As CASANOVA can control DNA binding of both, the catalytically active Cas9 as well as dCas9, it can be applied on dCas9–fluorophore fusions, thereby facilitating inducible CRISPR labeling experiments [15] (Fig. 1).

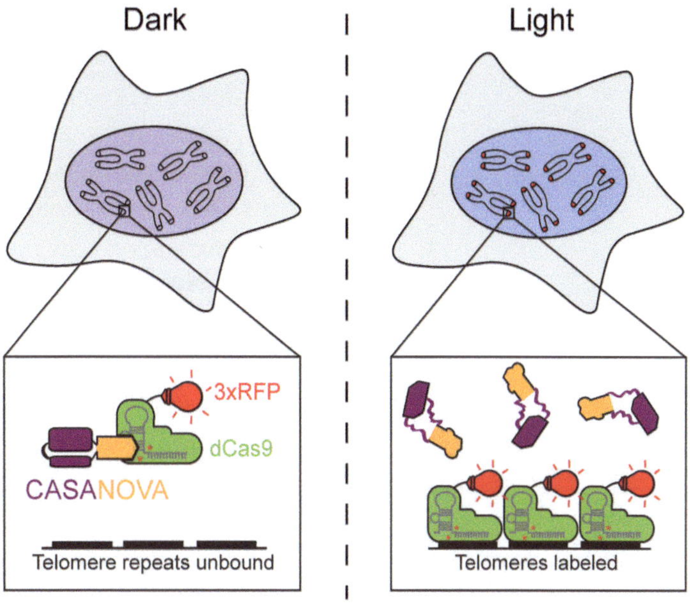

Fig. 1 Light-dependent telomere labeling with CASANOVA. Cells are cotransfected with dCas9–3×RFP, a gRNA targeting the telomere repeats, and CASANOVA, an engineered, light-dependent anti-CRISPR protein. In the dark state, CASANOVA traps dCas9–3×RFP molecules, thereby preventing CRISPR labeling. Upon irradiation with blue light, however, the CASANOVA conformation changes, resulting in release of dCas9–3×RFP molecules and subsequent recruitment to telomeres. Labeled telomeres can then be detected as red fluorescent spots in the cell nucleus by microscopy

We here detail the reagents needed and experimental steps required to establish light-inducible CRISPR labeling of telomeres with CASANOVA in U2OS cells. In addition, we provide instructions on how to analyze the corresponding imaging data using a fully automated analysis workflow in KNIME.

2 Materials

2.1 Cell Culture and Transient Transfection

1. Humidified cell culture incubator at 5% CO_2 and 37 °C.
2. Laminar flow hood.
3. Water bath (37 °C).
4. Standard cell culture microscope.
5. U2OS cells cultured in 75 cm^2 tissue culture flasks (*see* **Note 1**).
6. U2OS culture medium: Phenol red-free DMEM medium supplemented with 4.5 g/L glucose, 10% v/v fetal calf serum, 2 mM L-glutamine, 1 mM sodium pyruvate, 1% penicillin–streptomycin.

7. 1× Phosphate buffered saline (PBS).
8. 0.05% Trypsin–EDTA solution.
9. Cell counting device (e.g., Neubauer counting chamber).
10. Glass-bottom cell culture dishes suitable for fluorescence microscopy (e.g., one-compartment CELLview cell culture dishes, Greiner Bio-One) (*see* **Note 2**).
11. jetPRIME transfection reagent and jetPRIME buffer (PolyPlus Transfection; *see* **Note 3**).
12. 1.5 mL reaction tube.
13. CASANOVA expression vector (Addgene #113035).
14. AcrIIA4 expression vector (Addgene #113037).
15. dCas9–3×RFP/gRNA telo expression vector (Addgene #85717).
16. dCas9–3×RFP-P2A-CASANOVA/gRNA telo expression vector (Addgene #137711).
17. dCas9–3×RFP-P2A-wtAcrIIA4/gRNA telo expression vector (Addgene #137712).
18. gRNA telo/GFP expression vector (Addgene #137713) (*see* **Note 4**).
19. Stuffer DNA (e.g., pBluescript).
20. Vortex mixer.
21. Microcentrifuge (for 1.5 mL Eppis).
22. 4% PFA solution in 1× PBS (*see* **Note 5**).
23. Fume hood.
24. Mounting medium with DAPI for fluorescence imaging of fixed cells (e.g., SlowFade™ Diamond Antifade Mountant with DAPI, ThermoFisher).

2.2 Blue Light Setup

1. Six blue light high-power LEDs (e.g., type CREE XP-E D5-15; emission peak ~460 nm; emission angle ~130°) mounted onto aluminum cooling blocks and connected in series (*see* **Note 6** and Fig. 2).
2. Switching Mode Power Supply (e.g., Manson HCS-3102).
3. Raspberry Pi.
4. Transparent acrylic glass table (*see* **Note 6** and Fig. 2).
5. Light meter (e.g., LI-COR LI-250A).

2.3 Confocal Microscopy

1. Leica SP8 confocal laser scanning microscope equipped with 405 nm diode, 488 nm laser, 552 nm laser, and an HC PL APO 40×/1.3-NA (numerical aperture) oil objective (*see* **Note 7**).

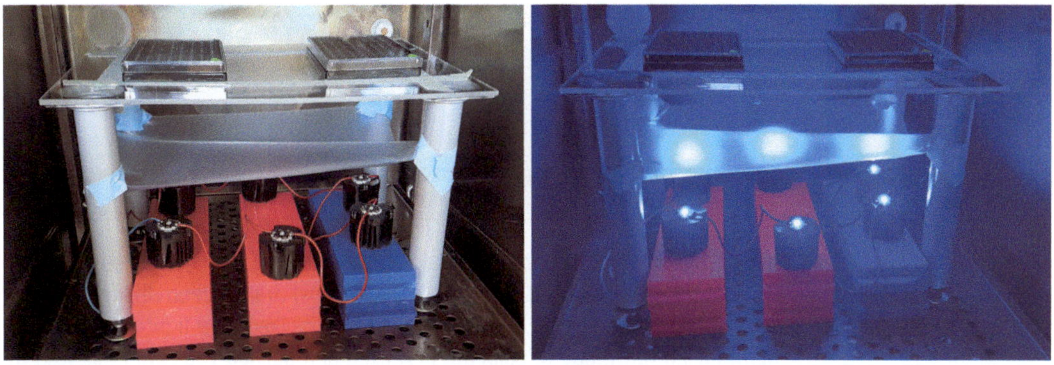

Fig. 2 Blue light setup. Six high-power LEDs emitting blue light are placed below a custom made acrylic glass table located in a cell culture incubator. The translucent foil positioned above the LEDs scatters the light, thereby enabling a more even illumination of samples located on top of the table. A Raspberry Pi connected to a laboratory power supply (not shown) is used to create a pulsatile illumination pattern

2.4 Image Analysis

1. KNIME software (KNIME AG; open source: https://www.knime.com) with ImageJ2 Integration (https://www.knime.com/community/imagej).
2. PC with KNIME-compatible hardware.

3 Methods

The overall protocol takes three consecutive days, excluding the image analysis (Fig. 3). All steps are performed at room temperature, unless otherwise indicated.

3.1 Cell Culture, Cell Seeding, and Transient Transfection

Cell passaging and transfection is performed in a laminar flow under sterile working conditions. Maintain U2OS cells in 75 cm^2 flasks in U2OS culture medium and passage when reaching 70–90% confluency.

1. Warm trypsin solution, PBS, and U2OS culture medium to 37 °C in a water bath.
2. Aspirate medium of a 70–90% confluent U2OS cell culture flask.
3. Add 10 mL PBS, gently tilt the flask forward/backward and sideward to wash off remaining medium and cell debris.
4. Aspirate PBS and add 1–2 mL of 0.05% Trypsin–EDTA solution.
5. Incubate at 37 °C for ~5 min.
6. Gently tap against the side of the flask a few times to detach cells and confirm that cells are detached from the culture flask surface using a cell culture microscope.

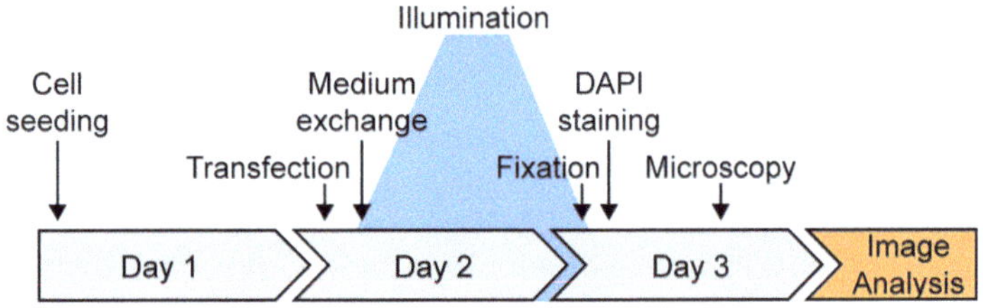

Fig. 3 Overview of the experimental steps and timing. Cells are seeded on day 1 and transfected 24 h later. Four hours post transfection the medium is exchanged and cells are either irradiated with pulsed blue light or kept in the dark for 20 h. On day 3, cells are PFA fixed and stained with DAPI followed by confocal fluorescence microscopy. Finally, the images are analyzed using an automated KNIME workflow

7. Add 8 mL of U2OS culture medium to stop trypsinization and gently pipet up and down several times to obtain a homogenous cell suspension.
8. Count cells and dilute to a cell density of 60,000 cells per mL using U2OS culture medium.
9. Seed six one-compartment CELLview cell culture dishes at a density of 120,000 cells per compartment by adding 2 mL of diluted cell suspension to each dish (*see* **Note 8**).
10. Gently tilt the dishes forward, backward, and sideward to evenly distribute the cell suspension within each dish.
11. Incubate cells in a humidified cell culture incubator at 37 °C and 5% CO_2 overnight.
12. Dilute 2.9 μg of plasmid DNA in 400 μL jetPRIME buffer in a 1.5 mL reaction tube. The plasmid DNA thereby comprises of the following (*see* **Note 9**).
    (a) CASANOVA samples (light-dependent telomere labeling): 2 μg gRNA telo/GFP expression vector, 300 ng of CASANOVA expression vector and 600 ng of dCas9–3×RFP-P2A-CASANOVA/gRNA telo vector.
    (b) Negative control samples: 2 μg of gRNA telo/GFP expression vector, 300 ng of AcrIIA4 expression vector and 600 ng of dCas9–3×RFP-P2A-wtAcrIIA4/gRNA telo vector (*see* **Note 10**).
    (c) Positive control samples: 2 μg of gRNA/GFP expression vector, 300 ng of stuffer DNA and 600 ng of dCas9–3×RFP/gRNA telo vector.
13. Vortex DNA mixes for 10 s and briefly spin down in a microcentrifuge.
14. Add 12 μL of jetPRIME reagent to each DNA mix.

15. Vortex for 10 s, spin down and incubate transfection mixes for 10–15 min at room temperature.
16. Transfect two cell culture dishes for each, the CASANOVA, negative control and positive control samples by adding 200 μL of transfection mix dropwise to each dish.
17. Gently tilt the cell culture dishes forward, backward, and sideward several times to evenly distribute the transfection mixes. Place samples into the incubator.
18. Replace medium 4 h post transfection (*see* **Note 11**) and directly proceed with **step 19** (*see* **Note 12**).
19. For each, the CASANOVA, negative and positive control samples, illuminate one dish with pulsatile blue light (5 s ON, 10 s OFF; *see* **Notes 6** and **13**) at a light intensity of ~3 W/m^2 (*see* **Note 14**) directly in the cell culture incubator (Fig. 2). Place the second dish (dark control) into the same incubator but shielded from light, for example by wrapping in aluminum foil (*see* **Note 15**).
20. Incubate samples in the light or dark for 20 h (*see* **Note 16**).

3.2 Sample Fixation

Perform the following steps in a fume hood while shielding samples from light as good as possible (*see* **Note 17**). Note, that PFA requires appropriate disposal.

1. Aspirate medium from the one-compartment dishes.
2. Wash the cells with PBS. To this end, add 2 mL 1× PBS to each dish, incubate for a few seconds and aspirate PBS.
3. Add 1 mL of 4% PFA solution and incubate for 20 min at room temperature.
4. Aspirate the 4% PFA solution and wash dishes 3× with 2 mL 1× PBS to remove residual PFA.
5. Following PBS aspiration, add a few drops of DAPI-containing mounting medium so that the glass surface is covered.
6. Incubate for ~2 h at 4 °C (*see* **Note 18**).

3.3 Microscopy

1. Start the Leica Application Suite X (LAS X) Software.
2. Place sample dish into the appropriate sample holder mounted onto the microscopy stage.
3. Select the HC PL APO 40×/1.3-NA oil objective. Set laser power to 1% for the UV laser (DAPI excitation), 2% for the 488 nm laser (GFP excitation) and to 1% for the 552 nm laser (RFP excitation). Set resolution to 1024 × 1024 pixel and line average to 5 (*see* **Note 19**). For imaging of DAPI (nuclei), GFP (transfection control), and RFP (dCas9–3×RFP), use detection wavelengths of 410–490 nm, 493–578, and 578–789 nm, respectively.

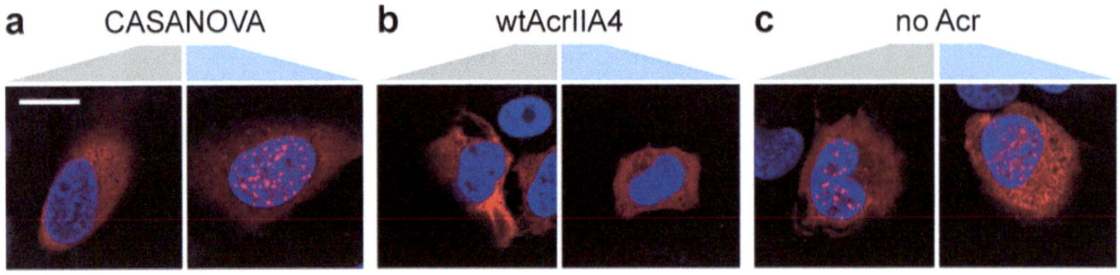

Fig. 4 Exemplary microscopy images. U2OS cells expressing dCas9–3×RFP, a telomere targeting gRNA, and either CASANOVA (**a**), wild-type (wt)AcrIIA4 (**b**, negative control), or no Acr (**c**, positive control) were incubated in the dark or irradiated with pulsed blue light (indicated by grey and blue coloring, respectively) for 20 h, followed by PFA fixation, DAPI staining and fluorescence microscopy. The CASANOVA samples show light-dependent formation of spots in the nucleus, while in the negative (wtAcrIIA4) and positive (no Acr) samples, spots are either absent or present independent of light, respectively. Shown images are overlays of the RFP and DAPI signals. Scale bar, 20 μm

4. Acquire images (DAPI, GFP, and RFP signal) for 30—100 nuclei per samples (*see* **Note 20**). Make sure to record the different fluorescence channels always in the same order (DAPI, GFP, RFP) as, otherwise, they have to be resorted before image analysis. Representative fluorescence microscopy images of nuclei for each, the CASANOVA, positive and negative control samples in the light and dark are shown in Fig. 4.
5. Export all data as lif files.

3.4 Image Analysis

Settings and items in the software are capitalized in the following section. Figure 5 summarizes the major steps in the image analysis pipeline.

1. Open the Telomere_Quantification_Pipeline_Bubeck_et_al.knmf workflow file in KNIME. Either double-click on the workflow or select IMPORT KNIME WORKFLOW in the FILE menu. The workflow is available as *Supplementary Data* in Bubeck et al. [15] (https://www.nature.com/articles/s41592-018-0178-9#Sec20).
2. Load the lif file(s) to be processed. To this end, right-click on the IMAGE READER node and select CONFIGURE. In the OPTIONS menu select the lif files to be processed by double-clicking on them (*see* **Note 21**). Selections can be removed by double-clicking on selected files.
3. Scroll to the right end of the workflow. Right-click on the CSV WRITER node and select CONFIGURE. Then, in the SETTINGS menu, denote the output location for the analyzed data.
4. Execute the workflow by right-clicking on the LOOP END node and selecting EXECUTE (*see* **Note 22**). If necessary,

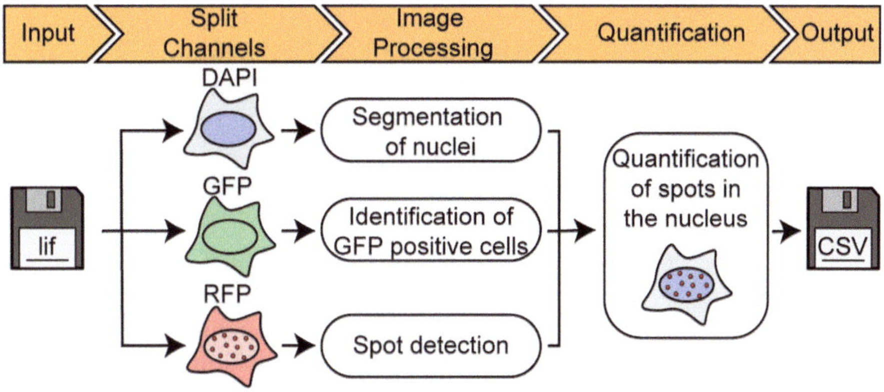

Fig. 5 Automated image analysis workflow. The KNIME workflow receives microscopy images (lif files) as input. First, channels are split into DAPI, GFP, and RFP signal and nuclei are segmented using the DAPI signal. GFP negative cells, that is, cells lacking the transfection marker, are excluded from the analysis. Fluorescent spots are detected in the RFP channel and spots located inside nuclear segments are considered labeled telomeres. The workflow output is a CSV table comprising the nuclei segment IDs, spots detected within each nuclear segment as well as the area (in pixels) and mean intensity of each detected spot

adapt the parameters in the image processing and spots detection metanodes before executing the workflow (*see* **Note 23**).

4 Notes

1. U2OS cells are commonly used for CRISPR labeling experiments [17, 18]. However, CRISPR labeling can also be performed in other cell lines (e.g., HEK293T, UMUC3, HeLa, and RPE cells) [12].
2. One-compartment CELLview cell culture dishes have a large growth area of 8.7 cm^2, which allows to acquire images from a number of different areas in one dish. Nonetheless, other glass bottom dishes or chambered slides that can be used in combination with the sample holders of the confocal microscope are suitable as well.
3. U2OS cells can be efficiently transfected with jetPRIME. When using other cell lines, transfection reagents and conditions have to be adapted accordingly.
4. The GFP reporter present on the gRNA telo/GFP vector serves as transfection control. The gRNA spacer sequence in this construct is 5′-GTTAGGGTTAGGGTTAGGGTT-3′ [19] and targets the telomere repeats.
5. 4% PFA solution can be obtained from commercial suppliers. It is, however, also rather simple to make it from 1× PBS and PFA powder. Note, that the procedure has to be performed under a fume hood. To make 1 L of 4% PFA in PBS, heat 800 mL PBS

in a glass beaker to ~60 °C. Add 40 g PFA powder and slowly raise the pH by adding 1 N NaOH dropwise. Once the solution clears (i.e., PFA is completely dissolved), adjust pH to 6.9 by adding 1 N HCL dropwise. Then, top up the volume to 1 L using PBS, sterile filter, aliquot, and freeze PFA solution at −20 °C until use. The PFA solution can also be stored at 2–8 °C for a few weeks.

6. Our blue light setup comprises a self-made acrylic glass table (length × width × height = 35 × 25 × 20 cm), below which the LEDs are positioned in a rectangle (Fig. 2). Samples are located on top of the table and are illuminated from below, that is, through the acrylic glass. Translucent foil located in between the LEDs and the table surface can be optionally used to scatter the light to reach a more even illumination of samples.
7. Other confocal microscopes that are suitable for imaging DAPI, GFP and RFP can be used, provided the objective magnification is at least 40× and the overall image quality and sensitivity is comparable to a Leica SP8.
8. Six one-compartment dishes are needed in total, that is, two for each, the positive control, the negative control, and the CASANOVA samples. One dish will be illuminated, while the second one will serve as the dark control.
9. The given volumes and amounts are for transfection of two one-compartment CELLview cell culture dishes, that is, a light and corresponding dark control sample. We recommend a vector ratio gRNA/GFP to dCas9–3×RFP to Acr/CASANOVA expression construct of 20:6:3. It is important to note, that for CRISPR labeling, it is desired that dCas9–3×RFP is expressed only at mild/intermediate levels. When strongly overexpressing dCas9–3×RFP, the RFP background fluorescence will be too high and mask the locus-specific labeling. Moreover, an excess of the telomere targeting gRNA construct is required as otherwise, sub-saturating concentrations of the gRNA are expressed, again reducing the labeling efficiency [19]. When adapting the protocol to genomic loci which are not repetitive, it will be necessary to cotransfect multiple gRNAs that target proximate sites around the target locus, so that a sufficient number of fluorescent Cas9 molecules will accumulate at the genomic site of interest.
10. We recommend performing an additional negative control, in which no telomere-targeting gRNA is expressed. To this end, dilute 2 μg of a gRNA expression vector lacking the telomere-targeting spacer (Addgene #113039), 600 ng of dCas9–3×RFP only vector (Addgene #64108; this vector is not coencoding a gRNA) and 300 ng of stuffer DNA in 400 μL jetPRIME buffer and proceed as with the other samples.

11. Refreshment of the medium 4 h post transfection is recommended to reduce the cell toxicity.
12. In the protocol described here, all samples and controls are processed in parallel, which simplifies establishing the method and as well as troubleshooting it, if necessary. Illumination thereby starts right after medium exchange (i.e., 4 h posttransfection) to guarantee the maximum induction of telomere labeling, while microscopy is eventually performed on fixed samples. However, it is also possible to perform illumination of samples directly under a confocal microscope, that is, to follow the recruitment of fluorescent dCas9 molecules to telomeres in real time [15]. If this is desired, samples should be incubated for ~24 h shielded from light. This provides sufficient time to allow expression of all constructs. Following this incubation step, samples are placed into a microscope equipped with temperature (37 °C) and CO_2 (5%) control. A field-of-view with cells of interest is repeatedly scanned (every ~30 s) using a 488 nm laser to release dCas9–3×RFP from CASANOVA, while—in parallel—the RFP signal is recorded to observe formation of fluorescent spots (dCas9–3×RFP molecules recruited to telomeres) in the nucleus.
13. The pulsatile illumination regime can be created by running a simple script available on GitHub (https://github.com/zanderharteveld/anti-CRISPR-designs/blob/master/light_pulse_control/Illumination_script.py) on the Raspberry Pi controlling the power supply. The current and voltage can be adjusted by replacing the default parameters ("VOLT174," i.e., 17.4 Volt and "CURR017," i.e., 0.17 Ampere) in the script.
14. Use a light meter to measure the blue light intensity. Adjust the voltage/current so that a light intensity of ~3 W/m^2 is observed right above the table, that is, where the samples are located. Illumination with light intensities higher than ~3 W/m^2 is not recommended due to potential light-related cell toxicity.
15. Do not tightly seal samples in foil, as otherwise gas exchange might be hindered, eventually resulting in diminished cell growth or cell death.
16. 20 h incubation time posttransfection is sufficient to enable modest dCas9–3×RFP expression desired for CRISPR labeling. However, it is possible to extend the incubation period to 48 h in case no or only a very faint RFP signal is observed upon 20 h of incubation.
17. Before the PFA fixation is completed, light exposure (also ambient room light) could induce CRISPR labeling in the CASANOVA dark control samples, which is undesired.

18. Fixed samples can be stored at 2–8 °C for several days. However, it is important to shield samples from light at all times, to avoid bleaching of the fluorescence signal.
19. High resolution images are obligate to visualize and spatially separate labeled telomeres.
20. Recording a large enough number of nuclei per sample is necessary to enable comparisons between samples with sufficient statistical power.
21. The default workflow configuration expects fluorescent channels to be recorded in the order DAPI–GFP–RFP. If the channel order differs from that, the outputs of the ROW SPLITTER nodes have to be reconfigured accordingly. To do so, simply delete the connections between the ROW SPLITTER outputs and the inputs of the following nodes by clicking on the lines connecting these nodes and pressing "delete." Then, reconnect the ROW SPLITTER outputs as follows: The DAPI channel enters the second input in the IMAGE PROCESSING node; the GFP channel enters the third input in the IMAGE PROCESSING node; the RFP channel enters the ROWID node as well as the first input in the IMAGE PROCESSING node.
22. The KNIME workflow automatically processes the images as follows (Fig. 5). First, the image stacks are split into the three different channels DAPI, GFP, and RFP. The DAPI signal is used to segment the nuclei. The GFP signal is used to detect cells that received the gRNA expression vector. GFP negative cells are excluded from the analysis. The RFP channel is used to identify and segment fluorescent spots, that is, dCas9–3×RFP labeled telomeres, in the nucleus. Spots outside the nucleus are excluded from the analysis. The software generates a CSV file as output. The columns 1–3 in the CSV output file represent (1) the ID of the detected nucleus, (2) the area (in pixels) of the detected spot and (3) the mean RFP fluorescence intensity in this spot, respectively. Each row represents a single fluorescent spot, that is, if the same nucleus contained multiple fluorescent spots these are listed in individual rows.

    The *interactive segmentation view* nodes implemented in this workflow can be used to visualize detected spots and thus verify that the workflow functions as anticipated. To do so, right-click on the node named "Spots in Nuclei" and select *view: interactive segmentation view*. Then, click on the *loop end* node and select *step loop execution*. This way, one can run the workflow image by image and manually inspect the performance of the image analysis pipeline.
23. When using this pipeline, several parameters in the image analysis pipeline might need to be adjusted, depending on the

image quality, used objective and zoom factors, image format (e.g., 8 bit vs. 32 bit), cell line, DAPI, GFP, and RFP signal intensity as well as overall labeling efficiency (i.e., signal-to-noise ratio).

(a) In the default workflow configuration, only spots with an average fluorescence intensity at least 1.7 times the average RFP fluorescence in the nucleus are counted as labeled telomeres, as otherwise, minor fluctuations of the fluorescence signal within the nucleus would lead to false positives. This parameter can be changed by double-clicking on the "Detect spots within nuclei & exclude RFP saturated cells" metanode. Then right-click on the *row filter* named "Filter spots with signal/background > 1.7" and select *configure*. Then, in the *filter criteria* menu, adjust the value for *lower bound* (default is "1.7"). Press *apply* to accept change.

(b) In the default configuration, nuclei with an average RFP fluorescence signal higher than 170, which is approximately two-thirds of the maximum (when using 8-bit images), are excluded from the analysis, as in these cells, background fluorescence is too high to enable reliable detection of labeled telomeres. This parameter can be changed by configuring the *row filter* node named "Filter nuclei with saturated RFP" in the same metanode.

(c) Other important parameters are those filtering for the minimum/maximum area and signal intensity of nuclei and spots (labeled telomeres) as well as the minimum GFP signal (transfection positive control). These parameters can be customized by configuring the corresponding *row filter* nodes in the "Detect spots within nuclei & exclude RFP saturated cells," the "Identify GFP positive cells" as well as the "Image Processing" metanodes.

Acknowledgments

M.D.H. was supported by a Helmholtz International Graduate School for Cancer Research scholarship (DKFZ, Heidelberg).

References

1. Jinek M, Chylinski K, Fonfara I et al (2012) A programmable dual-RNA-guided DNA endonuclease in adaptive bacterial immunity. Science 337(6096):816–821. https://doi.org/10.1126/science.1225829
2. Mali et al (2013) Science, 2013: RNA-guided human genome engineering via Cas9. Full citation can be found here: https://www.ncbi.nlm.nih.gov/pubmed/23287722
3. Cong L, Ran FA, Cox D et al (2013) Multiplex genome engineering using CRISPR/Cas systems. Science 339(6121):819–823. https://doi.org/10.1126/science.1231143
4. Mali P, Aach J, Stranges PB et al (2013) CAS9 transcriptional activators for target specificity

screening and paired nickases for cooperative genome engineering. Nat Biotechnol 31 (9):833–838. https://doi.org/10.1038/nbt.2675

5. Gilbert LA, Larson MH, Morsut L et al (2013) CRISPR-mediated modular RNA-guided regulation of transcription in eukaryotes. Cell 154 (2):442–451. https://doi.org/10.1016/j.cell.2013.06.044
6. Cheng AW, Wang H, Yang H et al (2013) Multiplexed activation of endogenous genes by CRISPR-on, an RNA-guided transcriptional activator system. Cell Res 23 (10):1163–1171. https://doi.org/10.1038/cr.2013.122
7. Perez-Pinera P, Kocak DD, Vockley CM et al (2013) RNA-guided gene activation by CRISPR-Cas9-based transcription factors. Nat Methods 10(10):973–976. https://doi.org/10.1038/nmeth.2600
8. Hilton IB, D'Ippolito AM, Vockley CM et al (2015) Epigenome editing by a CRISPR-Cas9-based acetyltransferase activates genes from promoters and enhancers. Nat Biotechnol 33(5):510–517. https://doi.org/10.1038/nbt.3199
9. Liu XS, Wu H, Ji X et al (2016) Editing DNA methylation in the mammalian genome. Cell 167(1):233–247.e217. https://doi.org/10.1016/j.cell.2016.08.056
10. Komor AC, Kim YB, Packer MS et al (2016) Programmable editing of a target base in genomic DNA without double-stranded DNA cleavage. Nature 533(7603):420–424. https://doi.org/10.1038/nature17946
11. Gaudelli NM, Komor AC, Rees HA et al (2017) Programmable base editing of A*T to G*C in genomic DNA without DNA cleavage. Nature 551(7681):464–471. https://doi.org/10.1038/nature24644
12. Chen BH, Gilbert LA, Cimini BA et al (2013) Dynamic imaging of genomic loci in living human cells by an optimized CRISPR/Cas system. Cell 155(7):1479–1491. https://doi.org/10.1016/j.cell.2013.12.001
13. Tanenbaum ME, Gilbert LA, Qi LS et al (2014) A protein-tagging system for signal amplification in gene expression and fluorescence imaging. Cell 159(3):635–646. https://doi.org/10.1016/j.cell.2014.09.039
14. Qi LS, Larson MH, Gilbert LA et al (2013) Repurposing CRISPR as an RNA-guided platform for sequence-specific control of gene expression. Cell 152(5):1173–1183. https://doi.org/10.1016/j.cell.2013.02.022
15. Bubeck F, Hoffmann MD, Harteveld Z et al (2018) Engineered anti-CRISPR proteins for optogenetic control of CRISPR-Cas9. Nat Methods 15(11):924–927. https://doi.org/10.1038/s41592-018-0178-9
16. Rauch BJ, Silvis MR, Hultquist JF et al (2017) Inhibition of CRISPR-Cas9 with bacteriophage proteins. Cell 168(1–2):150–158.e110. https://doi.org/10.1016/j.cell.2016.12.009
17. Ma H, Naseri A, Reyes-Gutierrez P et al (2015) Multicolor CRISPR labeling of chromosomal loci in human cells. Proc Natl Acad Sci U S A 112(10):3002–3007. https://doi.org/10.1073/pnas.1420024112
18. Ma H, Tu LC, Naseri A et al (2016) Multiplexed labeling of genomic loci with dCas9 and engineered sgRNAs using CRISPRainbow. Nat Biotechnol 34(5):528–530. https://doi.org/10.1038/nbt.3526
19. Pawluk A, Amrani N, Zhang Y et al (2016) Naturally occurring Off-Switches for CRISPR-Cas9. Cell 167(7):1829–1838.e1829. https://doi.org/10.1016/j.cell.2016.11.017

Chapter 10

Optogenetic Control of Gene Expression Using Cryptochrome 2 and a Light-Activated Degron

Carmen N. Hernández-Candia and Chandra L. Tucker

Abstract

Optogenetic tools allow for use of light as an external input to control cellular processes. When applied to regulate the function of transcription factors, optogenetic approaches provide a tunable, reversible, and bidirectional method to control gene expression. Herein, we present a detailed method to induce gene expression in mammalian cells using the light dependent dimerization of cryptochrome 2 (CRY2) and CIB1 to complement a split transcription factor. We also describe a protocol to disrupt gene expression with light by fusing a dimeric transcription factor to CRY2. When combined with a light-induced degron attached to the gene product, this method allows for rapid modulation of target protein abundance.

Key words Optogenetics, CRY2, Degron, Light, Transcription

1 Introduction

Tools that provide bidirectional control over gene expression enable induction of cellular behaviors and allow for interrogation of the biological role of specific gene products. Strategies to control gene expression include the use of chemical compounds to allosterically induce or inhibit gene expression such as Tet-ON and Tet-OFF systems [1, 2]. Chemical control of gene expression is also possible using chemical dimerizers to induce the reassembly of split proteins such as transcription factors [3], Cre recombinase [4], CRISPR/dCas9 systems [5], and T7 RNA polymerase [6].

As an alternative to chemical approaches, engineered photoreceptor proteins have been used to control transcription with light. Optogenetic approaches provide additional benefits including spatial resolution, minimal side effects, fast reversibility, and tunable control. One tool used for optogenetic control, a light-induced dimerization system based on the plant photoreceptor cryptochrome 2 (CRY2) and its interacting partner CIB1, has been used to gain bidirectional control of gene expression in mammalian cells

Dominik Niopek (ed.), *Photoswitching Proteins: Methods and Protocols*, Methods in Molecular Biology, vol. 2173,
https://doi.org/10.1007/978-1-0716-0755-8_10,

by inducing the reconstitution of split transcription factors or by preventing localization of a transcription factor to the nucleus [7].

In yeast, CRY2-CIB1 dimerization was used to complement a split Gal4 transcriptional activator, allowing for light-induced expression of a protein of interest under control of a GalUAS-containing promoter [8, 9]. Application of the same approach within mammalian cells, with CRY2 fused to the Gal4 DNA binding domain, failed to show light dependence [7]. The lack of light regulation was determined to be due to the multivalent state of the light-activated CRY2 protein when fused to the dimeric Gal4BD, resulting in formation of clusters with light that were cleared from the nucleus. As a result, the CRY2-Gal4BD protein is excluded from the nucleus, and cannot complement its activation domain half to allow for induction of transcription [7]. To prevent clustering and nuclear exclusion in mammalian cells, CRY2 was fused to a monomeric DNA binding domain, for example a version of Gal4BD (residues 1–65) missing the dimerization domain. When combined with a second construct containing a VP16 or VP64 activation domain fused to CIB1 (CIB1-VP64 or CIB1-VP16), this resulted in a 101.8-fold induction of transcription after 18 h of light treatment [7].

While clustering and nuclear exclusion were problematic for inducing activity with light, we were able to constructively utilize induced clustering and nuclear exclusion to allow for light-dependent inhibition of gene expression. To accomplish this, we fused an intact multivalent transcription factor, Gal4BD-VP16 or Gal4BD-tTA, to CRY2. With Gal4BD-VP16, this approach led to a 28-fold reduction of transcription after 18 h of light treatment [7]. By replacing the DNA binding domain of Gal4 with a catalytically dead version of Cas9 (CRY2-dCas9-VP64), transcription could also be induced at endogenous genomic sites [7].

One aim of gene regulation is to control protein availability within the cell. Upon activation of transcription, new protein can be synthesized within minutes. However, shutoff of transcription does not induce a corresponding rapid decrease in intracellular protein levels, since existing gene products can remain in cells for hours to days. To overcome this limitation, we developed a dual optogenetic system where light is used both to disrupt gene transcription as well as activate protein degradation through exposure of a degron, a specific sequence recognized by the protein quality control machinery [7]. We used the CRY2-Gal4BD-VP16 light-blocked transcriptional system to induce a protein of interest fused to an *As*LOV2-caged degron [10]. With the addition of light, we were able to simultaneously inhibit gene expression and trigger the exposure of the degron tag [7]. This method resulted in significantly decreased protein levels within 4 h of light onset, providing an approach to rapidly deplete protein at specific times within cells.

In this chapter, we present two step-by-step protocols used to regulate transcription and protein abundance in mammalian cell culture with light: a system using CRY2–CIB1 interaction to reconstitute a split transcription factor and activate transcription in light, and a system using a CRY2-fused transcription factor and a light-exposed degron to block transcription and deplete protein levels in light.

2 Materials

2.1 Equipment

1. Tissue culture incubator at 37 °C with 5% CO_2.
2. Tissue culture biosafety cabinet, for performing manipulations on cells in a sterile environment, outfitted with a red LED safelight.
3. A computer-controlled LED device can be used to provide pulsed blue light to samples [11]. Alternatively, a LED lamp outfitted with a repeat cycle timer can be used.

2.2 Cell Transfection

1. Optogenetic constructs for bidirectional transcription regulation are described in Pathak et al. [7] and are available at Addgene (www.addgene.org), including CRY2(+NLS, FL)-Gal4BD(1–147)-VP16AD(short) (Addgene #92031), mCherry-IRES-CRY2(+NLS,FL)-Gal4BD(1–65) (Addgene #92035), and CIB1(+NLS, FL)-VP64 (Addgene #92037). A plasmid carrying a protein of interest under control of a Gal4 UAS-containing promoter is also required (e.g., pGL2-GAL4-UAS-Luc, Addgene #33020, or GAL4UAS-Luciferase Reporter, Addgene #64125). To induce degradation of a protein of interest with light, the *As*LOV2-RRRG degron domain can be amplified from the plasmid pBMN HAYFP-LOV24 (Addgene vector #49570) [10].
2. 12-well tissue culture plates or imaging dishes.
3. Aluminum foil.
4. A confluent plate of HEK293T cells or other cultured cells.
5. Trypsin solution.
6. Phosphate Buffered Saline (PBS, 1× concentrated).
7. Dulbecco's modified Eagle medium (DMEM), serum-free.
8. DMEM supplemented with 10% Fetal Bovine Serum (FBS) and 1× Penicillin-Streptomycin (PS).
9. Sterile water.
10. 2.5 M $CaCl_2$ (filtered through 0.45 μm filter).

11. 2× Hepes buffered saline (HBS): 50 mM HEPES, 280 mM NaCl, 2.2 mM NaH_2PO_4, 2.2 mM Na_2HPO_4, pH 7.05–7.14, sterile filtered.
12. Lipofectamine 2000 (ThermoFisher, 11668027) or similar transfection reagent.

3 Methods

3.1 Reconstitution of a Split Transcription Factor Using CRY2-CIB1 Light-Induced Dimerization

1. To regulate expression, the open reading frame of a protein of interest should be placed under control of a Gal4 UAS-containing promoter. When initially testing these methods, it is recommended to use a control reporter construct for example expressing GFP or luciferase under Gal4UAS control.
2. Split a confluent plate of cultured cells, such as HEK293T cells, into an appropriate cell culture dish. For live cell experiments, imaging dishes or #1.5 thickness glass coverslips (for use with an imaging chamber) are recommended. For imaging of fixed cells, cells can be seeded onto a #1.5 thickness glass coverslip placed at the bottom of each well of a 12-well dish (*see* **Note 1**). Incubate overnight in a 37 °C tissue culture incubator.
3. Transfection can be performed the next day if cells have achieved ~50–80% confluence. Here we describe protocols for transfection using calcium phosphate (**steps 4–7**) or Lipofectamine 2000 (**steps 8–10**), but another transfection method can be substituted (*see* **Note 2**).
4. For calcium phosphate transfection, for each well of a 12-well dish, prepare two microcentrifuge tubes, labeled "A" and "B." To tube "A," add 5 μL of 2.5 M $CaCl_2$, 0.5 μg of CRY2-Gal (1–65) (Addgene #92035), 0.5 μg of CIB1-VP64 (Addgene #92037), and 0.5 μg of Gal4UAS-target protein, and bring to 50 μL total with sterile water.
5. To tube "B," add 50 μL of 2× HBS solution. Mix the solution in tube "A," then add dropwise to tube "B" while concurrently vortexing or mixing by tapping (*see* **Note 3**).
6. Let the mixture sit at room temperature for 15–20 min, then add dropwise to cells, at the same time gently rotating the plate to mix.
7. To prevent unwanted light exposure, loosely wrap each plate in aluminum foil and place the samples back into the tissue culture incubator. After 4 h (up to overnight, depending on the cell type) wash with 1× PBS and replace media, using a red LED safelight to prevent unwanted photoactivation. Continue with **step 11**.

8. For Lipofectamine 2000 transfection (*see* **Note 4**), prepare two tubes labeled "A" and "B."To tube "A" add DNA (equimolar amounts of CRY2-Gal(1–65), CIB1-VP64, and Gal4UAS-target protein) according to the manufacturer's specification into 50 μL total volume of DMEM media containing no serum. To tube "B" add 1.3 μL of Lipofectamine 2000 into 50 μL of DMEM media (no serum).
9. Incubate for 5 min, then add DNA mix in tube A to the Lipofectamine 2000 mix in tube B. Incubate for 20 min at room temperature.
10. Add the solution dropwise to the cells while gently rocking the plate. Wrap the dishes in aluminum foil and place in the tissue culture incubator. The media can be replaced with prewarmed DMEM/10% FBS/PS prior to light treatment, using a red safelight.
11. Start the light treatment at least 4 h after transfection by positioning the LED light source above the culture plate. Loosely wrap the LED light source and plate in aluminum foil (or use a light-tight box) to prevent illumination of other cells in the incubator.
12. Illuminate cells with a 1–2 s pulse of light every 3 min for 18 h (*see* **Note 5**). For dark-treated samples, keep the plate wrapped in aluminum foil and shielded from light. If dark-treated samples must be imaged during the experiment it is recommended to use only a red light source to illuminate the cells (*see* **Note 6**).
13. After completing light treatment, cells can be imaged, fixed, or harvested for downstream applications. It is recommended to harvest the dark-treated samples under a red safelight.

3.2 Combined Light-Mediated Disruption of Transcription and Induced Degradation

1. Fuse the target protein open reading frame at the N-terminus of the *As*LOV2-RRRG degron tag (for example, from Addgene vector #49570) and place under control of a Gal4 UAS promoter element.
2. Seed cells into a 12-well or other dish to be used for transfection (*see* Subheading 3.1, **step 2**). Incubate overnight in a tissue culture incubator or until cells achieve ~70% confluence.
3. Transfect cells (using a calcium phosphate method (HEK293/HEK293T) or a cationic lipid-mediated method (*see* **Note 2** and Subheading 3.1, **steps 4–10**). For HEK293T cells in 12-well plates we use a calcium phosphate method to transfect 1 μg of CRY2-Gal4BD-VP16AD plasmid (Addgene #92031) and 1 μg of a Gal4UAS-target protein-*As*LOV-degron per well (*see* **Note 7**). Wrap plates loosely in foil and place in a 37 °C tissue culture incubator.

4. 4–6 h after calcium phosphate transfection, wash samples with 1× PBS and replace media. Perform washes under a red safe-light to avoid photostimulation of the samples.
5. Loosely wrap dark-treated samples in aluminum foil and place them in the tissue culture incubator. For light-treated samples, place the light source on top of the plate and loosely wrap them in aluminum foil.
6. Start light treatment after replacing the media, delivering 2 s blue light every 3 min for 12–18 h. When light treatment is complete, samples can be imaged or harvested for downstream applications.

4 Notes

1. Place one coverslip in each well, then add 1 mL of medium and use a pipette tip to press the coverslip against the plate to ensure that it is resting flat at the bottom of the well.
2. Calcium phosphate transfection of HEK293T cells can achieve close to 100% efficiency even when transfecting multiple plasmids, but is not compatible with all cultured cell types. Alternatively, a cationic lipidic-mediated transfection reagent such as Lipofectamine 2000 can be used, but this method is less efficient.
3. The mixing step and the pH of the 2× HBS solution are critical to form the DNA–calcium phosphate precipitates. Every time a drop of the sample containing the DNA is added to the second tube containing the 2× HBS solution mix the sample, flick by hand or gently mix on a vortexer.
4. The dynamic range, activation level and background activity reported by Pathak et al. [7] correspond to HEK293T cells transfected with a calcium phosphate method. Different DNA delivery approaches or cell types could provide different values and would likely require optimization.
5. CRY2 shows peak response to blue light, with maximum excitation at a wavelength of 450–460 nm. Blue light treatment can be toxic to cells, thus when extended light treatments are required, pulsed light treatment instead of constant light is recommended to avoid inducing cellular stress. We deliver a short pulse of light (1–2 s) at a frequency approximately equivalent to half of the half-life of the protein's photocycle, which is sufficient to maintain significant photoreceptor excitation without inducing light toxicity. For CRY2 with a mean photocycle lifetime of 5.5 min, providing 1–2 s of light every 2–3 min ensures that the majority of the CRY2 population remains photoexcited during light treatment, with minimal toxicity.

Besides wavelength, the intensity, frequency, and duration of the light treatment are important parameters that can tune the expression level of the target gene.

6. Any light, including white light, containing wavelengths within the range of 400–514 nm can stimulate the sample. Thus, do not use white light to visualize dark samples, or a GFP filter set, as these will stimulate both CRY2 and *As*LOV2. Long-term exposure to high-intensity light at wavelengths higher than 514 nm (e.g., 561 nm) also can stimulate CRY2 at very low levels. To avoid blue light excitation while examining dark samples, the transmitted light of a microscope source can be filtered to eliminate blue light. For our confocal microscope, we use a 45 mm Schott RG610 long pass light filter (available from Chroma) in the transmitted light path. Alternatively, 561 nm light (used to visualize a mCherry reporter for example) delivered using a fluorescent light source and appropriate filters is suitable.
7. Use of a lower concentration of CRY2-Gal4BD-VP16AD vector decreased gene expression in the dark, reducing the dynamic range [11]. On the other hand, if a higher concentration of CRY2 is expressed, cluster formation could occur in the dark, hindering light-dependent regulation of the system.

Acknowledgments

This work was supported by grants from the National Institutes of Health [R01GM100225, R21GM126253] to C.L.T.

References

1. Gossen M, Bujard H (1992) Tight control of gene expression in mammalian cells by tetracycline-responsive promoters. Proc Natl Acad Sci U S A 89:5547–5551. https://doi.org/10.1073/PNAS.89.12.5547
2. Gossen M, Freundlieb S, Bender G et al (1995) Transcriptional activation by tetracyclines in mammalian cells. Science 268:1766–1769
3. Rivera VM, Clackson T, Natesan S et al (1996) A humanized system for pharmacologic control of gene expression. Nat Med 2:1028–1032. https://doi.org/10.1038/nm0996-1028
4. Jullien N, Sampieri F, Enjalbert A, Herman J (2003) Regulation of Cre recombinase by ligand-induced complementation of inactive fragments. Nucleic Acids Res 31:131e. https://doi.org/10.1093/nar/gng131
5. Zetsche B, Volz SE, Zhang F (2015) A split-Cas9 architecture for inducible genome editing and transcription modulation. Nat Biotechnol 33:139–142. https://doi.org/10.1038/nbt.3149
6. Pu J, Zinkus-Boltz J, Dickinson BC (2017) Evolution of a split rna polymerase as a versatile biosensor platform. Nat Chem Biol 13:432–438. https://doi.org/10.1038/nchembio.2299
7. Pathak GP, Spiltoir JI, Höglund C et al (2017) Bidirectional approaches for optogenetic regulation of gene expression in mammalian cells using Arabidopsis cryptochrome 2. Nucleic Acids Res 45:e167. https://doi.org/10.1093/nar/gkx260
8. Kennedy MJ, Hughes RM, Peteya LA et al (2010) Rapid blue-light–mediated induction

of protein interactions in living cells. Nat Methods 7:973–975. https://doi.org/10.1038/nmeth.1524

9. Hughes RM, Bolger S, Tapadia H, Tucker CL (2012) Light-mediated control of DNA transcription in yeast. Methods 58:385–391. https://doi.org/10.1016/j.ymeth.2012.08.004

10. Bonger KM, Rakhit R, Payumo AY et al (2014) General method for regulating protein stability with light. ACS Chem Biol 9:111–115. https://doi.org/10.1021/cb400755b

11. Hernández-Candia CN, Wysoczynski CL, Tucker CL (2019) Advances in optogenetic regulation of gene expression in mammalian cells using cryptochrome 2 (CRY2). Methods 164-165:81–90. https://doi.org/10.1016/J.YMETH.2019.03.011

Chapter 11

Optogenetic Downregulation of Protein Levels to Control Programmed Cell Death in Mammalian Cells with a Dual Blue-Light Switch

Patrick Fischbach, Patrick Gonschorek, Julia Baaske, Jamie A. Davies, Wilfried Weber, and Matias D. Zurbriggen

Abstract

Optogenetic approaches facilitate the study of signaling and metabolic pathways in animal cell systems. In the past 10 years, a plethora of light-regulated switches for the targeted control over the induction of gene expression, subcellular localization of proteins, membrane receptor activity, and other cellular processes have been developed and successfully implemented. However, only a few tools have been engineered toward the quantitative and spatiotemporally resolved downregulation of proteins. Here we present a protocol for reversible and rapid blue light-induced reduction of protein levels in mammalian cells. By implementing a dual-regulated optogenetic switch (Blue-OFF), both repression of gene expression and degradation of the target protein are triggered simultaneously. We apply this system for the blue light-mediated control of programmed cell death. HEK293T cells are transfected with the proapoptotic proteins PUMA and BID integrated into the Blue-OFF system. Overexpression of these proteins leads to programmed cell death, which can be prevented by irradiation with blue light. This experimental approach is very straightforward, requires just simple hardware, and therefore can be easily implemented in state-of-the-art equipped mammalian cell culture labs. The system can be used for targeted cell signaling studies and biotechnological applications.

Key words Optogenetics, Protein downregulation, Blue-light degron, Blue-light gene repression, Blue-OFF, Dual optogenetic switch, Optogenetic apoptosis control

1 Introduction

The development of synthetic switches for the targeted manipulation of protein levels in animal cells has facilitated the study of signaling and metabolic pathways [1–6]. A common approach is to control the expression or stability of a protein of interest with chemically induced switches. These normally consist of engineered activators or repressors, the binding of which to synthetic promoters is regulated by the presence of a chemical. For protein degradation, chemically regulated degrons have been developed

Dominik Niopek (ed.), *Photoswitching Proteins: Methods and Protocols*, Methods in Molecular Biology, vol. 2173, https://doi.org/10.1007/978-1-0716-0755-8_11,

[7]. However, chemically based switches have limitations such as toxicity, irregular spatially controlled distribution of the trigger by diffusion in cell culture or tissues, and limited reversibility [7–10]. To overcome these limitations, light has begun to be used as an inducer, and numerous optogenetic switches have been developed recently [1, 2, 11]. However, only a few have been designed for the targeted downregulation or destabilization of a protein [11, 12] and existing designs suffer from residual protein levels due to constant protein neosynthesis. Here, we present a protocol for the application of an optogenetic tool that combines blue light inducible repression of transcription and the simultaneous degradation of the target protein in animal cells [13]. We show its applicability to control apoptosis by regulating the levels of two proapoptotic proteins in mammalian cells.

1.1 Molecular Layout and Mechanism of the Blue-OFF System

The dual system designed for the simultaneous transcriptional repression and degradation of a protein of interest (POI) consists of two switches [13]: (1) the photosensitive transcription factor EL222 from *Erythrobacter litoralis* fused to the KRAB transrepressor domain [14–17] (pKM565) to inhibit transcription of the POI from a target promoter (Fig. 1); (2) the POI, in this case is either the proapoptotic proteins PUMA or BID (pPF088, pPF092), fused to the B-LID module which mediates the proteasomal degradation upon illumination with blue light [11]. The B-LID is based on the LOV2 domain of *Avena sativa* phototropin I (*As*LOV2). The *As*LOV2 contains a C-terminal Jα helix which is bound to the core domain of LOV2 in the dark but unwinds after illumination with blue light. B-LID harnesses this photoswitching mechanism to expose a RRRG degron fused to the Jα C-terminus in a light-dependent manner (Fig. 1). The proapoptotic proteins PUMA and BID are targets of the transcription factor p53 keeping the balance of cell cycle arrest and cell death upon DNA damage or other cell death insults [18–20]. Overexpression of those proteins leads to cell death. In this work we show the applicability of the Blue-OFF system for the regulation of apoptosis by reducing the levels of proapoptotic proteins constitutively expressed from a transfected plasmid in HEK293-T cells. The POI-B-LID module is cloned under the control of a synthetic promoter comprising five copies of the DNA target sequence of EL222-KRAB, namely $(C120)_5$, placed downstream of a constitutive promoter (Fig. 1). Upon illumination with 460 nm light, EL222-KRAB homodimerizes and binds via its helix-turn-helix (HTH) DNA-binding domains to the C120 sequences, thereby repressing transcription. Simultaneously, the LOV2 domain of the B-LID module exposes the RRRG degron, leading to degradation of the POI [13].

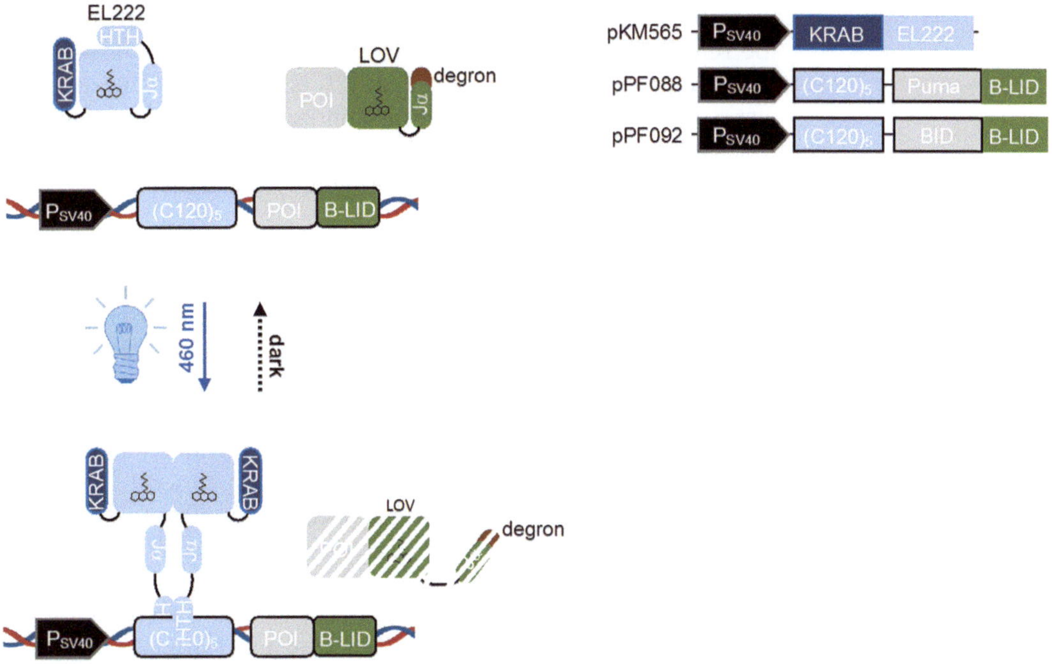

Fig. 1 Molecular design and mode of function of the Blue-OFF system for blue light-regulated downregulation of the levels of a POI. The pro-apoptotic proteins Puma and BID are fused to the B-LID system, and placed under the control of a constitutive SV40 promoter (pPF088, pPF092). The promoter sequence is followed by five copies of the EL222-binding sequence ($C120_5$). The photosensory transcription factor EL222 is fused to the inhibitory KRAB domain (pKM565). In the dark, the KRAB-EL222 fusion is not bound to the target sequence on the DNA and the B-LID system is inactive leading to accumulation of the POI. Upon blue light illumination, the Jα helix unwinds, exposing the docked degradation peptide (RRRG) which leads to proteasome mediated protein degradation. Simultaneously, the EL222 transcription factor dimerizes and binds to the $C120_5$ sequence inhibiting transcription via the fused KRAB repressor domain. Adapted from Baaske et al. [13]

1.2 Application and Experimental Considerations

This dual-controlled optogenetic switch shows highly efficient and rapid blue-light induced downregulation of protein expression and stability. These characteristics can be used to knock down essential genes in a cell, tissue or organism to study the effect of losing a given protein in an otherwise wild type context. We have previously shown a quantitative characterization of the system and its ability to control a synthetic caspase-based switch to induce programmed cell death [13]. Here, we describe a protocol to demonstrate further the applicability of the system by regulating the levels of ectopically overexpressed proapoptotic proteins such as PUMA or BID.

Blue light illumination can have toxic effects on cells. However, the intensity and time doses needed for full activation of the Blue-OFF system (20 $\mu mol/m^2/s$ for 8 h or 10 $\mu mol/m^2/s$ for 24 h) have no negative effect on the cells [13]. It is worth considering when designing an experiment that higher doses might have a negative influence on growth and health. The system represses transcriptional activity and targets the protein for degradation;

however, it has no effect over the mRNA, meaning that there might be remaining expression from previously synthesized messengers. An advantage of the Blue-OFF switch is that there is no need of extra addition of FMN, the chromophore of the LOV domains, to the growth media [21]. As the photoreceptors are activated also by daylight or room light, all work should be done under green or red safe light.

1.3 Experimental Design

In this protocol, HEK293T cells are transfected with plasmids encoding the proapoptotic proteins PUMA and BID engineered into the Blue-OFF system (pPF088, pPF092, pKM565). As a negative control, the plasmids encoding for the pro-apoptotic proteins fused to B-LID without the RRRG degron (pMZ1427, pTB505) under the control of a constitutive promoter were transfected. The light treatments were performed in closed LED boxes, with a wavelength of 460 nm and an intensity of 10 μmol/m^2/s for 24 h as described below. Control cells are kept in dark for the same incubation period. Transfections for microscopy are done in duplicates. After 24 h of treatment, the cells can be directly observed under the microscope or be fixed for long-term storage.

2 Materials

2.1 Reagents, Consumables, and Kits

1. Plasmids (Fig. 1):
   (a) pMZ1203: P_{SV40}-C120$_5$-Firefly-B-LID-pA.
   (b) pMZ1427: P_{SV40}-RFP-2A-Puma-B-LIDΔRRRG.
   (c) pTB505: P_{SV40}-RFP-2A-BID-B-LIDΔRRRG.
   (d) pPF088:P_{SV40}-C120$_5$-Puma-B-LID-pA.
   (e) pPF092: P_{SV40}-C120$_5$-BID-B-LID-pA.
2. Top10 chemically competent cells.
3. Plasmid isolation kit (e.g., M&N NucleoBond Xtra Midi Kit).
4. Ampicillin.
5. LB agar.
6. LB liquid medium.
7. HEK293T cells.
8. DMEM growth medium: DMEM supplemented with 10% (V/V) FBS and 1.4% (V/V) penicillin–streptomycin.
9. Trypsin–EDTA.
10. CASY ton buffer (OLS).
11. Opti-MEM.
12. Polyethyleneimine (PEI) linear molecular weight (MW) 25 kDa (Polyscience).

13. DMSO.
14. Bottle-top filter, 500 mL, 0.2 μM pore size (e.g., VWR).
15. CASY cups (OLS).
16. 10 cm cell culture dishes.
17. 24-well cell culture plates.
18. Glass cover slides.
19. Microscopy slides.
20. Paraformaldehyde (PFA).
21. 0.2 M Tris–HCl, pH 8.5
22. PBS (10× solution): Dissolve 26.82 mM KCl, 14.7 mM KH_2PO_4, 80.34 mM $Na_2HPO_4 \cdot 2H_2O$ and 1.37 M NaCl in 1 L ddH_2O. Dilute the PBS to 1×, sterile filter and aliquot it in 50 mL Falcon tubes.
23. Mowiol (Roth).
24. Dabco (Roth).

2.2 Equipment

1. Bacterial incubator with shaking function.
2. Spectrophotometer.
3. Tissue culture hood.
4. Tissue culture incubators.
5. CASY cell counter and analyzer (OLS, CasyTT).
6. Spectroradiometer (Avatec, Avaspec-2048).
7. LED band deco Flex RGB plug and light set (Prisma Leuchten).
8. LED Boxes (LEDs: Roithner LED450-series).
9. Confocal microscope (Nikon Eclipse Ti+ C2+ confocal upgrade).

2.3 Equipment Setup

2.3.1 Safe-Light Setup in the Lab

Stick the LED band to the internal surface/walls of the cell culture hood. The safe-light (green or red) can be turned on from the outside via remote control. If required, additional LED stripes can be installed all over the room. Cover all ambient light sources such as windows and/or doors with curtains or black adhesive vinyl foil to achieve full darkness.

2.3.2 LED Boxes

LED boxes were constructed and used as described in Müller et al. [8] and Ochoa-Fernandez et al. [22]. In brief, we use custom-made light boxes built out of PVC (20 cm × 20 cm × 20 cm) and equipped with an LED panel. The light boxes are additionally equipped with fans for gas exchange. Irradiation wavelength and intensity control is achieved with an Arduino microcontroller installed in the aluminum LED panel with a USB port for programming irradiation time and pulsing. In this protocol, boxes containing blue LEDs (460 nm) were used.

3 Methods

3.1 Plasmid DNA Preparation

1. Prepare LB agar plates by mixing LB agar (40 g/L) with H_2O according to manufacturer's instructions and autoclave it.
2. Add 100 μg/mL ampicillin (from a 100 mg/mL stock solution in H_2O, sterile filtered) to the cooled-down LB agar and pour it in 100 mm petri dishes and let it solidify. The plates can be stored at 4 °C for 1 month.
3. Transform chemically competent *Escherichia coli* TOP10 cells according to Beyer et al. [23] and plate 10 and 50 μL on LB agar plates supplemented with ampicillin. Incubate at 37 °C for 24 h.
4. Inoculate 120 mL of autoclaved LB medium supplemented with 100 μg/ml ampicillin with a single colony by using a sterile pipette tip, and incubate at 37 °C for 24 h at 150 rpm.
5. Centrifuge 100 mL of the overnight culture and isolate the DNA with the plasmid isolation kit (*see* **Note 1**) according to the manufacturer's instructions. Determine the DNA concentration with a spectrophotometer.

3.2 Reagent Setup

3.2.1 PEI Solution (1 mg/mL)

1. Dissolve 200 mg of PEI in 160 mL H_2O in a glass beaker and stir. For faster dissolution heat up to 50 °C.
2. Adjust the pH to 7 with HCl until PEI is completely dissolved and fill in with ddH_2O to 200 mL.
3. Filter the PEI solution through a 0.2 μm filter in a cell culture hood and divide it into 1 mL aliquots. The aliquots can be stored at −80 °C for at least 1 year.

3.2.2 Mowiol-DABCO-Solution

1. Mix 6 g of glycerol with 2.4 g Mowiol in a 50 mL Falcon tube, incubate for 30 min while vortexing every 10 min.
2. Add 6 mL H_2O and stir solution for 2 h.
3. Add 12 mL 0.2 M Tris–HCl, pH 8.5 and heat up to 53 °C in a water bath until dissolution (approximately 2 h). Stir every 30 min with a magnetic stirring bar.
4. Centrifuge the solution for 20 min at 5000 × *g*, transfer the supernatant to a fresh tube and add 25 mg DABCO for each mL of solution.
5. Stir until complete dissolution. Prepare 500 μL aliquots and freeze them at −20 °C.

3.3 Seeding of HEK293T Cells

Estimated duration 1 h (mid-day or afternoon of day 1)

1. Start with a HEK293T 80–90% confluent cell culture in 10 cm petri dishes maintained in 10 mL DMEM growth medium. Using healthy cells is essential (*see* **Note 2**).
2. For collecting the cells, remove the culture medium, add 2 mL of trypsin–EDTA solution and incubate at 37 °C for 5 min.
3. During the incubation time prepare a 15 mL tube with 8 mL fresh DMEM growth medium (to get a final total volume of 10 mL) (*see* **Note 3**).
4. Wash away the cells from the plate by rinsing the trypsin–EDTA cell suspension 2–3 times, and pipet the suspension up and down to resuspend the cells. Transfer the cell suspension into the 15 mL tube prepared in **step 3**.
5. Sediment the cells by centrifugation (3 min, 300 × *g*, RT). Discard the supernatant and resuspend the cells in 10 mL fresh DMEM growth medium.
6. Determine the cell concentration with the CASY cell counting system or with a Neubauer cell-counting chamber.
7. Seed HEK293T cells in 10 wells of each of two 24-well plates at a density of 40,000–50,000 cells per well in 500 μL DMEM growth medium. Incubate the cells for 24 h at 37 °C with 5% CO_2 in an incubator.

3.4 Transfection of HEK293T Cells

Estimated duration: 1.5 h in the morning and 30 min in the afternoon (day 2).

1. Inspect the seeded cells under the microscope. They should be uniformly and evenly distributed. The confluency should be about 30–50%.
2. Five different DNA mixes need to be prepared in Opti-MEM in a total volume of 250 μL as follows (*see* **Note 4**):
   (a) Mix 1: (the negative control) pMZ1203 = 1.875 μg and pKM565 = 1.875 μg.
   (b) Mix 2: (Blue-OFF controlled Puma) pPF088 = 1.875 μg and pKM565 = 1.875 μg.
   (c) Mix 3: (Blue-OFF controlled BID) pPF092 = 1.875 μg and pKM565 = 1.875 μg.
   (d) Mix 4: (unregulated Puma) pMZ1427 = 3.75 μg.
   (e) Mix 5: (unregulated BID) pTB505 = 3.75 μg.
3. In an additional 15 mL tube, prepare the PEI mix as follows: add 96 μL of 1 mg/mL PEI solution to 1829 μL Opti-MEM (calculate a 10% excess, in case of pipetting deviation/mistakes).

4. To produce the DNA–PEI complexes, add 250 μL of the PEI mix in two separate steps to the DNA mixes 1–5, and mix the tubes by vortexing for 10 s after each addition of the PEI mix. Incubate the tubes at RT for 10 min.
5. Add 100 μL of the transfection mix dropwise to each well of the plate seeded in Subheading 3.3, **step** 7. Transfect with each transfection mix 2 wells on the 24-well plate that will be kept in darkness (as control), and 2 wells on the 24-well plate that will be illuminated. Finally, distribute the added transfection mix evenly by gently moving the plates in "8-shape- or up-down/ left-right movements," and then incubate them in a CO_2 incubator at 37 °C, 5% CO_2.
6. Four to five hours after transfection, replace the culture medium carefully in both 24-well plates with 0.5 mL of pre-warmed DMEM growth medium per well (*see* **Note 5**).
7. From now on every step should be carried out in the absence of blue/room light, that is, "darkness." Use green (530 nm) or red (660 nm) safe-light to avoid activation of the system. Directly after changing the medium (**step 6**), illuminate one 24-well plate with 460 nm light with 10 μmol/m^2/s intensity and keep the other plate in "darkness" for 16 h at 5% CO_2 and 37 °C (*see* **Note 6**).

3.5 Fixing Cells for Long-Term Storage

Estimated duration: 1 h (day 3)

1. After illumination, aspirate the medium of the wells with the transfected cells. Wash the cells once with 500 μL PBS and add 200 μL PFA (*see* **Note** 7).
2. Incubate the cells covered with PFA for 10 min on ice, and an additional 10 min at RT.
3. Remove the PFA (*see* **Note 8**).
4. Add 500 μL PBS. The cells are now fixed and the following steps can be carried in normal room light.
5. Prepare microscopy slides and add 8 μL Mowiol/Dabco to the microscopy slide.
6. Use forceps to transfer the glass slides (remove carefully the excess liquid from the glass slide with a tissue paper) with cells upside down on the Mowiol/Dabco droplet on the microscopy slide. After 30 min incubation at 37 °C the slides can be stored for more than 1 month at 4 °C (*see* **Note 9**).

3.6 Analysis of Apoptosis

Estimated duration: 0.5 h (day 3)

1. Check for cell growth under the microscope. Observe the formation of a confluent monolayer (if it does form) (*see* **Note 10**).

2. Perform quantification and statistics accordingly/as needed. Note that, instead of using apoptosis/survival as readout, the rapid downregulation of protein levels can also be directly monitored (*see* **Note 11**).

4 Notes

1. For high transfection efficiencies it is recommended to use RNA-free, supercoiled DNA. For best results we use the NucleoBond Xtra MIDI Kit for DNA preparations.
2. Healthy cells are essential (viability and morphology)! Low cell viability leads to low expression levels. Ideally cells should be neither too young (passage number < 5) nor too old (passage number > 30) for best expression results.
3. Cells of a maximum of three plates can be pooled in one tube to speed up the process. In this case use only 4 mL fresh DMEM medium.
4. For each well of a 24-well plate mix a total of 0.75 μg of plasmid DNA in 50 μL of Opti-MEM, and 2.5 μL of 1 mg/mL PEI solution in 50 μL of Opti-MEM.
5. Incubation with PEI for more than 5 h might lead to decreased cell viability. However, very short incubation times with PEI decrease transfection efficiency. Additionally, the PEI solution has to be kept at pH = 7; this is essential for high transfection efficiency.
6. The system is very light-sensitive. After transfection, every step should be carried out in the absence of blue/room light, that is, always in "darkness." Use green (530 nm) or red (660 nm) safe-light to avoid activation of the system.
7. PFA is toxic! All work with PFA should be performed following the manufacturer's guidelines for proper handling. Use gloves and dispose of the liquid waste and all consumables/material which had contact with PFA under the toxic waste instructions of your institute.
8. Use a waste tube and dispose of PFA according to the toxic waste handling guidelines at your institute.
9. Fixed cells can be stored for at least 1 month at 4 °C. Expressed fluorescent proteins are still detectable with a fluorescence microscope.
10. The protocol described here was implemented for the optogenetic regulation of programmed cell death via the control of the levels of proapoptotic proteins. HEK293T cells were transfected with the Blue-OFF optogenetic switch engineered to control PUMA or BID. Incubation of the cells in the dark led

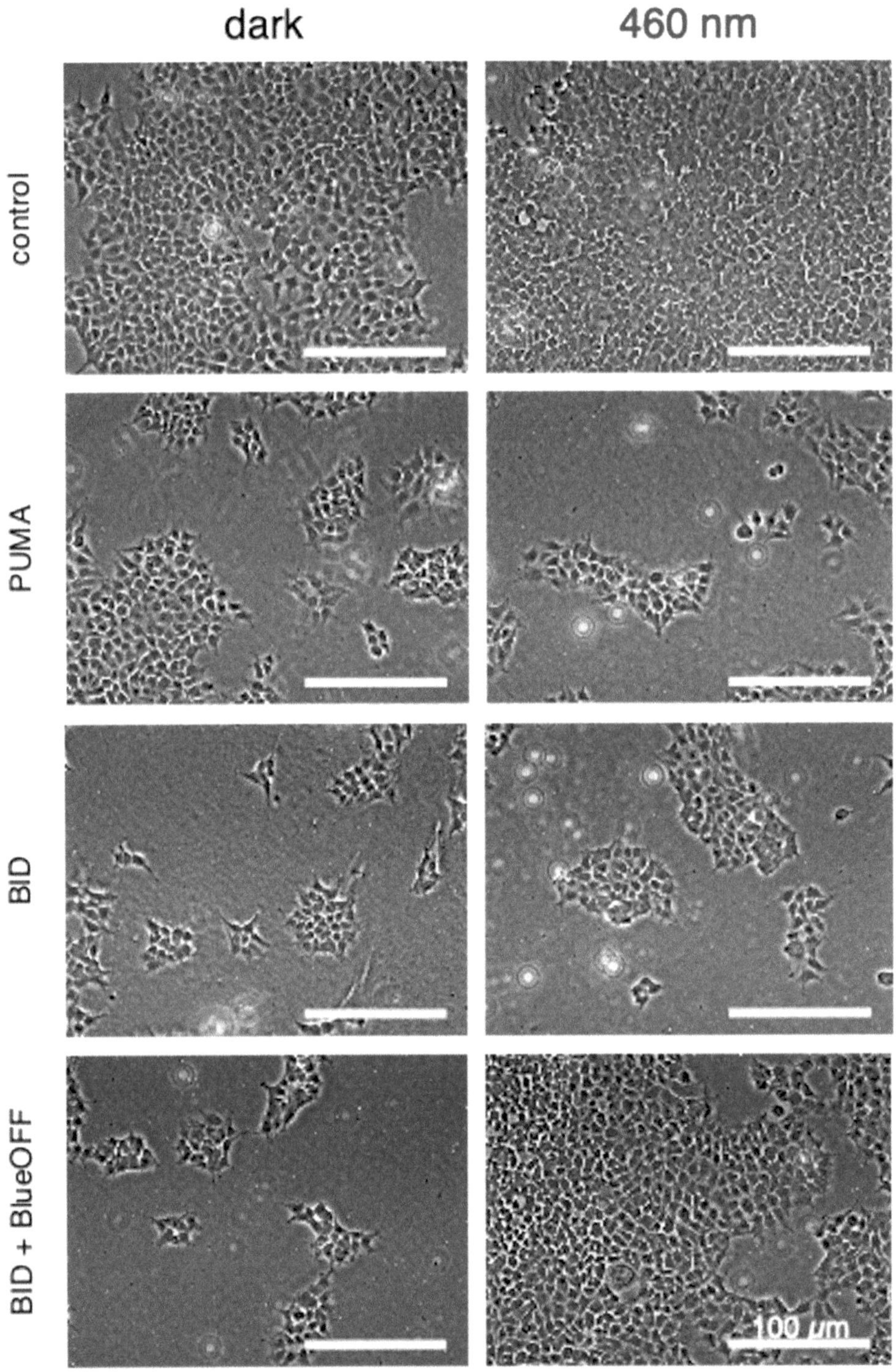

Fig. 2 Control of programmed cell death. Representative results of HEK293T cells transfected with control (pMZ1203) forming a uniform monolayer in darkness and under blue illumination. Constitutive expression of PUMA or BID (pMZ1427; pTB505) in darkness and blue light, and of PUMA-Blue-OFF and BID-Blue-OFF (pPF088 + pKM565; pPF092 + pKM565) in darkness leads to increased cell death. In contrast to this, cells transfected with PUMA-Blue-OFF or BID-Blue-OFF show a higher survival rate, observed as a uniform monolayer, upon blue light illumination

to a high cell death rate, in contrast to control cells transfected with pMZ1203 (Blue-OFF controlling FLuc) which formed a uniform monolayer (Fig. 2). Cells transfected with the light-dependent PUMA or BID systems showed as expected a higher survival rate (uniform monolayer) when illuminated with blue light. The results open up novel perspectives for the targeted regulation of programmed cell death in animal cells with applications in fundamental research such as the study of apoptotic and carcinogenic cellular mechanisms [24, 25]. Additionally, the high spatiotemporal resolution of the system might be of advantage for the establishment of cellular patterns in tissue engineering approaches.

11. To analyze the rapid downregulation of protein levels in a quantitative manner, one can use reporter genes (e.g., luciferases, phosphatases, or fluorophores) instead of proapoptotic proteins [13].

Acknowledgments

We thank J. Andres, L. Koch, and T. Blomeier for experimental support and fruitful discussions, and R. Wurm, M. Gerads, and J. Müller for valuable experimental support. This work was supported by the German Research Foundation (DFG) (grant ZU259/2-1 to M.D.Z. and under Germany's Excellence Strategy CEPLAS—EXC-2048/1—Project ID 390686111 to M.D.Z., BIOSS—EXC-294 and CIBSS—EXC-2189—Project ID 390939984 to W.W.) and the European Commission—Research Executive Agency (H2020 Future and Emerging Technologies (FET-Open) Project ID 801041 CyGenTiG to J.A.D. and M.D. Z.).

Author contributions: P.F. designed the system and performed the experiments, analyzed the data, and wrote the protocol. P.G. and J.B. designed the system. J.D. designed experiments and analyzed the data. W.W. designed the system and experiments, and analyzed the data. M.D.Z. designed the system and experiments, analyzed the data, and wrote the protocol.

References

1. Wend S, Wagner HJ, Muller K et al (2014) Optogenetic control of protein kinase activity in mammalian cells. ACS Synth Biol 3 (5):280–285. https://doi.org/10.1021/sb400090s
2. Beyer HM, Juillot S, Herbst K et al (2015) Red light-regulated reversible nuclear localization of proteins in mammalian cells and zebrafish. ACS Synth Biol 4(9):951–958. https://doi.org/10.1021/acssynbio.5b00004
3. Toettcher JE, Gong D, Lim WA et al (2011) Light-based feedback for controlling intracellular signaling dynamics. Nat Methods 8 (10):837–839. https://doi.org/10.1038/nmeth.1700
4. Levskaya A, Weiner OD, Lim WA et al (2009) Spatiotemporal control of cell signalling using a light-switchable protein interaction. Nature 461(7266):997–1001. https://doi.org/10.1038/nature08446

5. Andres J, Blomeier T, Zurbriggen MD (2019) Synthetic switches and regulatory circuits in plants. Plant Physiol 179(3):862–884. https://doi.org/10.1104/pp.18.01362
6. Kolar K, Knobloch C, Stork H et al (2018) OptoBase: a web platform for molecular optogenetics. ACS Synth Biol 7(7):1825–1828. https://doi.org/10.1021/acssynbio.8b00120
7. Bonger KM, Chen LC, Liu CW et al (2011) Small-molecule displacement of a cryptic degron causes conditional protein degradation. Nat Chem Biol 7(8):531–537. https://doi.org/10.1038/Nchembio.598
8. Muller K, Zurbriggen MD, Weber W (2014) Control of gene expression using a red- and far-red light-responsive bi-stable toggle switch. Nat Protoc 9(3):622–632. https://doi.org/10.1038/nprot.2014.038
9. Frey AD, Rimann M, Bailey JE et al (2001) Novel pristinamycin-responsive expression systems for plant cells. Biotechnol Bioeng 74 (2):154–163. https://doi.org/10.1002/bit.1105
10. Gossen M, Bujard H (1992) Tight control of gene-expression in mammalian-cells by tetracycline-responsive promoters. Proc Natl Acad Sci USA 89(12):5547–5551. https://doi.org/10.1073/pnas.89.12.5547
11. Bonger KM, Rakhit R, Payumo AY et al (2014) General method for regulating protein stability with light. ACS Chem Biol 9(1):111–115. https://doi.org/10.1021/cb400755b
12. Pathak GP, Spiltoir JI, Hoglund C et al (2017) Bidirectional approaches for optogenetic regulation of gene expression in mammalian cells using &ITArabidopsis &ITcryptochrome 2. Nucleic Acids Res 45(20):e167. https://doi.org/10.1093/nar/gkx260
13. Baaske J, Gonschorek P, Engesser R et al (2018) Dual-controlled optogenetic system for the rapid down-regulation of protein levels in mammalian cells. Sci Rep 8:15024. https://doi.org/10.1038/s41598-018-32929-7
14. Nash AI, McNulty R, Shillito ME et al (2011) Structural basis of photosensitivity in a bacterial light-oxygen-voltage/helix-turn-helix (LOV-HTH) DNA-binding protein. Proc Natl Acad Sci U S A 108(23):9449–9454. https://doi.org/10.1073/pnas.1100262108
15. Rivera-Cancel G, Motta-Mena LB, Gardner KH (2012) Identification of natural and artificial DNA substrates for light-activated LOV-HTH transcription factor EL222. Biochemistry 51(50):10024–10034. https://doi.org/10.1021/bi301306t
16. Moosmann P, Georgiev O, Thiesen HJ et al (1997) Silencing of RNA polymerases II and III-dependent transcription by the KRAB protein domain of KOX1, a Kruppel-type zinc finger factor. Biol Chem 378(7):669–677. https://doi.org/10.1515/bchm.1997.378.7.669
17. Motta-Mena LB, Reade A, Mallory MJ et al (2014) An optogenetic gene expression system with rapid activation and deactivation kinetics. Nat Chem Biol 10(3):196–202. https://doi.org/10.1038/Nchembio.1430
18. Nakano K, Vousden KH (2001) PUMA, a novel proapoptotic gene, is induced by p53. Mol Cell 7(3):683–694. https://doi.org/10.1016/S1097-2765(01)00214-3
19. Sax JK, Fei PW, Murphy ME et al (2002) BID regulation by p53 contributes to chemosensitivity. Nat Cell Biol 4(11):842–849. https://doi.org/10.1038/ncb866
20. Deng J (2017) How to unleash mitochondrial apoptotic blockades to kill cancers? Acta Pharm Sin B 7(1):18–26. https://doi.org/10.1016/j.apsb.2016.08.005
21. Muller K, Engesser R, Metzger S et al (2013) A red/far-red light-responsive bi-stable toggle switch to control gene expression in mammalian cells. Nucleic Acids Res 41(7):e77. https://doi.org/10.1093/nar/gkt002
22. Ochoa-Fernandez R, Samodelov SL, Brandl SM et al (2016) Optogenetics in plants: red/far-red light control of gene expression. Methods Mol Biol 1408:125–139. https://doi.org/10.1007/978-1-4939-3512-3_9
23. Beyer HM, Gonschorek P, Samodelov SL et al (2015) AQUA cloning: a versatile and simple enzyme-free cloning approach. PLoS One 10 (9):e0137652. https://doi.org/10.1371/journal.pone.0137652
24. Lee JH, Soung YH, Lee JW et al (2004) Inactivating mutation of the pro-apoptotic gene BID in gastric cancer. J Pathol 202 (4):439–445. https://doi.org/10.1002/path.1532
25. Giotopoulou N, Valiakou V, Papanikolaou V et al (2015) Ras suppressor-1 promotes apoptosis in breast cancer cells by inhibiting PINCH-1 and activating p53-upregulated-modulator of apoptosis (PUMA); verification from metastatic breast cancer human samples. Clin Exp Metastasis 32(3):255–265. https://doi.org/10.1007/s10585-015-9701-x

Chapter 12

Tracing Reversible Light-Induced Chromatin Binding with Near-infrared Fluorescent Proteins

Anne Rademacher, Fabian Erdel, Jorge Trojanowski, and Karsten Rippe

Abstract

Blue light-induced chromatin recruitment (BLInCR) is a versatile optogenetic tool to enrich effector proteins at specific loci within the nucleus using illumination in the 400–500 nm range. The resulting chromatin binding reaction is reversible on the time scale of minutes. BLInCR is advantageous over ligand-binding induced methods since it does not require a change of growth medium for the relatively slow depletion of the inducer from the nucleus. However, applying BLInCR for reversibility experiments is challenging because of the need to spectrally separate light-induced activation from visualization of the chromatin locus and effector and/or reader proteins by light microscopy. Here, we describe an improved BLInCR protocol for light-dependent association and dissociation of effectors using the near-infrared fluorescent protein iRFP713. Due to its spectral properties, iRFP713 can be detected separately from the red fluorescent protein mCherry. Thus, it becomes possible to trace two proteins labeled with iRFP713 and mCherry independently of the light activation reaction. This approach largely facilitates applications of the BLInCR system for experiments that test the reversibility, persistence, and memory of chromatin states.

Key words Optogenetics, Transcription activation, Automated microscopy, Image quantification, Chromatin binding

1 Introduction

Optogenetic proteins switch their conformation when illuminated with light of a certain wavelength range and revert to their original conformation in the absence of this trigger [1]. When exploiting the light-dependent reversibility—an inherent advantage of optogenetic systems—for microscopy studies, the repertoire of usable fluorescent protein tags is limited to excitation wavelengths outside the range that induces photoswitching. Two widely used blue light-induced photoswitches are the PHR domain from *Arabidopsis thaliana* [2, 3] and the LOV2 domain from *Avena sativa* [4, 5], which change their interaction properties in response to blue light. Fluorescently tagged proteins used in combination with these photoswitches need to be excited with green or longer wavelength

Dominik Niopek (ed.), *Photoswitching Proteins: Methods and Protocols*, Methods in Molecular Biology, vol. 2173, https://doi.org/10.1007/978-1-0716-0755-8_12,

light to visualize them without inducing photoswitching [3, 6]. Thus, commonly used fluorescent proteins such as CFP, GFP or YFP are not suited. Red fluorescent proteins such as mCherry ($\lambda_{ex} = 587$ nm, $\lambda_{em} = 610$ nm) [7] are compatible with the excitation wavelength requirements. However, it is often desirable to monitor a second protein independently of the photoswitch trigger, for example, to determine the location of the chromatin locus of interest or to include a functional readout for the recruited effector. To address this issue, fluorescent proteins with near-infrared emission [8] such as iRFP713 ($\lambda_{ex} = 690$ nm, $\lambda_{em} = 713$ nm) [9] can be used (*see* **Note 1**). By sequential image acquisition on a confocal fluorescence microscope with $\lambda_{ex1} = 633$ nm and $\lambda_{ex2} = 561$ nm and appropriately selected detection windows, the iRFP713 and mCherry signals can be spectrally separated and recorded without inducing blue light-dependent photoswitching (Fig. 1).

BLInCR relies on the PHR domain that switches to a conformation that is permissive for interaction with CIBN when illuminated with blue light [10]. In this system, CIBN is fused to a nuclear protein that adopts a specific localization in the nucleus (e.g., reporter gene arrays, telomeres, nucleoli, PML nuclear bodies or the nuclear lamina). Blue-light illumination in the 400–500 nm range results in high local concentration of PHR-fused effector proteins at the target site. This approach was applied previously to dissect gene expression kinetics by fusing CIBN to TetR and PHR to the transcriptional activator VP16 or a nuclear localization signal (NLS, mock effector) [10]. A detailed protocol for this application can be found elsewhere [11]. The U2OS 2-6-3 cell line [12] was used for these experiments. It contains an array consisting of repetitive, promoter-proximal binding sites for LacI and TetR and a reporter cassette comprising a gene with MS2 loops that codes for peroxisome-targeted CFP (Fig. 2a). Previously, we used a GFP array marker to visualize the reporter array in this cell line. Since GFP-based imaging is incompatible with reversibility experiments, we recorded *z*-stacks of the mCherry-tagged effector to ensure that we capture the entire signal from the array [10]. Here, we provide a detailed protocol for recording effector and localizer fluorescence independently of the BLInCR blue light trigger via mCherry and iRFP713 fusions. This approach reduces the imaging time (and thereby photobleaching) and simplifies the image analysis (Fig. 2b, c). In addition, recruitment and reversibility can be characterized with the same constructs in the same cell. This is beneficial for rigorously resolving differences among individual cells during complex activation-deactivation patterns.

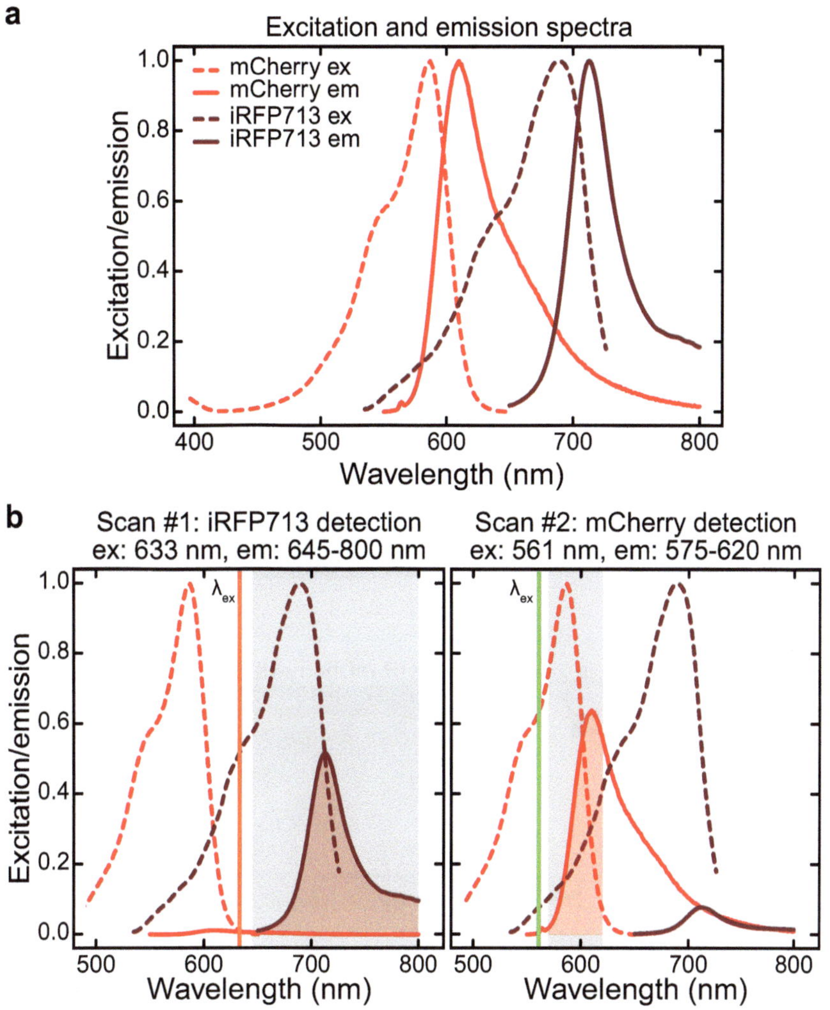

Fig. 1 Fluorescent proteins used for BLInCR reversibility experiments. (**a**) Excitation (dashed line) and emission (solid line) spectra of the red fluorescent protein mCherry and the near-infrared fluorescent protein iRFP713. The spectra are normalized to their respective maxima. (**b**) For the experiments described here, fluorescence is recorded in two sequential scans with the excitation wavelength (λ_{ex}) marked by a solid vertical line and the emission detection window marked by grey boxes. The emission spectra are scaled to the relative excitation at the excitation wavelength λ_{ex}. ex, excitation; em, emission

2 Materials

2.1 Cell Culture

1. Reporter cell line U2OS 2-6-3 [12] or other cell line containing repetitive arrays of *tet*O or *lac*O that can be bound by TetR or LacI, respectively (*see* **Note 2**).
2. Low glucose (1.0 g/L) DMEM without phenol red (ThermoFisher Scientific Inc., USA) supplemented with 10% tetracycline-free fetal bovine serum (FBS), 2 mM stable glutamine, 1× Penicillin/Streptomycin.

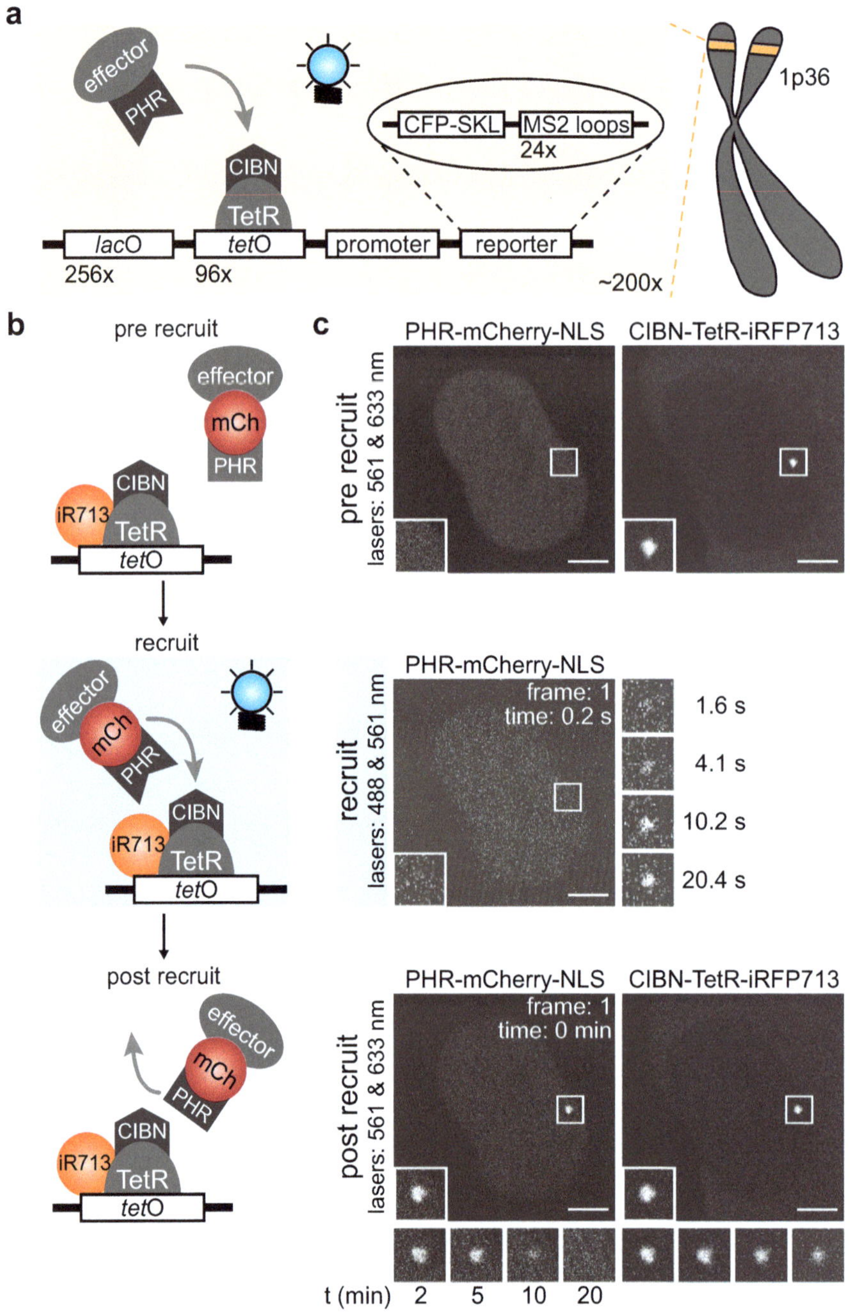

Fig. 2 Setup for BLInCR experiments with iRFP713. (**a**) The U2OS 2-6-3 cell line [12] contains promoter-proximal *tet*O and *lac*O arrays for TetR and LacI binding, respectively, that can be targeted by corresponding localizers fused to CIBN. In addition, it contains reporters for RNA (MS2) and protein (peroxisome-targeted CFP) readouts. (**b**) Schematic representation of pre recruit, recruit, and post recruit events for effector BLInCR to *tet*O arrays. The fluorophores can be visualized at every step independently of blue light-induced effector

3. Phosphate-buffered saline (PBS).
4. 0.05% trypsin/0.02% EDTA in PBS.
5. 8-well Lab-Tek chambers (ThermoFisher Scientific Inc.) or any other cell culture dish that is suitable for live cell microscopy (*see* **Note 3**).
6. Xtreme Gene 9 (Roche, Germany) or any other suitable transfection reagent (*see* **Note 4**) as well as transfection media, if needed (e.g., Opti-MEM for Xtreme Gene 9).
7. BLInCR constructs: CIBN-TetR-iRFP713 (localizer, *see* **Note 5**), PHR-mCherry-VP16 (effector, Addgene #103821), PHR-mCherry-NLS (mock effector, Addgene #103819) (*see* **Note 6**).
8. Nontransparent Styrofoam box (*see* **Note 7**).
9. Red flashlight (*see* **Note 7**).

2.2 Microscope

1. A confocal laser scanning microscope. Here, a Leica SP5 equipped with an HCX PL APO lambda blue 63.0 × 1.40 OIL UV objective is used.
2. Lasers for excitation at 488 nm, 561 nm and 633 nm.
3. Optional: Excitation/emission filter sets (e.g., Leica TX2 (*see* **Note 8**).
4. A chamber for incubation at 37 °C/5% CO_2 that can be protected from light. A microscope incubator box and control device from EMBLEM Technology Transfer GmbH, Germany is used in our work.

2.3 Software

1. LAS AF microscopy software (Leica).
2. Fiji distribution [13] of ImageJ [14], version 2.0.0.
3. R, version 3.5.2.
4. R package *EBImage*, version 4.24.0 [15].
5. R package *nlme*, version 3.1–137 [16].

Fig. 2 (continued) recruitment. iR713, iRFP713; mCh, mCherry. (**c**) Representative confocal laser scanning microscopy images for reversibly recruiting an mCherry-labeled mock effector (PHR-mCherry-NLS). The array location is marked by CIBN-TetR-iRFP713 (right, top). Prior to light induction, PHR-mCherry-NLS is evenly distributed throughout the cell (left, top). Upon illumination at 488 nm, it accumulates at the array within seconds (middle). After recruitment, it dissociates again in the absence of blue light (left, bottom) within 10-20 min. Notably, the iRFP713-tagged localizer construct can be detected continuously (right, bottom). The time points for the localizer constructs (right, bottom) are the same as for the effector (left, bottom). Note that iRFP713 is susceptible to photobleaching (*see* **Note 1**). Image intensities have been adjusted nonlinearly (gamma transformation) to improve the visibility. Scale bars: 5 μm; insets: 2× magnification

3 Methods

The protocols provided here were applied to the characterization of reversible association to and dissociation from a gene array in a reporter cell line (Fig. 2b). The use of iRFP713 and mCherry in these assays can be easily adapted for tagging other loci (telomeres, nucleoli, nuclear lamina, PML bodies, etc.) or readouts (e.g., MS2 coat protein to detect RNA carrying MS2 loop sequences) [11].

3.1 Cell Culture

U2OS 2-6-3 cells are cultured at 37 °C in 5% CO_2. They should be passaged every 3–4 days and can be frozen in DMEM containing 10% DMSO and 40% FBS.

1. Seed U2OS 2-6-3 cells in a dish that is compatible with live cell microscopy (e.g., Lab-Tek slides). To this end, passage the cells according to standard cell culture protocols. They should be 50–90% confluent on the next day for transfection.
2. On the next day, transfect cells with a localizer and an effector construct (e.g., CIBN-TetR-iRFP713 and PHR-mCherry-VP16 or -NLS) according to the manufacturer's protocol (*see* **Note 9**). Take care to protect transfected cells from light by placing them in a nontransparent Styrofoam box and by minimizing handling. It is recommended to transfect two wells and use one for searching the right focal plane using the ocular and one for the actual measurements.

3.2 Microscopy

The microscopy part of the experiment can be carried out 18–48 h after transfection for the constructs described here. Shorter transfection times might lead to incomplete maturation of the constructs whereas dilution or loss of the constructs will increase at longer transfection times. The imaging parameters given here provide reasonable time resolution for rapid recruitment and sufficient spatial resolution to reliably locate the array. Depending on the application and the used constructs, the temporal and spatial resolution should be adjusted. We have recorded high-resolution images before and after recruitment and after the reversibility series to be able to assess the amount of prerecruitment and residual enrichment of the effector construct at the array. All multicolor images are recorded using the "between lines" sequential scan modus to ensure that the images from the different scans are aligned.

1. Preheat the incubation chamber about 1 h prior to imaging to ensure that the microscope optics are at a constant temperature, thereby limiting drifting during the acquisition.
2. Place the cells on the microscope using the red flashlight. During the transfer, switch off all other (blue/white) light sources including computer screens.

3. Search for transfected cells using the ocular and an appropriate filter cube (e.g., Leica TX2 or Y5, *see* **Note 8**).
4. Set up sequential imaging in the LAS AF software with two scans. Use the following excitation/emission parameters (Fig. 1b): scan #1 (iRFP713), excitation at 633 nm, emission detection at 645–800 nm; scan #2 (mCherry), excitation at 561 nm, emission detection at 575–620 nm.
5. Use laser excitation and PMT detection in the microscope software to find a transfected cell (*see* **Note 10**). This should be done in a well that has not been used to initially search for transfected cells (in **step 3**) to minimize premature light exposure for the cell of interest.
6. Zoom in (e.g., zoom factor 9 corresponding to 53.5 nm per pixel) and adjust the excitation intensities, so that no pixels are saturated (*see* **Note 11**). Use the localizer channel (i.e., scan #1, iRFP713) to make sure that the array is in focus.
7. Record prerecruitment images (*see* **Note 12**). High quality: 512×512 px, 400 Hz line scan frequency, 4× line average (Fig. 2c). Low quality (optional): 256×256 px, 1400 Hz line scan frequency, 1× line average.
8. Exclude scan #1 and switch on the 488 nm laser (in addition to the 561 nm laser) in scan #2. Record a recruitment series with the same low-quality parameters as in **step** 7 (*see* **Note 13**, Fig. 2c). The frame time should be minimized (204 ms for the abovementioned imaging parameters).
9. Switch off the 488 nm laser in scan #2 and include scan #1 (iRFP713).
10. Optional: Record a high-quality image (parameters as listed in **step** 7) immediately after the recruitment.
11. Reduce the line average to 1× to limit photobleaching and record a dissociation series of both channels (90 frames, 15 s frame time, 512×512 px, 400 Hz line scan frequency, *see* **Note 14**, Fig. 2c).
12. Optional: Record a high-quality image with the parameters listed in **step** 7 (end-point).
13. Move on to the next cell and repeat **steps 5** through **12**.

3.3 Recruitment Analysis

The analyses of the recruitment time series were carried out essentially as described previously [10] (*see* Table 1). Image analysis was done in ImageJ and additional analyses were done using R.

1. Load the recruitment stack into ImageJ.
2. Make a maximum intensity projection of the time series stack (*Image* → *Stacks* → *Z Project...*). This projection serves to determine the area that is occupied by the array over the entire time course.

Table 1
BLInCR association and dissociation kinetics determined with different fluorophores or imaging settings

Experiment	Effector	Localizer[a]	*z*-axis image[b]	Data fit approach[c]	Characteristic time[d]
Association	PHR-mCherry-VP16	CIBN-TetR-iRFP713	Single	Individual	8.2 ± 6.1 s
				Global	8.0 ± 5.3 s
	PHR-YFP-VP16 [e]	CIBN-TetR-tagRFP-T	Single	Individual	11.9 ± 5.6 s
				Global	10.9 ± 3.5 s
	PHR-mCherry-NLS	CIBN-TetR-iRFP713	Single	Individual	8.6 ± 4.0 s
				Global	9.2 ± 3.9 s
	PHR-YFP-NLS [e]	CIBN-TetR-tagRFP-T	Single	Individual	25.9 ± 12.3 s
				Global	19.4 ± 3.2 s
Dissociation	PHR-mCherry-VP16	CIBN-TetR-iRFP713	Single	Individual	4.4 ± 0.8 min
	PHR-mCherry-VP16[f]	CIBN-TetR	Stack	Individual	4.9 ± 0.8 min
	PHR-mCherry-NLS	CIBN-TetR-iRFP713	Single	Individual	5.1 ± 0.5 min
	PHR-mCherry-NLS [f]	CIBN-TetR	Stack	Individual	4.8 ± 0.6 min

[a]To reliably localize the reporter array either a fluorescently labeled CIBN-LacIR or CIBN-LacI construct can be used
[b]Kinetics were recorded either by imaging one optical section ("single") or by recording image stacks (typically 4-6 optical sections, spaced 0.4–0.5 μm apart)
[c]For the individual fit procedure, all parameters in the fit equation given in **step 10** were determined for each single cell. For the global fit procedure, the rates k_1 and k_2 were fitted globally considering all cells from all association experiments listed here ($n = 61$). This assumes that the underlying rate processes, which recapitulate PHR photoswitching as well as PHR-CIBN and PHR-PHR association, are identical for the different constructs. The plateau value a as well as the contributions of the different rate processes to the overall binding kinetics b and c were fitted individually for each cell also in the global fit procedure
[d]The characteristic time refers to the time at which 50% of the effector has associated to ($\tau_{1/2}$) or dissociated from ($t_{1/2}$) the localizer protein bound to the array. The mean and standard deviation ($n > 10$ for all cases) is displayed
[e]Data from ref. [10] were analyzed to retrieve averaged results from individual fits. The global fit results differ slightly from the previously published data since they were analyzed with the new data from recruitment to CIBN-TetR-iRFP713 described here
[f]Values reported previously in ref. [10]

3. Select a circular region around the array and add the selection to the ROI manager (*Analyze → Tools → ROI Manager...*).
4. Optional: create a mask from the selection and save it for documentation purposes (*Edit → Selection → Create Mask*).
5. Repeat **steps 3** and **4** for a nuclear reference (*see* **Note 15**) and background region (Fig. 3a).
6. Select the recruitment series, then select the three ROIs in the ROI manager and measure their mean fluorescence intensity (*ROI Manager → More >> → Multi Measure*, Fig. 3a).
7. Save the data and load it into R.
8. Subtract the mean background intensity $I_{back}(t)$ from the mean intensity of the nuclear reference region $I_{nuc}(t)$ and fit the resulting (decaying) background-corrected reference intensity

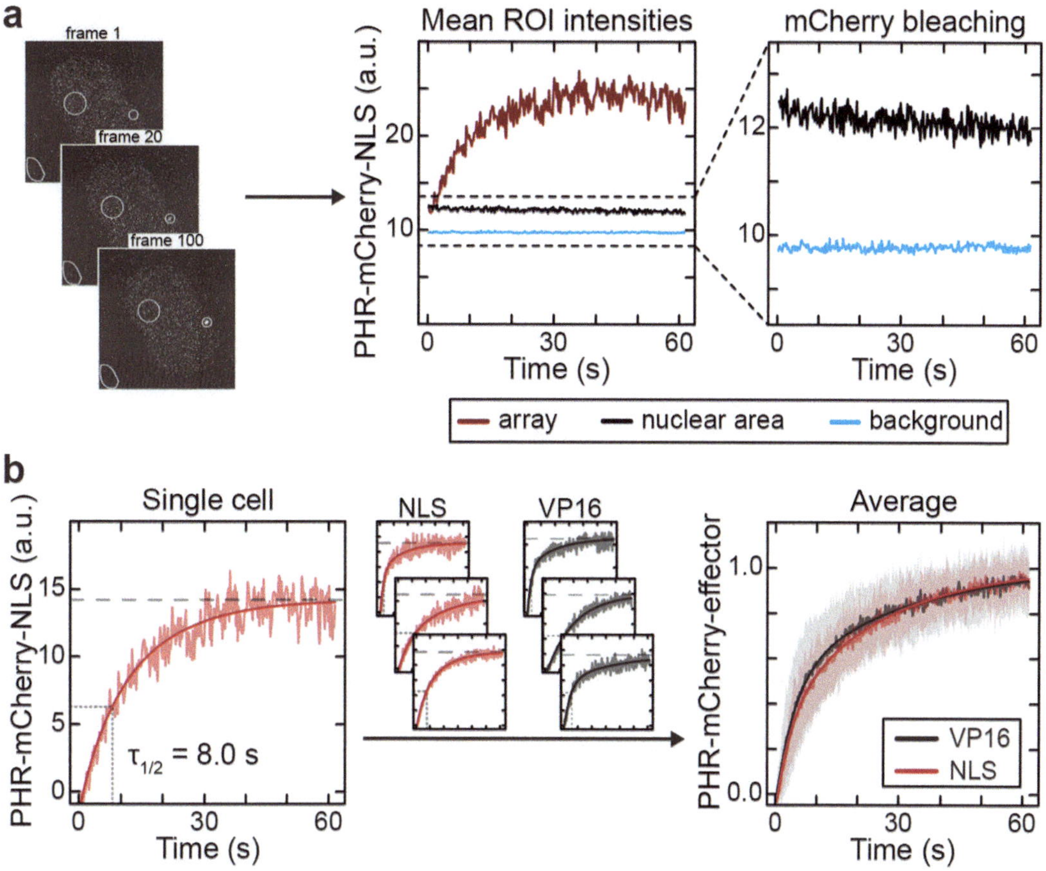

Fig. 3 Measuring BLInCR association kinetics. (**a**) Quantification of images. Regions of interest (ROIs) for array, nuclear reference region and background were selected manually on a maximum intensity projection of the 300 recruitment frames. Mean intensities in the three ROIs are measured for each frame (left), yielding the mean intensity traces (center). Photobleaching of mCherry was estimated from the nuclear area and the background mean intensity traces (right). (**b**) Data analysis. Single bleach- and nuclear background-corrected array intensities (left) are fitted to a model assuming two parallel first-order reactions (*see* **Note 17**). Measurements of several cells are normalized to account for different transfection efficiencies and averaged (right) to be able to compare the recruitment kinetics of different effectors. The average and standard deviations of VP16 ($n = 12$) and the mock effector NLS ($n = 12$) as well as the respective fits of the average curves are displayed. The times to reach half-maximal intensities are listed in Table 1

$I_{\mathrm{ref}}(t) = I_{\mathrm{nuc}}(t) - I_{\mathrm{back}}(t)$ with a single exponential to estimate mCherry bleaching (*see* **Note 16**, Fig. 3a): $I_{\mathrm{ref}}(t) = a_{\mathrm{ref}} \cdot \exp(-k_{\mathrm{bleach}} \cdot t)$, for example, by using the *nls* function from the *nlme* package in R.

9. Subtract the mean intensity of the reference region $I_{\mathrm{nuc}}(t)$ from the mean intensity of the array $I_{\mathrm{array}}(t)$ and correct the resulting net array signal $I(t) = I_{\mathrm{array}}(t) - I_{\mathrm{nuc}}(t)$ for bleaching using the estimated bleach parameter from **step 8**: $E(t) = I(t)/\exp(-k_{\mathrm{bleach}} \cdot t)$.

10. Fit the bleach-corrected effector enrichment at the array $E(t)$ (Fig. 3b, left) with a double exponential $E(t) = a - b \cdot \exp(-k_1 \cdot t) - c \cdot \exp(-k_2 \cdot t)$ (*see* **Note 17**), for example, using the *nls* function from the *nlme* package in R.
11. Retrieve the characteristic time to reach half-maximal levels $\tau_{1/2}$ from the model in **step 10** and the half-maximal level $E(\tau_{1/2})$ (*see* **Note 18**) by solving $0 = b\,(1 - 2 \cdot \exp(-k_1 \cdot \tau_{1/2})) + c\,(1 - 2 \cdot \exp(-k_2 \cdot \tau_{1/2}))$, for example, using the *uniroot* function in R.
12. Optional: Normalize the enrichment $E(t)$ to the plateau value a to account for different transfection efficiencies: $E_{\text{norm}}(t) = E(t)/a$. The traces that are normalized in this way can be used to calculate an average enrichment curve as shown in Fig. 3b (right).
13. Repeat **steps 1** through **12** for all cells.
14. Remove cells that moved during image acquisition or had a very low signal-to-noise ratio.

3.4 Reversibility Analysis

While the light-induced association reaches a plateau within a minute (Fig. 3), the analysis of the dissociation kinetics requires image acquisition on a time scale that is at least an order of magnitude larger (Table 1). During this longer acquisition time period, cells frequently undergo considerable movements and deformation that need to be accounted for in the analysis. Hence, nuclei and array regions were selected semiautomatically in each time frame based on intensity thresholds. The R package *EBImage* was used for this purpose. Alternatively, this task can also be accomplished with ImageJ (*Image → Adjust → Threshold...*).

1. Concatenate a (high-quality) prerecruitment image with the reversibility series (e.g., using ImageJ: *Image → Stacks → Tools → Concatenate...*).
2. Load the image series into R using the *EBImage* package. Alternatively, load the image series into ImageJ.
3. Use the PHR-mCherry-effector channel to segment the nuclei in each frame of the time series (*see* **Note 19**, Fig. 4a).
4. Use the CIBN-iRFP713-localizer channel to segment the array in each frame of the time series (*see* **Note 20**, Fig. 4a).
5. Create a mask of the nucleus without the array by subtracting the array mask from the whole nucleus mask. This selection will be referred to as "nucleus" subsequently.
6. Measure the mean fluorescence intensity in the respective selections for each time point (Fig. 4a).
7. Optional: Load the intensity data into R if ImageJ was used for segmentation.

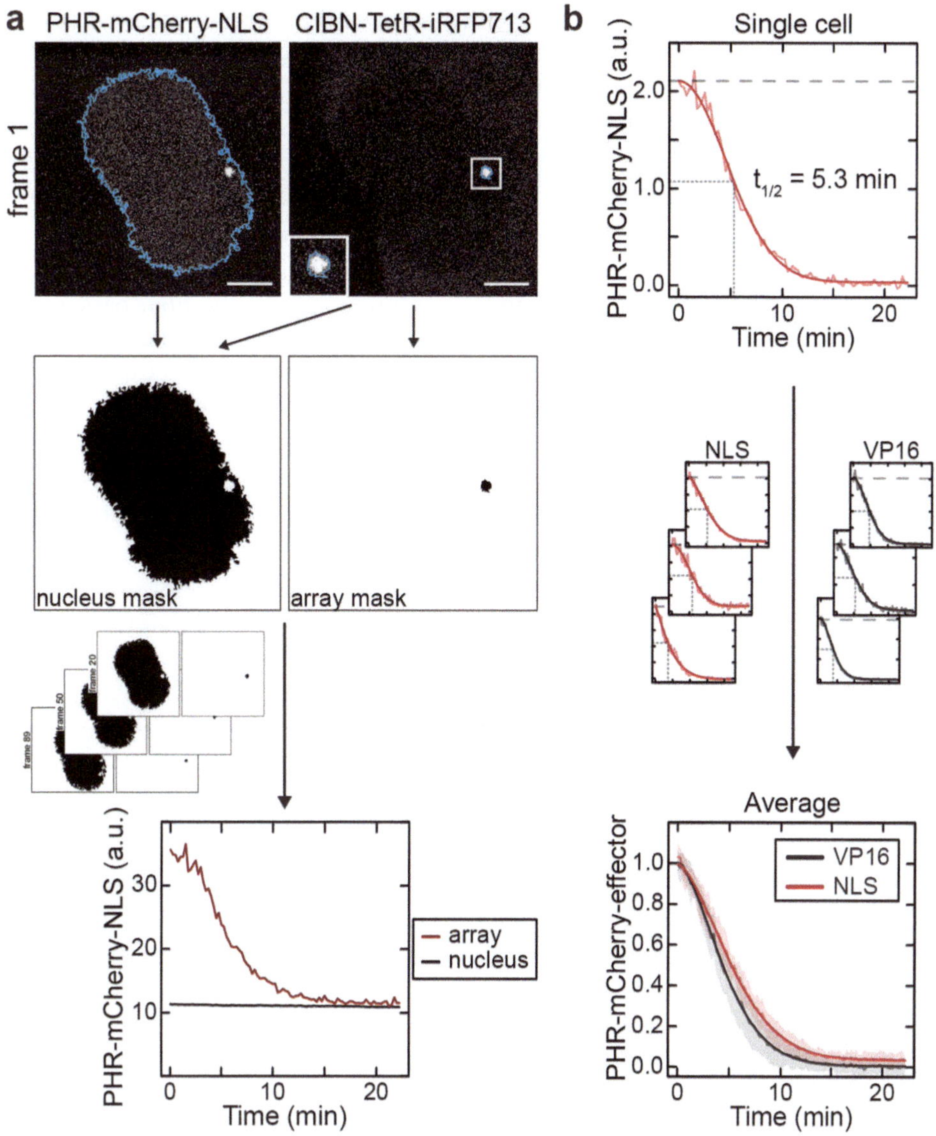

Fig. 4 Measuring BLInCR dissociation kinetics. (**a**) Quantification of images. The nucleus and array ROI were segmented based on intensity thresholds in the PHR-mCherry-NLS and the CIBN-TetR-iRFP713 image, respectively (top). The array mask was subtracted from the nucleus mask to yield the nucleus without array (middle). The masks were created for each frame and their mean fluorescence intensity was measured (bottom). Scale bars, 5 μm. (**b**) Data analysis. Bleach-, and nuclear background-corrected array intensities are corrected for the initial prerecruitment value (*see* **Note 22**) for individual cells and fitted to an exponential decay with a time-dependent (i.e., concentration-dependent) rate similar to the model described in ref. [17] (top). Multiple measurements are normalized to account for different effector expression and averaged to compare the recruitment kinetics of different effectors (bottom). The average and standard deviations of VP16 ($n = 11$) and the mock effector NLS ($n = 12$) are displayed. The resulting half-life times $t_{1/2}$ are listed in Table 1

8. Calculate the enrichment of fluorescence intensity at the array E_{array} from the array intensity $I_{array}(t)$ and the nucleus intensity (array excluded) $I_{nucleus}(t)$ and correct it for bleaching (*see* **Note 21**): $E_{array}(t) = (I_{array}(t) - I_{nucleus}(t))/I_{nucleus}(t)$.
9. Subtract the array enrichment of the "pre" image recorded before the initial recruitment from the array enrichment (*see* **Note 22**): $E(t) = E_{array}(t) - E_{array}(\text{"pre"})$.
10. Fit the corrected array enrichment $E(t)$ (Fig. 4b, top), for example using the *nls* function from the *nlme* package in R. A single exponential with a time-dependent (i.e., concentration-dependent) rate can be used $E(t) = a \cdot \exp(-k \cdot t^m) + c$ [17]. For alternative models *see* **Note 23**.
11. Determine the characteristic half-life $t_{1/2}$ from the model in **step 10** and the half-maximal level $E(t_{1/2})$ (*see* **Note 18**) by solving $0 = 2 \cdot \exp(-k \cdot t_{1/2}{}^m) - 1$, for example, using the *uniroot* function in R.
12. Optional: Normalize the cells to their respective initial levels ($E_{norm}(t) = E(t)/(a + c)$) to calculate an average curve as in Fig. 4b.

4 Notes

1. The main limitation of using iRFP713 for BLInCR applications is its relatively low brightness and high susceptibility to photobleaching. This is less of an issue for tracing the reporter array used here, which has a high number of fluorophore binding sites. However, for other applications where detection sensitivity is crucial, for example, for tracing RNA production, iRFP713 is not well-suited. Since the development and improvement of autofluorescent proteins with near-infrared emission is currently an active research area [8], we anticipate that brighter and more photostable constructs will become available. Recently, iRFP670 ($\lambda_{ex} = 643$ nm, $\lambda_{em} = 670$ nm) [18] and miRFP670nano ($\lambda_{ex} = 645$ nm, $\lambda_{em} = 670$ nm) [19] have been introduced but we have not tested them yet. We expect them to be applicable for BLInCR reversibility experiments with the protocol used here when adjusting for their somewhat different spectral properties.
2. In principle, any cell type that expresses the tagged protein constructs needed is suitable for BLInCR. We have chosen U2OS 2-6-3 for its transcription reporter, its easily detectable single array and its high transfection efficiency.

3. Alternatively, 10-well CELLview slides (Greiner Bio-One GmbH, Germany) can be used, which have black walls between the wells to limit stray light from imaging in a neighboring well. However, we did not observe a difference as compared to the transparent walls of the Lab-Tek slides, in particular when using laser excitation for imaging. In contrast, observing cells through the ocular with a mercury vapor lamp causes more stray light, which can cause premature photoswitching (depending on the filter sets and light intensities used).
4. To limit premature light exposure, it is advantageous to use a transfection reagent that does not need to be removed from the medium before imaging. We have also used lipofectamine 3000 for immortalized mouse embryonic fibroblasts (iMEFs) without changing the medium after transfection.
5. The localizer construct used here binds to *tet*O in the absence of doxycycline. We used doxycycline-free medium, resulting in high-affinity binding of CIBN-TetR-iRFP713 as soon as the construct was expressed. This caused no problems for transient transfections (24–48 h), but cells should be grown in doxycycline-containing medium for long-term experiments or when generating stable cell lines. Using reverse TetR [20], which binds *tetO* in the presence of doxycycline, is also a recommended option for the latter types of experiments (also *see* ref. [21] for a review on TetR properties and variants).
6. Other suitable constructs are also available: CIBN-TetR-tagRFP-T (localizer, Addgene #103809), PHR-iRFP713-VP16 (effector, Addgene #103823), PHR-iRFP713 (mock effector, Addgene #103818). For other effectors and localizers see the Addgene entries associated with [10].
7. To avoid premature light-induced binding of effectors, transfected cells should be kept in the dark at all times. White Styrofoam boxes are not suited for this purpose since they do transmit some light. If cells need to be handled after transfection, switch off all white light sources and use a red safelight (e.g., a removable bike tail light).
8. The Leica TX2 filter cube can be helpful to initially localize cells. It limits the excitation light using a bandpass filter (BP 560/40). Since it can also cause effector recruitment at high light intensities, it is recommended to only use the ocular to adjust the focal plane in the very beginning of the experiment. This can also be done in a well that is not used for the actual measurements later. Other filter cubes such as the Leica Y5 should work well but we have not tested them.
9. We have adapted the Xtreme Gene 9 (Roche, Germany) protocol for use in 8-well Lab-Tek slides as follows (amounts for transfecting one well corresponding to 0.8 cm^2): add 0.8 μl Xtreme Gene 9 reagent to 20 μl Opti-MEM in a suitable

reaction tube (e.g., 1.5 ml Eppendorf), add 300–400 ng plasmid DNA at a 1:1 ratio of localizer and effector constructs, flick the tube for mixing and incubate the mixture for 15 min at room temperature before adding the mix to the medium.

10. It is recommended to use cells that express the respective constructs at low levels. In our experience, high expression levels of PHR constructs lead to significant prerecruitment in the absence of light.

11. Keep in mind that the effector intensity at the array will be high upon recruitment. The intensity in the image prior to recruitment should thus be rather low to avoid oversaturated pixels at the array after recruitment. In addition, do not change the excitation during imaging to obtain comparable results.

12. The high-quality image is well-suited to ensure that there is no prerecruitment of PHR-mCherry-effector. It is also used in the analysis of the reversibility series (*see* Subheading 3.4). The low-quality image is recorded with the same imaging parameters as the recruitment series and can thus be used to assess the laser intensities. However, it cannot be included in the quantitative analyses of the recruitment series because the 488 nm laser is used for the recruitment but switched off during imaging prior to recruitment. Since mCherry has some absorbance at 488 nm (Fig. 1a), the absolute intensities of prerecruitment and recruitment images are not comparable.

13. For recruitment, the localizer was not imaged in order to increase the temporal resolution. We have recorded 300 frames corresponding to ~1 min total imaging time, during which the cells generally did not move or drift excessively. This time was generally sufficient to reach a plateau of PHR-mCherry-effector intensity at the array. The PHR switching and thus the recruitment kinetics also depend on the intensity of the blue light used. It should therefore be kept constant for all experiments to ensure comparability.

14. For the reversibility series, both localizer and effector images were recorded because the cells move, deform, and/or drift significantly over the recording time of 22.5 min. Hence, it is advantageous to record the localizer image to ensure that the array is still in focus. This is different from the approach used in our previous work [10] where image stacks were recorded to ensure that all signal from the array was detected without visualizing the array over time.

15. Here, we have used a nuclear reference region as opposed to the entire nucleus, because the fluorescently tagged (mock) effectors have a preference to be enriched or depleted from nucleoli. The mCherry-tagged effectors used here tend to be depleted from nucleoli. Hence, the mean intensity of the

nucleus will always be smaller than that of the array even if no effector is recruited. The nuclear reference region was placed in a region that had homogeneously distributed mCherry signal similar to that of the array; hence, $E(0) \approx 0$ if no construct is recruited without further normalization. Using a ring-shaped reference region around the array selection would also be feasible but is impractical since the array is often located close to the nuclear lamina.

16. We have fitted the data with a single exponential for practical reasons without any assumptions about the bleach process. For different time scales, a different model might be more appropriate. The background-corrected reference intensity I_{ref} can be fitted with any function that describes the data well and yields a constant value for the bleach-corrected reference intensity, which was in our case $I_{ref,cor} = (I_{nuc} - I_{back})/\exp(-k_{bleach} \cdot t) = a_{ref}$. Alternatively, the net array intensity $I(t) = I_{array}(t) - I_{nuc}(t)$ can simply be divided by the intensity of the nuclear reference region I_{nuc} yielding $E(t) = I(t)/I_{nuc}(t)$. The latter approach was used for the longer dissociation time series (Subheading 3.4). Using a (smooth) fit function for bleach correction was advantageous for the rather noisy association curves.

17. The fit model considers two parallel first-order reactions for binding to two subpopulations of binding sites. The two resulting rates could correspond to PHR photoswitching followed by (1) PHR-CIBN heterodimerization and by (2) PHR-PHR oligomerization and optodroplet formation [22, 23]. Note that the resulting times to reach half-maximal levels are shorter for the PHR-mCherry-effector constructs used here compared to the PHR-YFP-effector constructs used previously [10] (Table 1). This could be due to different propensities of the mCherry and YFP constructs to allow PHR-PHR oligomerization and optodroplet formation. A certain heterogeneity of binding kinetics between cells transfected with the same constructs is expected because the binding reaction rate depends on the (variable) expression level of the PHR ligand.

18. The half-maximal levels $E(\tau_{1/2})$ for the association and $E(t_{1/2})$ for the dissociation were calculated from the respective initial enrichment $E(0)$ and the respective plateau value $E(\infty)$ as $E(\tau_{1/2}) = E(0) + (E(\infty) - E(0))/2$ and, analogously, $E(t_{1/2}) = E(0) + (E(\infty) - E(0))/2$.

19. We have used the Otsu method [24] to determine a threshold and multiplied it with a factor of typically 0.7 (0.6–0.9). This factor was determined separately for each cell and was necessary because of different cytoplasmic or background signals.

The otsu() function in R calculates the threshold for each image in an image series. However, this histogram-based algorithm occasionally recognizes the cell as background and large arrays (compared to the cell size) as foreground resulting in poor segmentation of some of the nuclei in a series. Hence, we have determined the threshold on the last image of the effector series (where the effector construct has dissociated from the array) and used this value for segmenting the entire stack.

20. The threshold selection for the array segmentation was done as for the nucleus segmentation (*see* **Note 19**), but the factor with which the Otsu threshold was multiplied was typically 3.0 (2.5 5.0).

21. For the bleach correction of the dissociation curves, we have used the mean nucleus fluorescence intensity assuming that it bleaches in the same manner as the array intensity.

22. This normalization step is necessary because $E_{array}(0) \neq 0$ even if no construct is recruited (for an explanation, *see* **Note 15**).

23. The model in **step 10** described our data well (Fig. 4b), but alternative models can be used to describe the dissociation kinetics and interpret the fitted parameters. One model is represented by $E(t) = a \cdot \exp(-k \cdot t)/(b + \exp(-k \cdot t)) + c$. Here, b is a parameter related to the dissociation constant and concentration of binding-competent PHR molecules at the start of the dissociation time course. The parameter c accounts for a residual basal intensity level after dissociation. This equation is derived from the analytical solution to a model, in which a ligand can switch between a binding-competent and a noncompetent conformation with rate k. In the dark, the binding competent state B decays according to $B(t) = B_0 \cdot \exp(-k \cdot t)$. The binding to the target site S is assumed to be significantly faster than the conformational switch. Thus, at any time the concentration of the chromatin bound complex $BS(\mathrm{t})$ is in equilibrium: $BS(t) = S_{total} \cdot B(t)/(B(t) + k_{off}/k_{on})$ leading to $b = K_d/B_0$ and $a = S_{total}$ where $K_d = k_{off}/k_{on}$ is the dissociation constant. With this approach the conformational reversion rate k is obtained while a and b cannot be interpreted without knowledge about the relation of fluorescence intensity and absolute concentrations. Another possibility to represent the dissociation process would be to use a model with two sequential reactions with rate k according to $B_0\ (1 + k \cdot t) \cdot \exp(-k \cdot t) + c$ (*see* ref. [10] for the use of sequential reaction schemes to fit the transcription activation process).

Acknowledgments

We thank the DKFZ light microscopy core facility for technical support for imaging. This work was supported by the Deutsche Forschungsgemeinschaft (DFG grant RI 1283/14-1 to K.R.).

References

1. Losi A, Gardner KH, Moglich A (2018) Blue-light receptors for optogenetics. Chem Rev 118(21):10659–10709. https://doi.org/10.1021/acs.chemrev.8b00163
2. Liu H, Yu X, Li K, Klejnot J, Yang H, Lisiero D, Lin C (2008) Photoexcited CRY2 interacts with CIB1 to regulate transcription and floral initiation in Arabidopsis. Science 322(5907):1535–1539. https://doi.org/10.1126/science.1163927
3. Kennedy MJ, Hughes RM, Peteya LA, Schwartz JW, Ehlers MD, Tucker CL (2010) Rapid blue-light-mediated induction of protein interactions in living cells. Nat Methods 7 (12):973–975. https://doi.org/10.1038/nmeth.1524
4. Salomon M, Christie JM, Knieb E, Lempert U, Briggs WR (2000) Photochemical and mutational analysis of the FMN-binding domains of the plant blue light receptor, phototropin. Biochemistry 39(31):9401–9410. https://doi.org/10.1021/bi000585+
5. Huala E, Oeller PW, Liscum E, Han IS, Larsen E, Briggs WR (1997) Arabidopsis NPH1: a protein kinase with a putative redox-sensing domain. Science 278 (5346):2120–2123. https://doi.org/10.1126/science.278.5346.2120
6. Niopek D, Benzinger D, Roensch J, Draebing T, Wehler P, Eils R, Di Ventura B (2014) Engineering light-inducible nuclear localization signals for precise spatiotemporal control of protein dynamics in living cells. Nat Commun 5:4404. https://doi.org/10.1038/ncomms5404
7. Shaner NC, Campbell RE, Steinbach PA, Giepmans BN, Palmer AE, Tsien RY (2004) Improved monomeric red, orange and yellow fluorescent proteins derived from Discosoma sp. red fluorescent protein. Nat Biotechnol 22 (12):1567–1572. https://doi.org/10.1038/nbt1037
8. Karasev MM, Stepanenko OV, Rumyantsev KA, Turoverov KK, Verkhusha VV (2019) Near-infrared fluorescent proteins and their applications. Biochemistry (Mosc) 84(Suppl 1):S32–S50. https://doi.org/10.1134/S0006297919140037
9. Filonov GS, Piatkevich KD, Ting LM, Zhang J, Kim K, Verkhusha VV (2011) Bright and stable near-infrared fluorescent protein for in vivo imaging. Nat Biotechnol 29(8):757–761. https://doi.org/10.1038/nbt.1918
10. Rademacher A, Erdel F, Trojanowski J, Schumacher S, Rippe K (2017) Real-time observation of light-controlled transcription in living cells. J Cell Sci 130(24):4213–4224. https://doi.org/10.1242/jcs.205534
11. Trojanowski J, Rademacher A, Erdel F, Rippe K (2019) Light-induced transcription activation for time-lapse microscopy experiments in living cells. In: Shav-Tal Y (ed) Imaging gene expression: methods and protocols, methods in molecular biology, vol 2028. Springer Nature, New York, pp 251–270. https://doi.org/10.1007/978-1-4939-9674-2_17
12. Janicki SM, Tsukamoto T, Salghetti SE, Tansey WP, Sachidanandam R, Prasanth KV, Ried T, Shav-Tal Y, Bertrand E, Singer RH, Spector DL (2004) From silencing to gene expression: real-time analysis in single cells. Cell 116 (5):683–698. https://doi.org/10.1016/s0092-8674(04)00171-0
13. Schindelin J, Arganda-Carreras I, Frise E, Kaynig V, Longair M, Pietzsch T, Preibisch S, Rueden C, Saalfeld S, Schmid B, Tinevez JY, White DJ, Hartenstein V, Eliceiri K, Tomancak P, Cardona A (2012) Fiji: an open-source platform for biological-image analysis. Nat Methods 9(7):676–682. https://doi.org/10.1038/nmeth.2019
14. Schneider CA, Rasband WS, Eliceiri KW (2012) NIH image to ImageJ: 25 years of image analysis. Nat Methods 9(7):671–675
15. Pau G, Fuchs F, Sklyar O, Boutros M, Huber W (2010) EBImage—an R package for image processing with applications to cellular phenotypes. Bioinformatics 26(7):979–981. https://doi.org/10.1093/bioinformatics/btq046
16. Pinheiro J, Bates D, DebRoy S, Sarkar D, R_Core_Team (2018) nlme: linear and nonlinear mixed effects models. R package version 3. 1-137.https://CRAN.R-project.org/package=nlme
17. Sing CE, Olvera de la Cruz M, Marko JF (2014) Multiple-binding-site mechanism

explains concentration-dependent unbinding rates of DNA-binding proteins. Nucleic Acids Res 42(6):3783–3791. https://doi.org/10.1093/nar/gkt1327
18. Shcherbakova DM, Verkhusha VV (2013) Near-infrared fluorescent proteins for multicolor in vivo imaging. Nat Methods 10 (8):751–754. https://doi.org/10.1038/nmeth.2521
19. Oliinyk OS, Shemetov AA, Pletnev S, Shcherbakova DM, Verkhusha VV (2019) Smallest near-infrared fluorescent protein evolved from cyanobacteriochrome as versatile tag for spectral multiplexing. Nat Commun 10(1):279. https://doi.org/10.1038/s41467-018-08050-8
20. Gossen M, Freundlieb S, Bender G, Muller G, Hillen W, Bujard H (1995) Transcriptional activation by tetracyclines in mammalian cells. Science 268(5218):1766–1769. https://doi.org/10.1126/science.7792603
21. Berens C, Hillen W (2003) Gene regulation by tetracyclines. Constraints of resistance regulation in bacteria shape TetR for application in eukaryotes. Eur J Biochem 270 (15):3109–3121. https://doi.org/10.1046/j.1432-1033.2003.03694.x
22. Bugaj LJ, Choksi AT, Mesuda CK, Kane RS, Schaffer DV (2013) Optogenetic protein clustering and signaling activation in mammalian cells. Nat Methods 10(3):249–252. https://doi.org/10.1038/nmeth.2360
23. Shin Y, Berry J, Pannucci N, Haataja MP, Toettcher JE, Brangwynne CP (2017) Spatiotemporal control of intracellular phase transitions using light-activated optoDroplets. Cell 168(1-2):159–171.e114. https://doi.org/10.1016/j.cell.2016.11.054
24. Otsu N (1979) A threshold selection method from gray-level histograms. IEEE Trans Sys Man Cyber 9(1):62–66. https://doi.org/10.1109/tsmc.1979.4310076

Chapter 13

Construction of a Multiwell Light-Induction Platform for Traceless Control of Gene Expression in Mammalian Cells

Maysam Mansouri, Samson Lichtenstein, Tobias Strittmatter, Peter Buchmann, and Martin Fussenegger

Abstract

Mammalian cells can be engineered to incorporate light-responsive elements that reliably sense stimulation by light and activate endogenous pathways, such as the cAMP or Ca^{2+} pathway, to control gene expression. Light-inducible gene expression systems offer high spatiotemporal resolution, and are also traceless, reversible, tunable, and inexpensive. Melanopsin, a well-known representative of the animal opsins, is a G-protein-coupled receptor that triggers a Gαq-dependent signaling cascade upon activation with blue light (≈470 nm). Here, we describe how to rewire melanopsin activation by blue light to transgene expression in mammalian cells, with detailed instructions for constructing a 96-LED array platform with multiple tunable parameters for illumination of the engineered cells in multiwell plates.

Key words Optogenetics, Inducible gene expression, Cell engineering, Mammalian cells, Synthetic biology

1 Introduction

The ability to remotely control the expression of genes of interest in mammalian cells is a central goal in both basic and clinical research. Cell-based treatment strategies in translational medicine usually employ engineered cells that sense input signals and respond appropriately [1]. Available input signals include a variety of endogenous or exogenous stimuli, such as biomolecules [2], chemicals [3], temperature [4], pH [5], light [6], gas [7], and many others. Among them, light offers a promising opportunity for traceless control of cellular behavior with rapid response, good reversibility, and high spatiotemporal resolution [8]. The term optogenetics was coined to describe the use of light to achieve precise control of gene expression in engineered living cells [9, 10]. Originally designed and most widely employed to manipulate neural activities, optogenetic tools are being increasingly used in biomedical research in fields outside neuroscience [11–15].

Dominik Niopek (ed.), *Photoswitching Proteins: Methods and Protocols*, Methods in Molecular Biology, vol. 2173, https://doi.org/10.1007/978-1-0716-0755-8_13,

Photoactivatable proteins are at the core of the optogenetic tool kit. These proteins can be divided into two major groups, that is, opsins and nonopsins (light switches). Nonopsin photoactivatable proteins are widely used in nonneural optogenetics (i.e., non-ion-flux-based optogenetics) [8, 12]. Opsins are light-sensitive transmembrane proteins found in a variety of organisms ranging from microbes (Type I opsins) to primates (Type II opsins). Type II opsins are G-protein-coupled receptors (GPCR) that trigger a signaling cascade upon activation by light [16]. Melanopsin, a representative member of the animal opsins, is endogenously expressed in intrinsically photosensitive retinal ganglion cells (ipRGCs) of the inner retina [17]. It plays a crucial role in the adaptation of mammals to different light intensities, as well as in the circadian timing system [18]. Exposure to light induces isomerization of the retinal chromophore (11-cis-retinal to all-trans-retinal) of melanopsin, and this changes the conformation of the receptor, resulting in activation of the heterotrimeric $G\alpha\beta\gamma$ protein and inducing dissociation of the $G\alpha q$ subunit from the dimeric $G\beta\gamma$ complex. The $G\alpha q$ subunit then activates phospholipase C, which initiates calcium-dependent signaling within mammalian cells. Ultimately, this leads to activation of the phosphatase calcineurin, which dephosphorylates the nuclear factor of activated T-cells (NFAT). Dephosphorylated NFAT translocates to the nucleus and triggers expression of target genes [19] (Fig. 1).

Expression of melanopsin in mammalian cells provides functionality for blue-light dependent calcium influx into the cells. This feature allows us to rewire the signal transduction of light-activated melanopsin to a synthetic NFAT-responsive promoter, enabling the expression of gene(s) of interest to be triggered by blue light [20]. It is also possible to reprogram mammalian cells, including induced pluripotent stem cells (iPSCs) [21], for specific purposes by using blue light to alter endogenous pathways.

The following protocol describes in detail how to create a blue-light-inducible gene expression system for mammalian cells, including instructions for constructing a 96-LED array platform with multiple tunable parameters for stimulation of engineered cells in multiwell plates. Owing to the modularity of the LED array, the platform can easily be modified for use with other optogenetic receptors and light switches that operate at different wavelengths, simply by replacing the LEDs.

2 Materials

2.1 Construction of a Programmable LED Array

1. 60 Blue LEDs (475 nm, 20–30 cd, 50 mA, 5 mm, 15 o; Roithner Lasertechnik GmbH).
2. Electrical wire and tools (soldering iron, wire stripper, etc.).

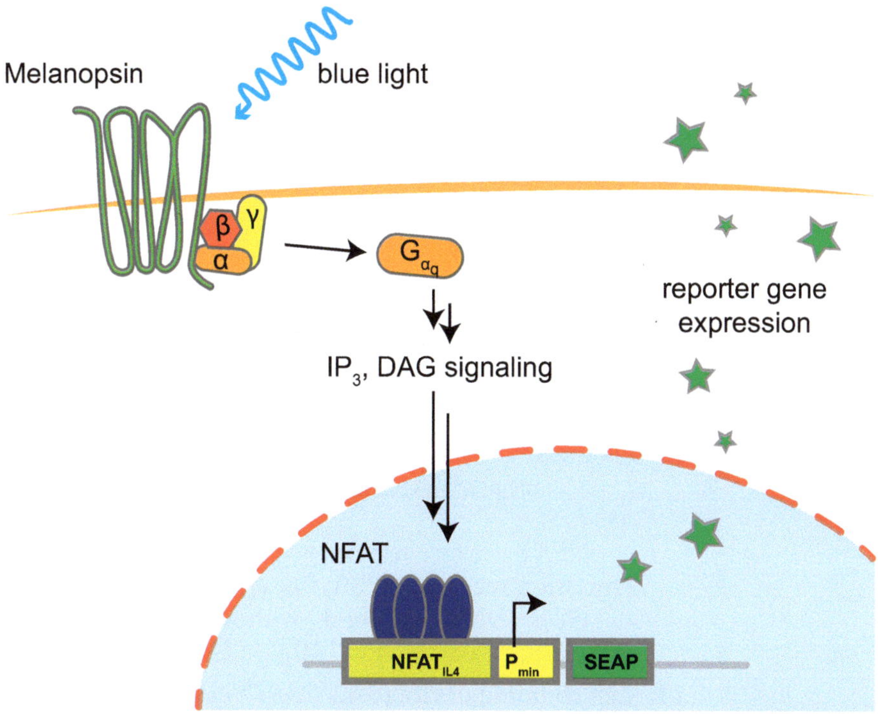

Fig. 1 Schematic of blue light-induced activation of endogenous signaling for gene expression in mammalian cells. Melanopsin signaling through the Gaq pathway leads to activation of the transcription factor NFAT (nuclear factor and activator of transcription). NFAT induces expression of the reporter gene, secreted placental alkaline phosphatase (SEAP), from a synthetic promoter equipped with binding sites derived from the interleukin 4 promoter

3. Copper-terminal printed circuit board.
4. USB cable.
5. Photometer (Ophir, NOVA, P/N7Z01500).
6. Arduino™ microcontroller and software.
7. Arduino drivers.
8. Step-up converter (SG3524) to achieve LED driving voltage.
9. Current mode switching circuit MLX10803 for LED current regulation.

2.2 Cell Culture

1. Dulbecco's modified Eagle's medium (DMEM) supplemented with 10% fetal bovine serum (FBS) and 1% penicillin/streptomycin.
2. Colorless DMEM without phenol red, supplemented with 10% fetal bovine serum (FBS) and 1% penicillin/streptomycin.
3. DMEM medium without FBS and penicillin/streptomycin antibiotics.
4. Trypsin–EDTA solution (0.05% (v/v)).

5. Cell line (HEK293T).
6. Black 96-well plates with transparent bottom (e.g., Greiner Bio-One).
7. Transparent 96-well plates.
8. 96 deep-well plate.
9. CO_2 tissue-culture incubator.
10. Sellotape.
11. Polyethyleneimine (PEI) HCl MAX, Linear, Mw 40,000 solution (1 mg/mL, Polysciences Inc.).
12. All-trans retinal (ATR) (Sigma-Aldrich, 1 mM stock in ethanol).
13. Plasmids:

    Melanopsin expression vector (pHY42; P_{CMV}-OPN4-pA) [19].

    NFAT-driven secretory alkaline phosphatase (SEAP) expression vector (pHY30; P_{NFATx3}-SEAP-pA) [19].

    Filler DNA (plasmid without mammalian promoter (pDF145; P_{T7}-SpA_H-Env140$_{ac}$)) [22].
14. Cell counting device (e.g., CASY® cell counter).
15. Multichannel pipette and tray (reservoir).
16. Cell culture plasticware (100 mm petri dishes).
17. Centrifuge.

2.3 SEAP Measurement

1. SEAP buffer (2×): 20 mM homoarginine (Acros Organics), 1 mM $MgCl_2$, 21% (v/v) diethanolamine, pH 9.8 (Acros Organics).
2. Substrate for SEAP: 120 mM para-nitrophenyl phosphate (pNPP) (Acros Organics) in 2× SEAP buffer.
3. 96-well plates.
4. Plate reader (e.g., TECAN AG, Maennedorf, Switzerland.

3 Methods

3.1 Construction of a Programmable LED Array

The 96-LED array is designed to allow for programming of different light intensities as well as illumination patterns (Fig. 2a–d). For this, the following steps are needed:

1. Design and print a circuit board consisting of a 10 × 6 LED holder. Space the LEDs to the dimensions of a 96-well plate so that one LED is located at the center of each well.
2. The LEDs are organized in two rows of five LEDs in series for each channel.

3. Based on the forward voltage of 4 V for the blue LEDs, the output voltage of the Boostcircuit has to be set to 24 V to achieve accurate current regulation.
4. Each channel contains one switch-mode LED current-regulating circuit MLX10803. The MLX10803 offers the opportunity to add a PWM Signal (from Arduino) to achieve dimming of the LED in 0–255 steps.
5. To program the LED array, download the Arduino™ software from www.arduino.cc. Write software to set the desired illumination pattern (constant or pulsating), illumination time and light intensity. Upload the program to the Arduino™ Microcontroller by connecting the Arduino™ board to the computer via a USB cable and initiate the transfer from the Arduino programming GUI (Fig. 2a–d).
6. To correlate Arduino™ values to the light intensity, program Arduino™ with different values (0 to 255) and measure the emitted light from LEDs with a photometer following the manufacturer's instructions.

3.2 Cell Culture

The following protocols for cell culture experiments have been optimized for HEK293T cells and would need to be modified for other cell lines.

1. Cultivate HEK293T cells in 12 mL of complete DMEM (supplemented with 10% FBS and 1% penicillin/streptomycin) at 37 °C in an atmosphere of 5% CO_2.
2. After the cells reach 70–80% confluence, they can be used for light-induction experiments. Aspirate the medium and expose the cells to 1 mL of 0.05% (v/v) trypsin–EDTA solution for 5 min at 37 °C in the incubator.
3. Remove the trypsinized cells from the incubator, and add 9 mL of complete DMEM to the cells. Detach the cells from the plate by gentle tapping.
4. Collect the cells from the plate, transfer them into a 15 mL conical tube, and centrifuge for 3 min at 450 × *g*.
5. Discard the supernatant, and resuspend the cells in 10 mL of complete DMEM.
6. Count the cells using a cell counting device.
7. Prepare 12 mL of seeding cell suspension in a 50-mL tube at a concentration of 1.25×10^5 cells/mL.
8. Close the 50 mL tube, invert several times to ensure proper mixing of the suspension, and pipette the cells into the multichannel tray reservoir.

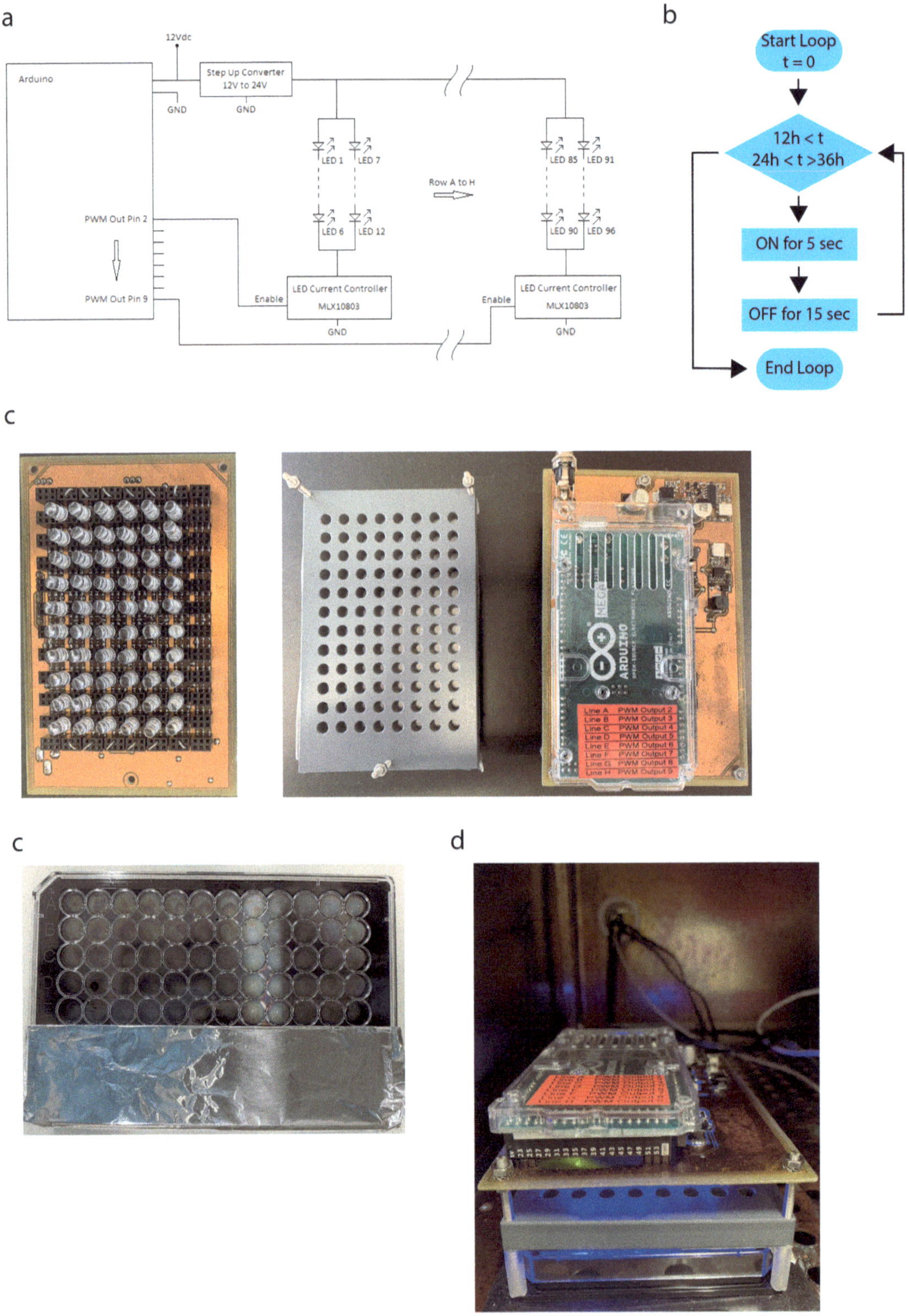

Fig. 2 Experimental setup. (**a**) Circuit diagram of the architecture of the LED array. (**b**) Flowchart of the program running on the Arduino™. Pulses of 5 s ON and 15 s OFF were applied twice for 12 h with a pause of 12 h in

9. Mix the cell suspension by pipetting up and down. Use a multichannel pipette to transfer 125 μL of the cell suspension into each well of a 96-well plate (*see* **Note 1**).
10. Incubate the plate with the seeded HEK293T cells at 37 °C under 5% CO_2 for 12–24 h prior to transfection.

3.3 Transfection

We typically co-transfect HEK293T cells with 5 ng of pHY42 (P_{CMV}-OPN4-pA), 90 ng of pHY30 (P_{NFATx3}-SEAP-pA; *see* **Note 2**), and 30 ng of pDF145 (P_{T7}-SpA_H-Env140$_{ac}$), making a total amount of 125 ng DNA per well of a 96-well plate (*see* **Notes 3** and **4**).

1. Prepare a master mix for 60 wells of a 96-well plate with pHY42, pHY30 and pDF145. Per well, add 50 μL of DMEM (without FBS and antibiotics) and 0.625 μL of PEI solution (1 mg/mL; *see* **Note 5**). This amounts to 60 × 50 μL = 3 mL of DMEM and 60 × 0.625 μL = 37.5 μL of PEI solution for the master mix. Vortex the PEI/DNA/DMEM mixture thoroughly for 20 s.
2. Incubate the PEI/DNA/DMEM mixture for 20 min at RT.
3. Transfer 300 μL of PEI/DNA/DMEM mixture to wells A2 to A11 of a deep well plate. Transfer 50 μL of the transfection cocktail with the multichannel pipette to each of the inner 60 wells of the 96-well plate (i.e., wells B2 to B11, C2 to C11, D2 to D11, E2 to E11, F2 to F11, and G2 to G11).
4. Row G will not be light-stimulated (dark) and serves as a control. It is protected from light by covering it with aluminum foil. Cut out two pieces of aluminum foil and glue them over rows F (which serves as a spacer between light-exposed and dark rows), G and H on the lid and the bottom of the 96-well plate.
5. Incubate transfected cells for 6–12 h at 37 °C under 5% CO_2. Then, pour the media out of the plate and gently tap the plate on a stack of paper towels to remove most of the remaining media. Replace with fresh complete DMEM and continue to incubate the cells for 24 h at 37 °C under 5% CO_2 (*see* **Note 6**).

3.4 Light Induction in Transfected Mammalian Cells

Illumination should be started 24 h after transfection to ensure good integration of cell membrane proteins. Program the LED array with Arduino software as follows:

Fig. 2 (continued) between. (**c**) View of disassembled parts. Left: bottom view showing LED holders; middle: fitting spacer to ensure correct orientation of LEDs; right: top view of the Arduino controller. (**c**) 96-well plates with the bottom rows covered with aluminum foil to protect control cells from light (control). (**d**) Operating the LED array on a 96-well plate inside the mammalian cell culture incubator

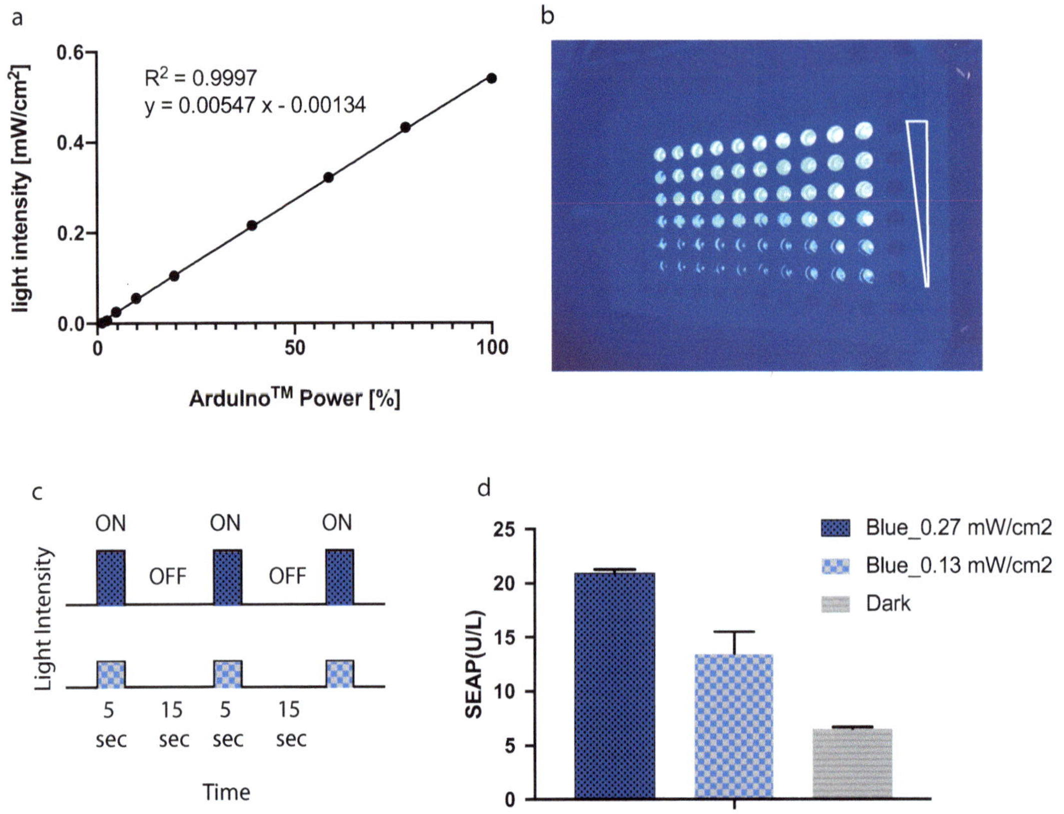

Fig. 3 Light intensity experiments. (**a**) Calibration curve showing light intensity as a function of Arduino units. The curve needs to be established for every construction, because many parameters, including LED type, current and filters, affect the light intensity. (**b**) Image of an operating LED array programmed with a light intensity gradient. (**c**) Light irradiation pattern with 5 s ON and 15 s OFF at different light intensities. (**d**) Light intensity-dependent SEAP expression levels

1. Select the desired illumination pattern (i.e., constant or pulsating) and illumination time, as well as the light intensity (Fig. 3). We employ a pulse pattern (5 s ON and 15 s OFF) for successive periods of 12 h on and 12 h off (total: 48 h) at 0.27 mW/cm^2 or 0.13 mW/cm^2.
2. Take 12 mL prewarmed colorless medium (+10% FBS +1% Pen/Strep), add 5 μM ATR, and transfer into a pipetting reservoir.
3. Gently remove the medium from the transfected cell plate by tapping it upside-down onto a stack of paper towels.
4. Gently add 120 μL of the medium supplemented with ATR to each well of the 96-well plate using the multichannel pipette.
5. Put the LED array on top of the plate (*see* **Note 7**) and connect it to the power supply. Illumination will start automatically as soon as the Arduino boots up.

3.5 Reporter Expression Analysis

We typically measure the reporter gene expression after 24 h and 48 h (Fig. 3; *see* **Note 8**), but SEAP analysis can be done at any desired time point (*see* **Note 9**). The fold induction is calculated as the ratio of SEAP level of stimulated cells to that of unstimulated cells. Measurement is done according to the following protocol.

1. Transfer 10 μL of the SEAP-containing cell supernatant to a fresh 96-well assay plate using a multichannel pipette.
2. Add 90 μL of ddH_2O to the SEAP plate using a multichannel pipette.
3. Seal the SEAP plate with Sellotape and incubate the plate for 30 min at 65 °C and 350 rpm to inactivate endogenous phosphatases present in the cells.
4. Centrifuge the plate for 2 min at 14,000 × *g*.
5. During the last 5 min of incubation, prepare the SEAP assay master mix for 60 wells as follows: For each well, use 80 μL of 2× SEAP buffer and 20 μL of pNPP solution (substrate).
6. Measure the absorbance at 405 nm every 30 s for 30 min with a plate reader.
7. Quantify SEAP levels according to Lambert–Beer's law:

$$E = \varepsilon \cdot c \cdot d \rightarrow c = \frac{E}{\varepsilon \cdot d}$$

E = increase in absorbance per minute (absorbance/min), ε = molar extinction coefficient of the product, p-nitrophenolate (ε = 18,600 M^{-1} cm^{-1}), c = increase in concentration of p-nitrophenolate (M/min), d = length of the light path in the sample (cm) (usually 0.625 cm for a 96-well plate).

8. Calculate the activity of SEAP (U/L) using the formula: SEAP activity = $c \times 10^6 \times (200/10)$ (dilution factor).

Illumination with blue light activates the melanopsin receptor on the cell surface and triggers the Ca^{2+} signaling pathway, leading to transcription from the NFAT-responsive promoter and ultimately expression of the reporter gene (SEAP) (Fig. 1). The level of SEAP expression can be fine-tuned by adjusting the light intensity (Fig. 3c) and amount of transfected reporter (*see* **Note 10**).

4 Notes

1. Treatment of the 96-well plate with poly-L-lysine for 30 min at room temperature prior to cell seeding is recommended to avoid cell detachment during changes of media.

2. The reporter gene utilized in the experiment (SEAP) can in principle be exchanged for any other desired reporter gene, such as genes encoding fluorescent proteins or luciferases.
3. Unwanted mutations can occur in the plasmid during bacterial propagation. Transfection of DNA from different bacterial clones is a useful way to find the best clone.
4. The ratio of receptor to reporter influences the fold induction. Thus, it may be necessary to optimize this ratio. The quality of the extracted DNA ($A_{260/280}$) should always be checked before transfection to confirm that the plasmid is RNA-free.
5. Transfection efficiency is highly dependent on the transfection reagent and cell type. A constitutively active reporter gene that can be multiplexed with the SEAP reporter (e.g., a fluorescent protein) can be coexpressed to monitor transfection efficiency.
6. Working with cells in ambient light may activate melanopsin. After transfection, cells should be handled under a hood with a safe red light, and all other lighting should be switched off.
7. Inserting a diffuser between the LEDs and cell plate generates a spatially more homogeneous illumination source.
8. Long-term blue-light illumination is toxic to mammalian cells. To monitor the cytotoxicity, it may be necessary to include a suitable control. For example, pSEAP2-Control (P_{SV40}-SEAP-pA) from Clontech can be used to evaluate the cytotoxicity by comparing the expression levels in light-stimulated and unstimulated cells.
9. Illumination with blue light can upregulate some endogenous transcription factors, leading to unwanted SEAP expression (Tyssowski and Gray, *bioRxiv*, 2019). To monitor this type of artefact, it is necessary to have a control containing only the reporter (without receptor) to track nonspecific induction.
10. A high level of reporter gene expression in control cells increases the leakiness of the system. Reducing the reporter amount and also culturing light-protected control cells in a separate plate can ameliorate this kind of leakiness.

Acknowledgments

This work was supported by the European Research Council (ERC) advanced grant (ElectroGene; grant no. 785800) and in part by the National Centre of Competence in Research (NCCR) for Molecular Systems Engineering.

References

1. Scheller L, Fussenegger M (2019) From synthetic biology to human therapy: engineered mammalian cells. Curr Opin Biotechnol 58:108–116. https://doi.org/10.1016/j.copbio.2019.02.023
2. Bacchus W, Lang M, El-Baba MD et al (2012) Synthetic two-way communication between mammalian cells. Nat Biotechnol 30 (10):991–996. https://doi.org/10.1038/nbt.2351
3. Weber W, Fux C, Daoud-el Baba M et al (2002) Macrolide-based transgene control in mammalian cells and mice. Nat Biotechnol 20 (9):901–907. https://doi.org/10.1038/nbt731
4. Weber W, Marty RR, Link N et al (2003) Conditional human VEGF-mediated vascularization in chicken embryos using a novel temperature-inducible gene regulation (TIGR) system. Nucleic Acids Res 31(12): e69. https://doi.org/10.1093/nar/gng069
5. Auslander D, Auslander S, Charpin-El Hamri G et al (2014) A synthetic multifunctional mammalian pH sensor and CO2 transgene-control device. Mol Cell 55(3):397–408. https://doi.org/10.1016/j.molcel.2014.06.007
6. Redchuk TA, Omelina ES, Chernov KG et al (2017) Near-infrared optogenetic pair for protein regulation and spectral multiplexing. Nat Chem Biol 13(6):633–639. https://doi.org/10.1038/nchembio.2343
7. Weber W, Rimann M, Spielmann M et al (2004) Gas-inducible transgene expression in mammalian cells and mice. Nat Biotechnol 22 (11):1440–1444. https://doi.org/10.1038/nbt1021
8. Mansouri M, Strittmatter T, Fussenegger M (2019) Light-controlled mammalian cells and their therapeutic applications in synthetic biology. Adv Sci (Weinh) 6(1):1800952. https://doi.org/10.1002/advs.201800952
9. Repina NA, Rosenbloom A, Mukherjee A et al (2017) At Light speed: advances in optogenetic systems for regulating cell signaling and behavior. Annu Rev Chem Biomol Eng 8:13–39. https://doi.org/10.1146/annurev-chembioeng-060816-101254
10. Tischer D, Weiner OD (2014) Illuminating cell signalling with optogenetic tools. Nat Rev Mol Cell Biol 15(8):551–558. https://doi.org/10.1038/nrm3837
11. Ma G, Wen S, He L et al (2017) Optogenetic toolkit for precise control of calcium signaling. Cell Calcium 64:36–46. https://doi.org/10.1016/j.ceca.2017.01.004
12. Kolar K, Weber W (2017) Synthetic biological approaches to optogenetically control cell signaling. Curr Opin Biotechnol 47:112–119. https://doi.org/10.1016/j.copbio.2017.06.010
13. Muller K, Naumann S, Weber W et al (2015) Optogenetics for gene expression in mammalian cells. Biol Chem 396(2):145–152. https://doi.org/10.1515/hsz-2014-0199
14. Endo M, Ozawa T (2017) Strategies for development of optogenetic systems and their applications. J Photoch Photobio C 30:10–23. https://doi.org/10.1016/j.jphotochemrev.2016.10.003
15. Muller K, Weber W (2013) Optogenetic tools for mammalian systems. Mol BioSyst 9 (4):596–608. https://doi.org/10.1039/c3mb25590e
16. Guru A, Post RJ, Ho YY et al (2015) Making sense of optogenetics. Int J Neuropsychopharmacol 18(11). https://doi.org/10.1093/ijnp/pyv079
17. Hattar S, Lucas RJ, Mrosovsky N et al (2003) Melanopsin and rod-cone photoreceptive systems account for all major accessory visual functions in mice. Nature 424(6944):76–81. https://doi.org/10.1038/nature01761
18. Mure LS, Hatori M, Zhu QS et al (2016) Melanopsin-encoded response properties of intrinsically photosensitive retinal ganglion cells. Neuron 90(5):1016–1027. https://doi.org/10.1016/j.neuron.2016.04.016
19. Ye HF, Daoud-El Baba M, Peng RW et al (2011) A synthetic optogenetic transcription device enhances blood-glucose homeostasis in mice. Science 332(6037):1565–1568. https://doi.org/10.1126/science.1203535
20. Zhao BX, Wang YC, Tan XH et al (2019) An optogenetic controllable T cell system for hepatocellular carcinoma immunotherapy. Theranostics 9(7):1837–1850. https://doi.org/10.7150/thno.27051
21. Khamo JS, Krishnamurthy VV, Chen QX et al (2019) Optogenetic delineation of receptor tyrosine kinase subcircuits in PC12 cell differentiation. Cell Chem Biol 26(3):400–410.e3. https://doi.org/10.1016/j.chembiol.2018.11.004
22. Auslander S, Fuchs D, Hurlemann S et al (2016) Engineering a ribozyme cleavage-induced split fluorescent aptamer complementation assay. Nucleic Acids Res 44(10). https://doi.org/10.1093/nar/gkw117

Chapter 14

Dual Activation of cAMP Production Through Photostimulation or Chemical Stimulation

Nyla Naim, Jeff M. Reece, Xuefeng Zhang, and Daniel L. Altschuler

Abstract

cAMP is a crucial mediator of multiple cell signaling pathways. This cyclic nucleotide requires strict spatiotemporal control for effective function. Light-activated proteins have become a powerful tool to study signaling kinetics due to having quick on/off rates and minimal off-target effects. The photoactivated adenylyl cyclase from *Beggiatoa* (bPAC) produces cAMP rapidly upon stimulation with blue light. However, light delivery is not always feasible, especially in vivo. Hence, we created a luminescence-activated cyclase by fusing bPAC with nanoluciferase (nLuc) to allow chemical activation of cAMP activity. This dual-activated adenylyl cyclase can be stimulated using short bursts of light or long-term chemical activation with furimazine and other related luciferins. Together these can be used to mimic transient, chronic, and oscillating patterns of cAMP signaling. Moreover, when coupled to compartment-specific targeting domains, these reagents provide a new powerful tool for cAMP spatiotemporal dynamic studies. Here, we describe detailed methods for working with bPAC-nLuc in mammalian cells, stimulating cAMP production with light and luciferins, and measuring total cAMP accumulation.

Key words cAMP, Adenylyl cyclase, Optogenetics, bPAC, Nanoluciferase

1 Introduction

cAMP (cyclic adenosine monophosphate) is a ubiquitous signaling molecule found in all prokaryotic and eukaryotic cells. It was first identified as a second messenger in G-protein coupled receptor (GPCR) signaling and is now recognized as a common node in multiple pathways. cAMP is integral to basic cell function as it regulates proliferation, differentiation, migration, attachment, and many other processes. Hence, irregular cAMP signaling has been linked to several diseases. For example, activating mutations that lead to constitutive cAMP accumulation have been reported in thyroid hyperplasia and adenoma [1, 2]. Diminished cAMP signaling has been found in certain mood disorders, Alzheimer's disease, and multiple sclerosis [3, 4]. Furthermore, a growing body of

Dominik Niopek (ed.), *Photoswitching Proteins: Methods and Protocols*, Methods in Molecular Biology, vol. 2173,
https://doi.org/10.1007/978-1-0716-0755-8_14,

literature suggests dysregulation of localized cAMP signaling can elicit physiological consequences [5].

Steady-state intracellular cAMP levels are primarily determined by the rate of synthesis by adenylyl cyclases and degradation by phosphodiesterases. Several reports have shown that the basal concentration of cellular cAMP is between 0.1 and 1 μM depending on the cell type [6, 7]. Upon hormonal stimulation, this can increase by more than ten-fold [8, 9]. Superphysiologic levels of cAMP, up to 90-fold above baseline, can be reached by stimulating adenylyl cyclases with forskolin and inhibiting phosphodiesterases with IBMX (3-isobutyl-1-methylxanthine) [10]. Cellular cAMP concentration fluctuates following three temporal patterns: transient, sustained, and oscillating (*see* Fig. 1a). Transient spikes of cAMP production after GPCR stimulation only last on the order of minutes due to phosphodiesterase activity [11–14] and cAMP-dependent protein kinase A (PKA)-driven negative feedback loops. Sustained cAMP production has been reported during GPCR endocytosis [15–20] and activating mutations along the pathway can lead to constitutive cAMP production [21–24].

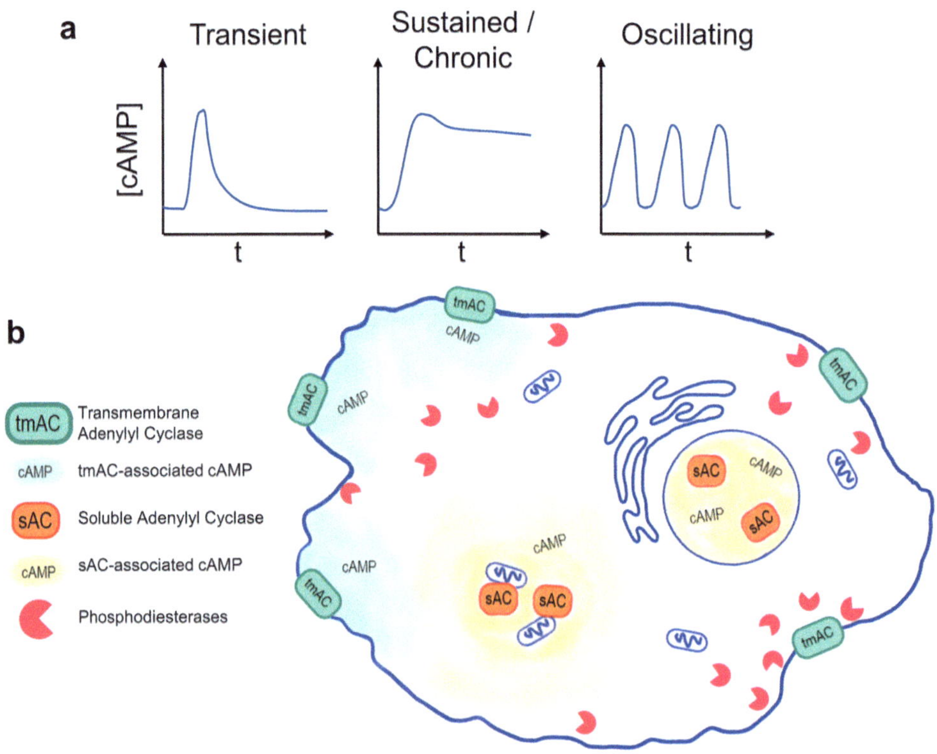

Fig. 1 (**a**) Diagram of cAMP concentration over time (*t*) following three temporal patterns. (**b**) Schematic of cellular cAMP compartmentalization or localized cAMP signaling from transmembrane adenylyl cyclases (tmAC) and soluble adenylyl cyclases (sAC). cAMP diffuses unevenly through the cell and gradients are largely shaped by phosphodiesterase activity

Additionally, cAMP oscillations have been reported in pancreatic β cells [25, 26], neurons [27, 28], pituitary cells [29], and myocardiac cells [30].

The concentration and temporal patterns of cAMP accumulation alone do not fully account for how cAMP activates distinct pathways [7, 31]. Over the last few decades, mounting evidence suggests cAMP is spatially regulated, forming regions of high and low concentrations (*see* Fig. 1b). This uneven distribution, or compartmentalization, allows preferential activation of specific signalosomes [32]. Under this model, studies suggest local cAMP signaling is regulated by adenylyl cyclase proximity to specific cAMP effectors [33], cAMP efflux [34], buffering by effector proteins [35, 36], phosphodiesterase barriers or sinks [37], restricted diffusion due to steric hindrance from intracellular structures [38, 39], and overall cell geometry [40, 41]. Hence to further validate this model and mimic cAMP signaling from transmembrane and soluble adenylyl cyclases, tools with high spatial and temporal control are needed.

While there are several pharmacological agents available to modulate cAMP levels, chemical activation lacks cellular specificity and has limited spatiotemporal control. Genetically encoded tools can help overcome these issues. For example, the optogenetic photoactivated adenylyl cyclase from *Beggiatoa* (bPAC) has been successfully used for tissue specific expression [42, 43] and subcellular localization [44, 45] of cAMP synthesis. bPAC is a ~40 kDa monomer containing a BLUF (sensors of *b*lue-*l*ight *u*sing *F*AD) domain and Type III adenylyl cyclase domain [46, 47]. As a dimer, it is activated by 435–455 nm light exposure within milliseconds and can produce up to a 300-fold increase in cAMP [47, 48]. bPAC returns to its dark-state quickly ($\tau_{\text{off}} = 12$ s) although residual cAMP accumulation is reported ($\tau = 23 \pm 2$ s) [47]. bPAC offers unique optogenetic strategies to study cAMP; however, there are limitations to this technique. For example, delivering light into deep tissues for in vivo experiments can be difficult and involve invasive procedures. Moreover, light can be phototoxic if not regulated. To circumvent issues with light delivery and expand the methods of activation for optogenetic proteins, Hochgeschwender and colleagues created a fusion protein of the light-sensitive channelrhodopsin ion channel and Gaussia luciferase. This "luminopsin" represented a new class of proteins offering both optogenetic and chemical control [49–51].

Following this concept, our group created a luminescence-activated adenylyl cyclase by fusing bPAC to nanoluciferase (nLuc). nLuc was selected for its small size (19 kDa) and blue-shifted luminescence in response to the luciferins furimazine (Fz) and h-coelenterazine (h-CTZ) [52]. A myc-tag was included between the two moieties to aid identification and visualization. The resulting bPAC-nLuc fusion protein (~62 kDa) can be

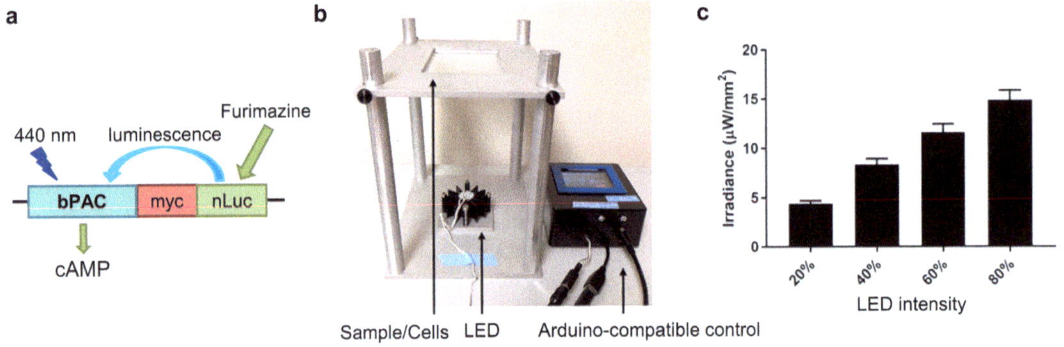

Fig. 2 (**a**) Schematic of luminescence activated cyclase, bPAC-nLuc, which can be activated by blue light and luciferins. (**b**) An image of an Arduino controlled LED mounted on a stage where cell culture dishes can be placed above. (**c**) The measured irradiance at 440 nm of a royal blue LED at varying intensities (mean ± SD of 54 measurements across the area of a 6-well dish). Data were originally published in the Journal of Biological Chemistry. Naim et al. Luminescence-activated nucleotide cyclase regulates spatial and temporal cAMP synthesis. (J. Biol. Chem. 2018; 294:1095–1103. © the American Society for Biochemistry and Molecular Biology)

activated by blue light or chemical stimulation of luminescence [53] (*see* Fig. 2a). This protocol describes the general working conditions and optimization process for using bPAC-nLuc in mammalian cell lines. bPAC-nLuc expressing cells must be kept away from light <500 nm. Upon stimulation with luciferins, such as furimazine, luminescence can be measured and used to confirm expression. For effective photo-activation using a blue LED light source, it is important to have even illumination across a sample. cAMP accumulation resulting from either chemical activation or photoactivation can be measured using an enzyme linked immunosorbent assay (ELISA) or live-cell imaging with a cAMP biosensor. Here, a rat hepatoma cell line (HC1) [54, 55] was selected for most assays because they exhibit low basal cAMP levels and high transfection efficiency. For live-cell imaging, rat follicular thyroid cells (PCCL3) [56] were used since they are more physiologically relevant to our interests. These techniques can be adopted to other cells lines, including human embryonic kidney cells (HEK293) [53]; however, the expression level of bPAC-nLuc and light/chemical dosage may need to be tuned to match endogenous cAMP levels.

Dual regulation of bPAC-nLuc using light and chemical activation provides flexibility in experimental design. Transient and sustained cAMP synthesis can be tested using the same construct, creating a broad range of temporal signatures. Interestingly, new analogs of furimazine, such as Endurazine™ and Vivazine™ (Promega), have been created for extended luminescent lifetimes (>48 h) and lower cytotoxicity. Furthermore, we found bPAC-nLuc could be easily targeted to various subcellular locations by adding targeting motifs to the N- and C-termini [53]. Hence, bPAC-nLuc can be used to study location-biased cAMP signaling or compartmentalization with a high degree of temporal control, thus expanding the available tool kit to artificially regulate local cAMP synthesis.

2 Materials

2.1 Cell Culture Materials

1. HC1 cells.
2. Dulbecco's Modified Eagle's Medium (DMEM).
3. PCCL3 cells.
4. Complete Coon's Media: Nutrient Mixture F-12 Ham (Coon's modification, Sigma-Aldrich) supplemented with 2.68 g/L sodium bicarbonate, 5% fetal bovine serum (FBS), 1% Penicillin/Streptomycin, 2 mM L-glutamine, and 4 Hormone solution: insulin (1 μg/mL), apo-transferrin (5 μg/mL), hydrocortisone (1 nM), and thyroid stimulating hormone (1 IU/L).
5. Starvation Coon's media: Complete Coon's Media (still containing 5% FBS) supplemented with 0.2% bovine serum albumin but lacking insulin, hydrocortisone, and thyroid stimulating hormone.
6. Phosphate Buffered Solution (PBS) without Ca^{2+} or Mg^{2+}.
7. 0.25% Trypsin–22.1 mM EDTA (Corning).
8. Cell dissociation solution, nonenzymatic 1× (Sigma-Aldrich).
9. 0.1% gelatin (prepared in PBS).
10. 10 cm and 6-well dishes.
11. 15 mL conical tubes.
12. 96-well dishes, white/opaque.
13. 25-mm glass coverslips.
14. Opti-MEM, phenol-red free (Invitrogen).
15. X-tremeGENE HP (Roche).
16. Lipofectamine 3000 (Thermo Fisher Scientific).
17. pcDNA3.1+ bPAC-myc-nLuc DNA [53].
18. Red Dimerization Dependent Sensor for cAMP (Montana Molecular).
19. Laminar Flow Hood for sterile cell culture.
20. 37 °C, 5% CO_2 incubator.
21. 37 °C hot water bath.
22. Hemocytometer.

2.2 bPAC-nLuc Activation and Lighting Conditions

1. Thor Labs Laser Power Meter (ThorLabs PM1100D, detector S130C).
2. Blackout curtains and/or heavy-weight, opaque paper (optional for blocking light from doors/windows).
3. Aluminum foil.
4. A heavy-duty cutting blade.
5. 13 W amber compact fluorescence bulb (Low Blue Lights, Photonic Developments LLC).
6. Red safelight lamp (e.g., Kodak GBX-2 Safelight Filter).
7. NanoGlo® Luciferase Substrate (Furimazine, Promega).
8. Arduino-controlled LED system (full description and parts list in Naim et al. 2018) [53].
9. High-power LED (royal blue CREE XTE Tri-Star LED, LED Supply).

2.3 cAMP ELISA

1. Monoclonal Anti-cAMP Antibody Based Direct cAMP ELISA Kit (nonacetylated, NewEast Biosciences).
2. 0.1 M HCl (prepared in ultrapure water).
3. BCA Protein Assay Kit (e.g., Pierce™).
4. TeccanSpark 20 M plate reader (SparkControl V1.2 software).
5. 3-Isobutyl-1-methylxanthine (IBMX, stock prepared in DMSO, Sigma-Aldrich).

2.4 Live Cell Imaging

1. Imaging chamber.
2. Olympus IX70 microscope.
3. Till Polychrome V monochromator.
4. 60×/1.4 NA oil objective
5. Hamamatsu CCD camera (Photonics Model C4742-80-12AG).
6. Slidebook software (Intelligent Imaging Innovations Inc.).
7. Emission filter FF01-620/52 (Semrock).
8. Emission filter S470/30 m (Chroma Technology Corp).
9. 560 nm long pass dichroic (HQ570 LP, Chroma Technology Corp).
10. Forskolin (stock prepared in DMSO, Sigma-Aldrich).

3 Methods

3.1 LED Calibration

1. To approximate the amount of light through polystyrene cell culture dishes and ensure even light distribution, set up the LED light source mounted on a stage that fits a standard 6-well dish (*see* **Note 1**) (*see* Fig. 2b).

2. Prepare the lid of a 6-well dish by carefully removing the edges from one or two sides of the lid using a heavy-duty cutting blade so that the laser power meter can lay flat against the plastic.
3. Place the lid upside-down on the stage above the LED light source.
4. Turn on the LED (*see* **Note 2**).
5. Place the laser power meter wand flat against the lid and measure the 440 nm light intensity at multiple points across the lid, ensuring areas where wells would reside are well represented (i.e., 9 measurements per well of a 6-well dish).
6. Calculate the irradiance (power per area, $\mu W/mm^2$) using the measured intensity values and area of the probe sensor.
7. Repeat process for each LED intensity used (*see* Fig. 2c).

3.2 Cell Culture Environment

All cells and samples containing functional bPAC-nLuc (i.e., during cell culture but not necessarily after cell lysis or fixation) should be protected from light below 500 nm to prevent protein activation. This can be accomplished by covering windows and doorways using blackout curtains, aluminum foil, or heavy-weight opaque paper. Use a laser power meter to detect diffuse light at 430–460 nm wavelengths. Light sources should be above 500 nm. Combining an amber light bulb in a red safelight lamp provides sufficient light for performing cell culture while maintaining low basal activity of bPAC-nLuc. All cell culture should be performed in a sterile environment using prewarmed reagents. All buffers should be prepared in ultrapure water unless otherwise noted.

3.3 HC1 Cell Culture and Transfection

1. Culture HC1 cells in DMEM containing 10% FBS, penicillin, and streptomycin in a 10 cm dish at 37 °C, 5% CO_2. Passage every 2–3 days once cells are 80–90% confluent using 1:5 dilutions.
2. Aspirate media and wash the culture dish surface with 5 mL of PBS.
3. Aspirate PBS, add 1 mL of trypsin solution and incubate for 3–5 min in an incubator.
4. Once cells have detached, resuspend the cells in 4 mL of media and transfer to a 15 mL conical tube.
5. If seeding cells for an experiment, count the cell density using a hemocytometer.
6. Dilute the cell suspension in media to the desired final cell concentration.
7. Pipet the desired cell number into experimental dishes (e.g., 96-well plate or 6-well plate).

8. Incubate for 24 h and replace media afterward.
9. Transfect each well with bPAC-nLuc using X-tremeGENE HP, following the manufacturer's protocol using a ratio of 1 μg DNA–2 μL X-tremeGENE reagent–100 μL Opti-MEM. Use 10 μL of transfection mixture per well of a 96-well plate, and 100 μL per well of a 6-well plate (*see* **Note 3**).
10. Incubate transfected samples for 24 h protected from light (e.g., using aluminum foil to shield samples from light) (*see* **Note 4**).

3.4 Chemical Luminescence Activation Assay

1. Seed 8000 HC1 cells per well in a white, opaque 96-well dish using 100 μL culture volume per well.
2. Incubate for 24 h, transfect bPAC-nLuc as described above, and incubate samples for 24 h.
3. Prepare furimazine solution in phenol red-free Opti-MEM for a twofold serial dilution ranging from 1:20 to 1:2560. Prepare enough solution to run samples in triplicates.
4. Aspirate media and wash wells once in PBS.
5. For each dose, aspirate media from three wells and add 100 μL of furimazine to each well.
6. Measure luminescence on a plate reader immediately as the intensity decays within seconds.
7. Repeat **steps 5** and **6** for each dose (*see* Fig. 3a and **Note 5**).

3.5 cAMP ELISA of bPAC-nLuc Activity

1. Seed 1.5×10^5 HC1 cells per well in two 6-well dishes and incubate for 24 h.
2. Transfect cells with bPAC-nLuc following the protocol described above and incubate for 24 h (*see* **Note 6**).
3. Prepare stimulating conditions in 1 mL media containing either DMSO (vehicle), 100 μM IBMX (basal cAMP activity, *see* **Note** 7), 1:100 furimazine (activation), or 100 μM IBMX + 1:100 furimazine (max activation). Other dilutions of furimazine can be used after optimizing the protocol (*see* **Note 8**).
4. Aspirate media from one dish and add the stimulating reagents.
5. Incubate for 10 min in incubator.
6. Aspirate stimulating reagents and wash with PBS.
7. Aspirate PBS and lyse cells in 100 μL of 0.1 M HCl for 10 min at room temperature (*see* **Note 9**).
8. Collect samples by scraping cells with a rubber policeman. Store samples at −80 °C.
9. Treat a second dish with cells with DMSO (vehicle) or IBMX (max activation) for 10 min at 37 °C.

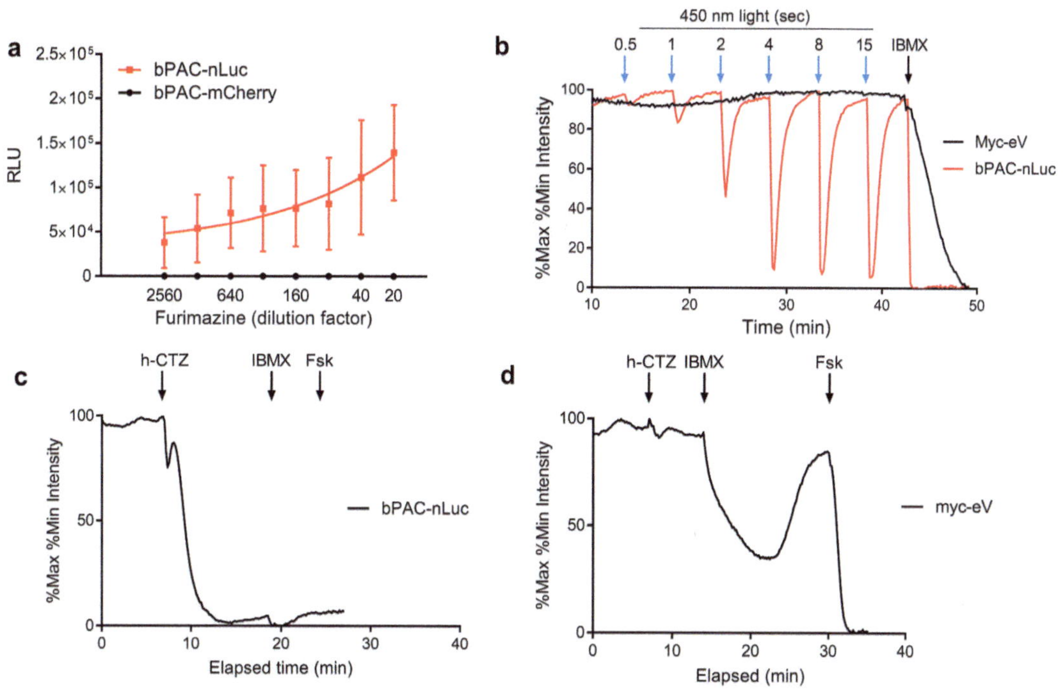

Fig. 3 (**a**) Measured relative luminescence (RLU) of HC1 cells expressing bPAC-nLuc or bPAC-mCherry stimulated with varying dilutions of Furimazine (mean ± SD, $n = 4$). (**b–d**) Live-cell, real-time monitoring of cAMP accumulation in PCCL3 cells coexpressing the Red Dimerization Dependent Sensor for cAMP with bPAC-nLuc or myc-empty vector (myc-eV) (representative cell traces shown). (**b**) Cells were stimulated with light pulses of varying duration at ~4 μW/mm^2 (250 μM IBMX indicates sensor saturation point) or (**c**, **d**) with 25 μM h-coelenterazine (h-CTZ), a luciferin capable of activating nLuc (100 μM IBMX and 10 μM forskolin (Fsk) indicate sensor saturation). Panel B was originally published in the Journal of Biological Chemistry. Naim et al. Luminescence-activated nucleotide cyclase regulates spatial and temporal cAMP synthesis. (J. Biol. Chem. 2018; 294:1095–1103. © the American Society for Biochemistry and Molecular Biology)

10. During the last minute of treatment, stimulate cells with 440 nm light for 1 min using the LED device (*see* **Note 10**).
11. Immediately wash samples with PBS and lyse cells as in **steps 6–8**.
12. Measure cAMP levels using an ELISA kit following the manufacturer's directions.
13. Quantify cAMP accumulation per amount of protein loaded (pmol cAMP/μg protein lysate) which can be measured using a bicinchoninic acid assay (BCA).

3.6 Live-Cell Monitoring of bPAC-nLuc Activation

1. Culture PCCL3 cells in "Complete Coon's Media" (*see* Subheading 2) in a 10 cm dish at 37 °C and 5% CO_2. Passage every 2–3 days once cells are 80–90% confluent using 1:3 dilutions.
2. Place sterilized 25 mm glass coverslips into a 6-well dish.
3. Coat coverslips by applying 0.1% gelatin solution for 30 min.

4. Aspirate gelatin solution and wash with PBS.
5. Remove a ~90% confluent 10 cm dish of PCCL3 cells from the incubator and aspirate media.
6. Wash with PBS.
7. Aspirate PBS and add a mixture of 2 mL of cell dissociation solution and 50 μL trypsin.
8. Incubate for approximately 5 min until cells begin to round up.
9. Aspirate the solution and gently dislodge cells by pipetting 10 mL of Coon's complete medium across the surface of the dish.
10. Count the cell density using a hemocytometer.
11. Dilute the cell suspension as needed to seed 1.5×10^6 cells per well of the 6-well dish.
12. Following seeding, incubate cells for 24 h.
13. Provide 1.5 mL of fresh media and cotransfect bPAC-nLuc and Red Dimerization Dependent Sensor for cAMP using Lipofectamine 3000 following the manufacturer's instructions, using a ratio of 0.25 μg bPAC-nLuc DNA–0.75 μg Red Dimerization Dependent Sensor for cAMP–2 μL P3000 reagent–3 μL Lipofectamine 3000 reagent for each well (*see* **Note 11**).
14. Incubate for approximately 24 h.
15. Aspirate media and add "Coon's Starvation Media" (*see* Subheading 2) to each well 3 h before imaging.
16. Start the monochromator, microscope components, Arduino-controlled LED, and imaging software.
17. Invert the LED-mounted stage and position the LED light source above the microscopy stage, using the same distance used for LED Calibration (*see* Subheading 3.2).
18. Wash a coverslip in PBS and transfer to an imaging chamber.
19. Add 0.9 mL of Opti-MEM (phenol red-free) to the chamber.
20. Transport the chamber, protected from light, to the microscope stage.
21. Setup imaging conditions to monitor red fluorescence (e.g., 60× oil objective, 20% monochromator intensity, 15 nm bandpass, 570 nm excitation, 560 nm long pass dichroic, 620 nm emission filter).
22. Identify red fluorescent cells (*see* **Note 12**) to find a field-of-view with cells that show diffuse localization of the sensor (*see* **Note 13**).
23. Take a test image to select an exposure time between 200 and 500 ms and adjust the gain as needed.

24. Allow cells to equilibrate without any light exposure for approximately 5 min.
25. Begin imaging the cells using a 10 s capture rate.
26. Select each cell as a region of interest and monitor the intensity in real-time.
27. Once a stable base line has been established (5–10 min), deliver varying pulses of blue light between 100 ms to 100 s and monitor decreases in red fluorescent intensity. Allow the intensity to return to basal levels between each pulse (*see* Fig. 3b).
28. At the end of each experiment, saturate the sensor by adding 10 μM forskolin and 100 μM IBMX directly to the chamber. This can be done by pipetting 100 μL of a 10× working solution prepared in Opti-MEM.
29. Finally, demonstrate bPAC-nLuc expression via luminescence. Take an image of the cell field luminescence using a long (10–15 s) exposure using a CFP emission filter (470/30 nm) but with no excitation.
30. Add a 1:1000 dose of furimazine and take another image of the luminescence.
31. To test the effect of furimazine in real-time, repeat **steps 18–24**.
32. Take a "before" picture of cell luminescence as in **step 29**.
33. Begin the imaging time course to monitor red fluorescence.
34. After establishing a baseline, stimulate with 1:1000 furimazine prediluted in Opti-MEM (*see* Fig. 3c, d).
35. Monitor red fluorescence for >30 min or until signal returns to baseline.
36. Saturate sensor as in **step 28**.
37. Take an "after" picture of luminescence as in **step 29**.
38. Save all files and export raw data to Excel.
39. Subtract the background fluorescence and plot the change in red fluorescent intensity over time. This can be normalized as a percentage of the maximum and minimum signal.
40. Quantify the change in luminescence between background-corrected "before" and "after" images.

4 Notes

1. To excite bPAC-nLuc with 440 nm light, we used a custom-built Arduino-compatible LED system (*see* Fig. 2b). A complete parts list, description of how it was built, and the code used to alter light pulse frequency, duration, and intensity was

previously described [53]. Other systems can be used as long as the LED emits blue wavelengths in the 440 nm range with an intensity of at least 4 μW/mm^2, which is the half maximal activation value reported [47]. For example, step-by-step guides to build LED arrays have been described by other groups [57, 58].

2. Do not look directly into the LED light source as it can cause damage to the eyes. Protective eyewear is recommended.
3. bPAC-nLuc constructs with subcellular targeting domains can also be used. Additionally, if transient transfections are not ideal, lentiviral version of bPAC-nLuc have been created and stable cell lines can be prepared [53].
4. In addition to the luminescence assay, expression can be tested in transfected cells using the myc-tag located between the bPAC and nLuc moiety. The anti-myc antibody, clone 9E10, can be used for both immunofluorescence and western blot analysis.
5. Other luciferins can be used to activate bPAC-nLuc as long as they emit strong blue light (~440 nm). For example, we have used h-coelenterazine (NanoLight Technology) and a prototype of the Nano-Glo Endurazine Live Cell Substrate (Furimazine-4377, Promega). When testing new luciferins, ensure the light emitted covers blue wavelengths by measuring the luminescent spectrum. Furimazine emitted maximum luminescence at 455 nm [53]. Luminescence can also be monitored over time to obtain kinetic data on the lifetime of the luminescence.
6. Controls for transfection must be included in luminescence assays, ELISAs, and live-cell imaging assays. For example, an myc-empty vector can be used to control for transfection efficiency while bPAC (not fused with nLuc but preferably myc-tagged, that is, bPAC-myc-mCherry) can be used demonstrate specificity of chemical activation.
7. bPAC has a reported dark activity of 33 ± 5 pmol/min/mg of protein [47]. Treating cells with IBMX helps estimate the basal activity of bPAC-nLuc and can help determine if sample treatment conditions are sufficiently dark.
8. To optimize activation conditions, compare cAMP accumulation to other stimulating agents of physiologically relevance, such as agents stimulating GPCR signaling.
9. Lysis conditions may be different depending on the ELISA kit protocol used. Follow the manufacturer's instructions.
10. 1 min of light activation at 4 μW/mm^2 is a useful starting point since it should stimulate very high levels of cAMP. The light intensity and duration of illumination can be reduced after establishing that the assay works as expected.

11. Other cAMP biosensors can be used as long as wavelengths below 500 nm are not required. Using a lower amount of bPAC-nLuc DNA helps prevent overexpression of the construct and thus saturation of the biosensor.
12. Do not use bright field imaging to identify cells as it can activate bPAC-nLuc activity.
13. Red Dimerization Dependent Sensor for cAMP should not form punctate dots, rather it should be expressed evenly throughout the cytoplasm.

Acknowledgments

This research was supported by National Institute of General Medical Sciences (NIGMS) of the US National Institutes of Health (NIH), and the Molecular Pharmacology Training Program of the University of Pittsburgh under grant Awards Number R01-GM09975, R01-GM130612, T32-GM00842419/20/21, and the Wistar Morris's Cotswold Foundation Fellowship.

References

1. Ledent C, Dumont JE, Vassart G, Parmentier M (1992) Thyroid expression of an A2 adenosine receptor transgene induces thyroid hyperplasia and hyperthyroidism. EMBO J 11 (2):537–542
2. Kosugi S, Shenker A, Mori T (1994) Constitutive activation of cyclic AMP but not phosphatidylinositol signaling caused by four mutations in the 6th transmembrane helix of the human thyrotropin receptor. FEBS Lett 356 (2–3):291–294. https://doi.org/10.1016/0014-5793(94)01286-5
3. Gould TD, Manji HK (2002) Signaling networks in the pathophysiology and treatment of mood disorders. J Psychosom Res 53 (2):687–697. https://doi.org/10.1016/S0022-3999(02)00426-9
4. Poppinga WJ, Muñoz-Llancao P, González-Billault C, Schmidt M (2014) A-kinase anchoring proteins: cAMP compartmentalization in neurodegenerative and obstructive pulmonary diseases. Br J Pharmacol 171(24):5603–5623. https://doi.org/10.1111/bph.12882
5. Gold MG, Gonen T, Scott JD (2013) Local cAMP signaling in disease at a glance. J Cell Sci 126(20):4537–4543. https://doi.org/10.1242/jcs.133751
6. Borner S, Schwede F, Schlipp A, Berisha F, Calebiro D, Lohse MJ, Nikolaev VO (2011) FRET measurements of intracellular cAMP concentrations and cAMP analog permeability in intact cells. Nat Protoc 6(4):427–438. https://doi.org/10.1038/nprot.2010.198
7. Koschinski A, Zaccolo M (2017) Activation of PKA in cell requires higher concentration of cAMP than in vitro: implications for compartmentalization of cAMP signalling. Sci Rep 7 (1):14090. https://doi.org/10.1038/s41598-017-13021-y
8. Terasaki WL, Brooker G (1977) Cardiac adenosine 3′:5′-monophosphate. Free and bound forms in the isolated rat atrium. J Biol Chem 252(3):1041–1050
9. Iancu RV, Ramamurthy G, Warrier S, Nikolaev VO, Lohse MJ, Jones SW, Harvey RD (2008) Cytoplasmic cAMP concentrations in intact cardiac myocytes. Am J Physiol Cell Physiol 295(2):C414–C422. https://doi.org/10.1152/ajpcell.00038.2008
10. Smith FD, Esseltine JL, Nygren PJ, Veesler D, Byrne DP, Vonderach M, Strashnov I, Eyers CE, Eyers PA, Langeberg LK, Scott JD (2017) Local protein kinase A action proceeds through intact holoenzymes. Science 356 (6344):1288–1293. https://doi.org/10.1126/science.aaj1669
11. Nikolaev VO, Bunemann M, Hein L, Hannawacker A, Lohse MJ (2004) Novel single chain cAMP sensors for receptor-induced signal propagation. J Biol Chem 279

(36):37215–37218. https://doi.org/10.1074/jbc.C400302200
12. Zaccolo M, De Giorgi F, Cho CY, Feng L, Knapp T, Negulescu PA, Taylor SS, Tsien RY, Pozzan T (2000) A genetically encoded, fluorescent indicator for cyclic AMP in living cells. Nat Cell Biol 2(1):25–29. https://doi.org/10.1038/71345
13. Cui W, Smith A, Darby-King A, Harley CW, McLean JH (2007) A temporal-specific and transient cAMP increase characterizes odorant classical conditioning. Learn Mem 14 (3):126–133. https://doi.org/10.1101/lm.496007
14. Vedel L, Brauner-Osborne H, Mathiesen JM (2015) A cAMP biosensor-based high-throughput screening assay for identification of Gs-coupled GPCR ligands and phosphodiesterase inhibitors. J Biomol Screen 20 (7):849–857. https://doi.org/10.1177/1087057115580019
15. Calebiro D, Nikolaev VO, Gagliani MC, de Filippis T, Dees C, Tacchetti C, Persani L, Lohse MJ (2009) Persistent cAMP-signals triggered by internalized G-protein-coupled receptors. PLoS Biol 7(8):e1000172. https://doi.org/10.1371/journal.pbio.1000172
16. Ferrandon S, Feinstein TN, Castro M, Wang B, Bouley R, Potts JT, Gardella TJ, Vilardaga J-P (2009) Sustained cyclic AMP production by parathyroid hormone receptor endocytosis. Nat Chem Biol 5(10):734–742. https://doi.org/10.1038/nchembio.206
17. Kuna RS, Girada SB, Asalla S, Vallentyne J, Maddika S, Patterson JT, Smiley DL, DiMarchi RD, Mitra P (2013) Glucagon-like peptide-1 receptor-mediated endosomal cAMP generation promotes glucose-stimulated insulin secretion in pancreatic β-cells. Am J Physiol Endocrinol Metab 305(2):E161–E170. https://doi.org/10.1152/ajpendo.00551.2012
18. Merriam LA, Baran CN, Girard BM, Hardwick JC, May V, Parsons RL (2013) Pituitary adenylate cyclase 1 receptor internalization and endosomal signaling mediate the pituitary adenylate cyclase activating polypeptide-induced increase in Guinea pig cardiac neuron excitability. J Neurosci 33(10):4614–4622. https://doi.org/10.1523/jneurosci.4999-12.2013
19. Inda C, Dos Santos Claro PA, Bonfiglio JJ, Senin SA, Maccarrone G, Turck CW, Silberstein S (2016) Different cAMP sources are critically involved in G protein-coupled receptor CRHR1 signaling. J Cell Biol 214 (2):181–195. https://doi.org/10.1083/jcb.201512075
20. Pavlos NJ, Friedman PA (2017) GPCR signaling and trafficking: the long and short of it. Trends Endocrinol Metab 28(3):213–226. https://doi.org/10.1016/j.tem.2016.10.007
21. Persani L, Lania A, Alberti L, Romoli R, Mantovani G, Filetti S, Spada A, Conti M (2000) Induction of specific phosphodiesterase isoforms by constitutive activation of the cAMP pathway in autonomous thyroid Adenomas1. J Clin Endocrinol Metabol 85 (8):2872–2878. https://doi.org/10.1210/jcem.85.8.6712
22. Weinstein LS, Shenker A, Gejman PV, Merino MJ, Friedman E, Spiegel AM (1991) Activating mutations of the stimulatory G protein in the McCune-Albright syndrome. N Engl J Med 325(24):1688–1695. https://doi.org/10.1056/nejm199112123252403
23. Boot AM, Lumbroso S, Verhoef-Post M, Richter-Unruh A, Looijenga LHJ, Funaro A, Beishuizen A, van Marle A, Drop SLS, Themmen APN (2011) Mutation analysis of the LH receptor gene in Leydig cell adenoma and hyperplasia and functional and biochemical studies of activating mutations of the LH receptor gene. J Clin Endocrinol Metab 96 (7):E1197–E1205. https://doi.org/10.1210/jc.2010-3031
24. Min KS, Liu X, Fabritz J, Jaquette J, Abell AN, Ascoli M (1998) Mutations that induce constitutive activation and mutations that impair signal transduction modulate the basal and/or agonist-stimulated internalization of the Lutropin/Choriogonadotropin receptor. J Biol Chem 273(52):34911–34919
25. Dyachok O, Isakov Y, Sagetorp J, Tengholm A (2006) Oscillations of cyclic AMP in hormone-stimulated insulin-secreting beta-cells. Nature 439(7074):349–352. https://doi.org/10.1038/nature04410
26. Tian G, Sagetorp J, Xu Y, Shuai H, Degerman E, Tengholm A (2012) Role of phosphodiesterases in the shaping of sub-plasma-membrane cAMP oscillations and pulsatile insulin secretion. J Cell Sci 125 (Pt 21):5084–5095. https://doi.org/10.1242/jcs.107201
27. Vitalis EA, Costantin JL, Tsai P-S, Sakakibara H, Paruthiyil S, Iiri T, Martini J-F, Taga M, Choi ALH, Charles AC, Weiner RI (2000) Role of the cAMP signaling pathway in the regulation of gonadotropin-releasing hormone secretion in GT1 cells. Proc Natl Acad Sci 97(4):1861–1866. https://doi.org/10.1073/pnas.040545197
28. Nicol X, Voyatzis S, Muzerelle A, Narboux-Neme N, Sudhof TC, Miles R, Gaspar P

(2007) cAMP oscillations and retinal activity are permissive for ephrin signaling during the establishment of the retinotopic map. Nat Neurosci 10(3):340–347. https://doi.org/10.1038/nn1842

29. Haisenleder DJ, Yasin M, Marshall JC (1992) Enhanced effectiveness of pulsatile 3′,5′-cyclic adenosine monophosphate in stimulating prolactin and alpha-subunit gene expression. Endocrinology 131(6):3027–3033. https://doi.org/10.1210/endo.131.6.1280210
30. Brooker G (1973) Oscillation of cyclic adenosine monophosphate concentration during the myocardial contraction cycle. Science 182 (4115):933–934
31. Rich TC, Fagan KA, Nakata H, Schaack J, Cooper DM, Karpen JW (2000) Cyclic nucleotide-gated channels colocalize with adenylyl cyclase in regions of restricted cAMP diffusion. J Gen Physiol 116(2):147–161
32. Ghigo A, Mika D (2019) cAMP/PKA signaling compartmentalization in cardiomyocytes: lessons from FRET-based biosensors. J Mol Cell Cardiol 131:112–121. https://doi.org/10.1016/j.yjmcc.2019.04.020
33. Agarwal SR, Gratwohl J, Cozad M, Yang PC, Clancy CE, Harvey RD (2018) Compartmentalized cAMP signaling associated with lipid raft and non-raft membrane domains in adult ventricular myocytes. Front Pharmacol 9:332. https://doi.org/10.3389/fphar.2018.00332
34. Sassi Y, Abi-Gerges A, Fauconnier J, Mougenot N, Reiken S, Haghighi K, Kranias EG, Marks AR, Lacampagne A, Engelhardt S, Hatem SN, Lompre A-M, Hulot J-S (2012) Regulation of cAMP homeostasis by the efflux protein MRP4 in cardiac myocytes. FASEB J 26(3):1009–1017. https://doi.org/10.1096/fj.11-194027
35. Chen C, Nakamura T, Koutalos Y (1999) Cyclic AMP diffusion coefficient in frog olfactory cilia. Biophys J 76(5):2861–2867
36. Agarwal SR, Clancy CE, Harvey RD (2016) Mechanisms restricting diffusion of intracellular cAMP. Sci Rep 6:19577. https://doi.org/10.1038/srep19577
37. Lohse C, Bock A, Maiellaro I, Hannawacker A, Schad LR, Lohse MJ, Bauer WR (2017) Experimental and mathematical analysis of cAMP nanodomains. PLoS One 12(4):e0174856. https://doi.org/10.1371/journal.pone.0174856
38. Richards M, Lomas O, Jalink K, Ford KL, Vaughan-Jones RD, Lefkimmiatis K, Swietach P (2016) Intracellular tortuosity underlies slow cAMP diffusion in adult ventricular myocytes. Cardiovasc Res 110(3):395–407. https://doi.org/10.1093/cvr/cvw080
39. Monterisi S, Favia M, Guerra L, Cardone RA, Marzulli D, Reshkin SJ, Casavola V, Zaccolo M (2012) CFTR regulation in human airway epithelial cells requires integrity of the actin cytoskeleton and compartmentalized cAMP and PKA activity. J Cell Sci 125(Pt 5):1106–1117. https://doi.org/10.1242/jcs.089086
40. Maiellaro I, Lohse MJ, Kittel RJ, Calebiro D (2016) cAMP signals in drosophila motor neurons are confined to single synaptic boutons. Cell Rep 17(5):1238–1246. https://doi.org/10.1016/j.celrep.2016.09.090
41. Bacskai BJ, Hochner B, Mahaut-Smith M, Adams SR, Kaang BK, Kandel ER, Tsien RY (1993) Spatially resolved dynamics of cAMP and protein kinase A subunits in Aplysia sensory neurons. Science 260(5105):222–226
42. Efetova M, Petereit L, Rosiewicz K, Overend G, Haussig F, Hovemann BT, Cabrero P, Dow JA, Schwarzel M (2013) Separate roles of PKA and EPAC in renal function unraveled by the optogenetic control of cAMP levels in vivo. J Cell Sci 126(Pt 3):778–788. https://doi.org/10.1242/jcs.114140
43. Jansen V, Alvarez L, Balbach M, Strunker T, Hegemann P, Kaupp UB, Wachten D (2015) Controlling fertilization and cAMP signaling in sperm by optogenetics. Elife 4. https://doi.org/10.7554/eLife.05161
44. Tsvetanova NG, von Zastrow M (2014) Spatial encoding of cyclic AMP signaling specificity by GPCR endocytosis. Nat Chem Biol 10 (12):1061–1065. https://doi.org/10.1038/nchembio.1665
45. Averaimo S, Assali A, Ros O, Couvet S, Zagar Y, Genescu I, Rebsam A, Nicol X (2016) A plasma membrane microdomain compartmentalizes ephrin-generated cAMP signals to prune developing retinal axon arbors. Nat Commun 7:12896. https://doi.org/10.1038/ncomms12896
46. Ryu MH, Moskvin OV, Siltberg-Liberles J, Gomelsky M (2010) Natural and engineered photoactivated nucleotidyl cyclases for optogenetic applications. J Biol Chem 285 (53):41501–41508. https://doi.org/10.1074/jbc.M110.177600
47. Stierl M, Stumpf P, Udwari D, Gueta R, Hagedorn R, Losi A, Gartner W, Petereit L, Efetova M, Schwarzel M, Oertner TG, Nagel G, Hegemann P (2011) Light modulation of cellular cAMP by a small bacterial photoactivated adenylyl cyclase, bPAC, of the soil bacterium Beggiatoa. J Biol Chem 286

(2):1181–1188. https://doi.org/10.1074/jbc.M110.185496

48. Lindner R, Hartmann E, Tarnawski M, Winkler A, Frey D, Reinstein J, Meinhart A, Schlichting I (2017) Photoactivation mechanism of a bacterial light-regulated adenylyl cyclase. J Mol Biol 429(9):1336–1351. https://doi.org/10.1016/j.jmb.2017.03.020
49. Berglund K, Birkner E, Augustine GJ, Hochgeschwender U (2013) Light-emitting channelrhodopsins for combined optogenetic and chemical-genetic control of neurons. PLoS One 8(3):e59759. https://doi.org/10.1371/journal.pone.0059759
50. Berglund K, Clissold K, Li HE, Wen L, Park SY, Gleixner J, Klein ME, Lu D, Barter JW, Rossi MA, Augustine GJ, Yin HH, Hochgeschwender U (2016) Luminopsins integrate opto- and chemogenetics by using physical and biological light sources for opsin activation. Proc Natl Acad Sci U S A 113(3): E358–E367. https://doi.org/10.1073/pnas.1510899113
51. Park SY, Song SH, Palmateer B, Pal A, Petersen ED, Shall GP, Welchko RM, Ibata K, Miyawaki A, Augustine GJ, Hochgeschwender U (2017) Novel luciferase-opsin combinations for improved luminopsins. J Neurosci Res. https://doi.org/10.1002/jnr.24152
52. Hall MP, Unch J, Binkowski BF, Valley MP, Butler BL, Wood MG, Otto P, Zimmerman K, Vidugiris G, Machleidt T, Robers MB, Benink HA, Eggers CT, Slater MR, Meisenheimer PL, Klaubert DH, Fan F, Encell LP, Wood KV (2012) Engineered luciferase reporter from a deep sea shrimp utilizing a novel imidazopyrazinone substrate. ACS Chem Biol 7(11):1848–1857. https://doi.org/10.1021/cb3002478
53. Naim N, White AD, Reece JM, Wankhede M, Zhang X, Vilardaga JP, Altschuler DL (2018) Luminescence-activated nucleotide cyclase regulates spatial and temporal cAMP synthesis. J Biol Chem 294(4):1095–1103. https://doi.org/10.1074/jbc.AC118.004905
54. Insel PA, Maguire ME, Gilman AG, Bourne HR, Coffino P, Melmon KL (1976) Beta adrenergic receptors and adenylate cyclase: products of separate genes? Mol Pharmacol 12(6):1062–1069
55. Ross EM, Howlett AC, Ferguson KM, Gilman AG (1978) Reconstitution of hormone-sensitive adenylate cyclase activity with resolved components of the enzyme. J Biol Chem 253 (18):6401–6412
56. Kimura T, Van Keymeulen A, Golstein J, Fusco A, Dumont JE, Roger PP (2001) Regulation of thyroid cell proliferation by TSH and other factors: a critical evaluation of in vitro models. Endocr Rev 22(5):631–656
57. Polstein LR, Gersbach CA (2014) Light-inducible gene regulation with engineered zinc finger proteins. Methods Mol Biol 1148:89–107. https://doi.org/10.1007/978-1-4939-0470-9_7
58. Dietler J, Stabel R, Möglich A (2019) Pulsatile illumination for photobiology and optogenetics. In: Deiters A (ed) Methods in enzymology, vol 624. Academic Press, Cambridge, Massachusetts, pp 227–248. https://doi.org/10.1016/bs.mie.2019.04.005

Chapter 15

Synthesis of a Light-Controlled Phytochrome-Based Extracellular Matrix with Reversibly Adjustable Mechanical Properties

Maximilian Hörner, Philipp Hoess, Ramona Emig, Balder Rebmann, and Wilfried Weber

Abstract

Synthetic extracellular matrices with reversibly adjustable mechanical properties are essential for the investigation of how cells respond to dynamic mechanical cues as occurring in living organisms. One interesting approach to engineer dynamic biomaterials is the incorporation of photoreceptors from cyanobacteria or plants into polymer materials. Here, we give an overview of existing photoreceptor-based biomaterials and describe a detailed protocol for the synthesis of a phytochrome-based extracellular matrix (CyPhyGel). Using cell-compatible light in the red and far-red spectrum, the mechanical properties of this matrix can be adjusted in a fully reversible, wavelength-specific, and dose-dependent manner with high spatiotemporal control.

Key words Biomaterials, Extracellular matrix, Hydrogels, Optogenetics, Phytochromes, Mechanosensing

1 Introduction

The mechanical properties of the extracellular environment are key in determining cell fate and function [1]. While the majority of studies investigating matrix–cell interactions were performed using materials with static mechanical properties, matrices with adjustable mechanical properties can more closely emulate the dynamic environment a cell is exposed to in a living organism [2–4]. For this reason a variety of materials has been developed which allow the modulation of the mechanical properties in response to chemical or optical stimuli [2–6]. One interesting approach to engineer remotely controllable materials is the use of photoreceptors from plants or cyanobacteria. Naturally occurring photoreceptors are well suited for the control of synthetic extracellular matrices as they are inherently functional in a biological environment and

Dominik Niopek (ed.), *Photoswitching Proteins: Methods and Protocols*, Methods in Molecular Biology, vol. 2173, https://doi.org/10.1007/978-1-0716-0755-8_15,

respond to light intensities and colors compatible with cellular systems. So far, light-responsive materials utilizing the photoreceptors UVR8 [7], LOV2 [8], Dronpa145N [9, 10], $CarH_C$ [11], and Cph1 [12] as molecular switches have been described [13] (Table 1).

In this chapter, we give a detailed protocol for the synthesis of a phytochrome-based light-responsive hydrogel, termed here CyPhyGel [12]. This material is responsive to light in the cell-compatible red/far-red spectrum and allows a fully reversible, wavelength-specific and dose-dependent adjustment of the mechanical properties under cell culture conditions in a spatiotemporally controlled manner.

The system is based on an engineered variant of the *cy*anobacterial *phy*tochrome Cph1 (termed Cph1∗) that was genetically modified with a tandem arginine–glycine–aspartic acid (RGD) motif to allow cell adhesion and a C-terminal cysteine for coupling to the polymer [12, 14]. Coupling of Cph1∗ to 8-arm polyethylene glycol (PEG) vinyl sulfone via Michael type addition resulted in the formation of a covalently cross-linked hydrogel, likely caused by binding of vinyl sulfones to the C-terminal cysteine and the surface-exposed Cph1-internal cysteine 371 (Fig. 1) [12]. Illumination with red light (~660 nm) induced Cph1∗ dimerization and thus increased the number of cross-links within the material and consequently its stiffness. Vice versa, far-red light (~740 nm)-induced Cph1∗ monomerization decreased the number of cross-links and thus the stiffness of the material.

In the following, we describe in detail the production of Cph1∗ in *E. coli* flask cultures (Subheading 3.1), the purification of the protein via immobilized metal ion affinity chromatography (IMAC, Subheading 3.2) and finally the synthesis of the CyPhyGels (Subheading 3.3). In contrast to our previous publication [12], Cph1∗ is now produced using an optimized expression plasmid that is compatible with high-cell-density fermentation [15]. In addition, we provide an advanced and simplified hydrogel synthesis protocol that circumvents the need of a glovebox for hydrogel synthesis under an oxygen-free atmosphere [16].

2 Materials

Prepare all solutions using analytical grade reagents and ultrapure water. If not stated otherwise, all components are stored at room temperature (RT).

2.1 Production of Cph1∗

1. Plasmid pMH48 (available via Addgene, ID 131861) encoding the protein Cph1∗ and the enzymes HO1 and PcyA for phycocyanobilin (PCB) synthesis under control of two T7 promoters [15]. Store at −20 °C.

Table 1
Overview of photoreceptor-based light-tunable biohybrid materials

Photoreceptor	Functionality	Light condition 1	Light condition 2	Coupling mechanism	Photoreceptor modification	Backbone
UVR8 [7]	Reversible gel–sol phase transition	UV light→ UVR8 monomers→ No cross-links→ sol	Dark→ UVR8 dimers→ Cross-links→ Gel	Protein–peptide interaction (TIP1/ WRESAI)	UVR8 genetically fused to WRESAI	Peptide Nap-GFFYGE based nanofibers with TIP1
LOV2 [8]	Reversible stiffness modulation	Blue light→ Displacement of Jα helix→ Longer cross-links→ Reduced stiffness	Dark→ Jα helix binding to LOV2→ Shorter cross-links→ Increased stiffness	Strain-promoted azide–alkyne cycloaddition (SPAAC)	LOV2-Jα enzymatically functionalized on N- and C-termini with reactive azides	4-arm PEG tetrabicyclononyne (20 kDa)
Dronpa145N [9]	Reversible gel–sol phase transition	Cyan light→ Dronpa145N monomers→ No cross-links→ Sol	Violet light→ Dronpa145N tetramers→ Cross-links→ Gel	Covalent protein–peptide interaction (SpyCatcher–SpyTag)	Dronpa145N genetically fused to SpyTag	Protein polymer (GB1-SpyCatcher)$_3$
Dronpa145N [10]	Reversible gel–sol phase transition	Cyan light→ Dronpa145N monomers→ No cross-links→ Sol	Violet light→ Dronpa145N tetramers→ Cross-links→ Gel	Michael-type addition	No modification, one Dronpa145N-internal solvent-exposed cysteine	4-arm PEG maleimide (20 kDa)
CarH$_C$ [11]	Unidirectional gel–sol phase transition	Green light→ CarH$_C$ monomers→ No cross-links→ Sol	Dark→ CarH$_C$ tetramers→ Cross-links→ Gel	Covalent protein–peptide interaction (SpyCatcher–SpyTag)	Protein polymer based on SpyTag-ELP-CarH$_C$-ELP-SpyTag and SpyCatcher-ELP-CarH$_C$-ELP-SpyCatcher	
Cph1 [12]	Reversible stiffness modulation	Red light→ Cph1 dimers→ Cross-links→ Increased stiffness	Far-red light→ Cph1 monomers→ No cross-links→ Reduced stiffness	Michael-type addition	Cph1 genetically fused to C-terminal cysteine	8-arm PEG vinyl sulfone (40 kDa)

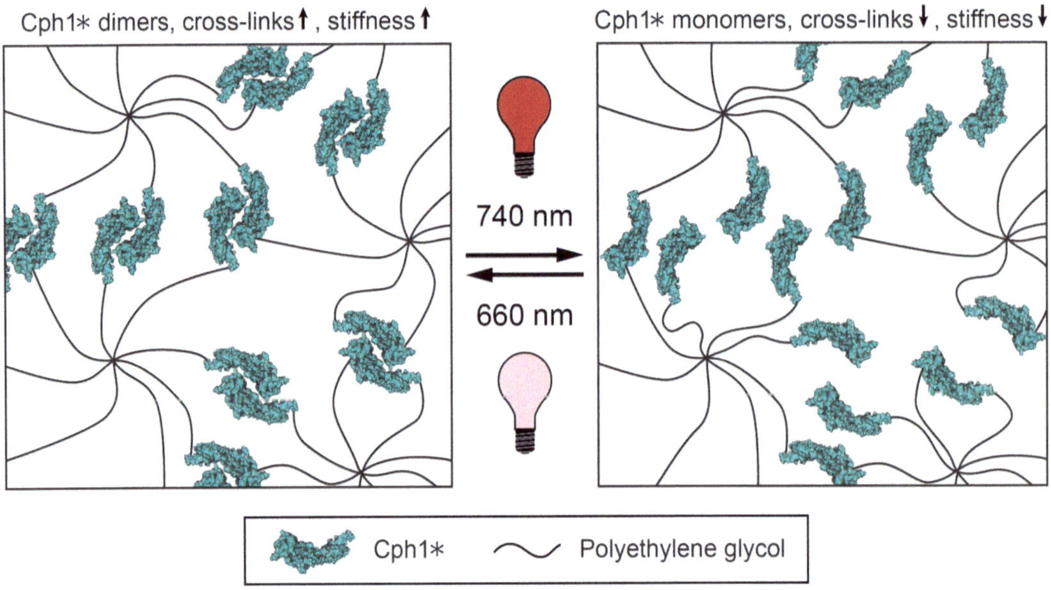

Fig. 1 Schematic overview of the phytochrome-based extracellular matrix (CyPhyGel). Red (660 nm) and far-red (740 nm) light induced dimerization and monomerization of polyethylene glycol-coupled Cph1∗ modulate the cross-link density and thus the stiffness of the material. Structure of Cph1: PDB ID 3ZQ5 [14]

2. Chemically competent BL-21 Star (DE3) cells (Thermo Fisher Scientific, Waltham, MA). Store at −80 °C.
3. Heat block or water bath at 42 °C.
4. Sterile lysogeny broth (LB) medium.
5. Streptomycin stock solution: 50 mg/mL streptomycin sulfate in water. Weigh 2.5 g of streptomycin sulfate and fill up to 50 mL with water. Filter through a 0.22 μm PES filter and store in aliquots at −20 °C.
6. 50 mL conical centrifuge tubes.
7. Heated and refrigerated shaking incubator for bacteria.
8. Autoclaved 0.5 L and 2 L baffled flasks containing 200 mL or 1 L of LB medium, respectively.
9. Spectrophotometer for cuvettes and absorbance measurement at 600 nm.
10. Isopropyl β-D-1-thiogalactopyranoside (IPTG) stock solution: 1 M IPTG in water. Weigh 11.9 g IPTG and fill up to 50 mL with water. Filter through a 0.22 μm PES filter and store in aliquots at −20 °C.
11. Centrifuge with 1 L centrifuge bottles for bacteria culture.
12. Ni-lysis buffer: 50 mM NaH_2PO_4, 300 mM NaCl, 10 mM imidazole, pH 8.0 (adjusted with NaOH). Store at 4 °C.

2.2 Purification of Cph1* by IMAC

1. Water bath at 37 °C.
2. French press or high-pressure homogenizer for bacteria lysis (e.g., APV 2000; SPX FLOW, Charlotte, NC).
3. Centrifuge (30,000 × *g*) for clarification of lysate.
4. Fast protein liquid chromatography (FPLC) system (e.g., ÄKTA system; GE Healthcare, Chicago, IL), or alternatively empty gravity flow columns.
5. Ni-NTA Superflow agarose (Qiagen, Venlo, Netherlands) for empty FPLC or gravity flow columns. Alternatively, prepacked Ni-NTA Superflow cartridges (Qiagen) can be used with an FPLC system.
6. Ni-wash buffer: 50 mM NaH_2PO_4, 300 mM NaCl, 20 mM imidazole, pH 8.0 (adjusted with NaOH). Store at 4 °C.
7. Ni-elution buffer: 50 mM NaH_2PO_4, 300 mM NaCl, 250 mM imidazole, pH 8.0 (adjusted with NaOH). Store at 4 °C.

2.3 Synthesis of CyPhyGels

1. Syringe and syringe filters, 0.22 μm PES.
2. Centrifugal concentrators, 10 kDa MWCO, PES (e.g., Spin-X UF 20, 10 K MWCO; Corning, Corning, NY).
3. Centrifuge (8000 × *g*) for centrifugal concentrators.
4. Reaction buffer: PBS (2.7 mM KCl, 1.5 mM KH_2PO_4, 8.1 mM Na_2HPO_4, 137 mM NaCl) with 2 mM EDTA, pH 8.0. Mix 100 mL of 10x PBS (pH 7.4) with 4 mL of 0.5 M EDTA, pH 8.0 stock solution and fill up to 1 L with water. Adjust pH to 8.0 with 5 M NaOH. Autoclave.
5. Dextran desalting columns, 5 K MWCO, 10 mL (Pierce, Thermo Fisher Scientific).
6. 15 mL conical centrifuge tubes, 1.5 mL microcentrifuge tubes.
7. Centrifuge (17,000 × *g*) for microcentrifuge tubes.
8. Bradford assay solution (e.g., protein assay dye reagent concentrate; Bio-Rad, Hercules, CA).
9. Bovine serum albumin (BSA) protein standard: 1 mg/mL BSA in water. Weigh 500 mg BSA and fill up to 500 mL with water. Store in single-use aliquots at −20 °C.
10. Semi-micro cuvettes, path length: 10 mm.
11. Spectrophotometer for semi-micro cuvettes and absorbance measurement at 672 nm.
12. Illumination panel with red light-emitting diodes (LEDs, peak wavelength: 660 nm (e.g., LH W5AM; Osram Opto Semiconductors, Regensburg, Germany)) and adjustable intensity.
13. Fiber-optic spectrometer (e.g., AvaSpec-ULS2048; Avantes BV, Apeldoorn, Netherlands) to measure red light intensity.

14. Tris(2-carboxyethyl)phosphine (TCEP) stock solution: 100 mM TCEP in 0.5 M $NaHCO_3$, pH 8.0. Freshly prepare 0.5 M $NaHCO_3$, pH 9.0 in water. Weigh 2.87 g TCEP hydrochloride and fill up to 100 mL with 0.5 M $NaHCO_3$, pH 9.0. Adjust pH to 8.0 with NaOH/HCl and filter solution through a 0.22 μm PES filter and store in single-use aliquots at −20 °C.
15. Triethanolamine (TEA) stock solution: 1 M TEA in reaction buffer. Weigh 6.0 g of TEA and add 4 mL of 10× PBS and 160 μL of 0.5 M EDTA, pH 8.0 stock solution. Fill up to ~35 mL with water and adjust pH to 8.0 with concentrated HCl. Fill up to 40 mL with water, recheck pH, and adjust if necessary. Filter solution through a 0.22 μm PES filter *see* **Note 1**.
16. 8-arm polyethylene glycol-vinyl sulfone (PEG-VS) stock solution: 100 mg/mL in reaction buffer. Weigh 1 g of PEG-VS (40 kDa, Sunbright HGEO-400VS (NOF Europe, Frankfurt, Germany)) and fill up to 10 mL with reaction buffer. Filter solution through a 0.22 μm PES filter and store in single-use aliquots at −80 °C.
17. Plastic box with transparent lid for hydrogel samples.
18. Cysteine quenching solution: 150 mM cysteine in reaction buffer, pH 8.0. Weigh 0.55 g of L-cysteine and fill up to 30 mL with reaction buffer. Adjust pH to 8.0 with 5 M NaOH. Always prepare this solution fresh.

3 Methods

If not stated otherwise, all steps are performed at RT.

3.1 Production of Cph1∗

In the following, the production of Cph1∗ in 12× 1 L LB medium is described. For different expression volumes adjust the protocol accordingly.

1. Transform BL-21 Star (DE3) cells with the Cph1∗ expression plasmid pMH48. To this aim thaw one aliquot of competent *E. coli* cells on ice. Add 1 μL of pMH48 plasmid DNA (10–500 ng/μL) and gently mix the tube by tapping (*see* **Note 2**). After incubation for 20 min on ice, incubate the tube for 45 s in a heat block or water bath at 42 °C. Immediately transfer the tube on ice again, incubate for 2 min and add 500 μL of LB medium. Following incubation for 1 h at 37 °C with shaking (750 rpm), inoculate 10 mL LB medium supplemented with 100 μg/mL streptomycin in a 50 mL conical centrifuge tube with the bacterial culture. Incubate the culture overnight at 37 °C with shaking (150 rpm).

2. On the next morning, inoculate 200 mL LB medium supplemented with 100 μg/mL streptomycin in a 500 mL flask with the bacterial overnight culture. Incubate at 37 °C with shaking (150 rpm) for 3–4 h.
3. Inoculate 12× 1 L LB medium supplemented with 100 μg/mL streptomycin with 15 mL of the bacterial culture each. Incubate at 30 °C with shaking (150 rpm).
4. Once the bacterial cultures are grown to an optical density at 600 nm (OD_{600}) of 0.6–0.8, induce expression by addition of 1 mM IPTG. To this aim, add 1 mL of 1 M IPTG stock solution to each flask. Incubate at 18 °C with shaking (150 rpm) in the dark. From now on, protect bacteria from bright light (*see* **Note 3**).
5. After protein production for 20–24 h, harvest bacteria by centrifugation with 6500 × *g* for 10 min at 4 °C. The bacterial pellets should have a blue-green color due to the produced Cph1∗ and the chromophore PCB (Fig. 2a). Resuspend the pellets in 30 mL ice-cold Ni-lysis buffer per 1 L of production culture. Shock freeze 30–40 mL aliquots of resuspended bacteria in 50 mL conical centrifuges tubes in liquid nitrogen and store the tubes at −80 °C (*see* **Note 4**). Alternatively, proceed directly with **step 2** for Cph1∗ purification (Subheading 3.2).

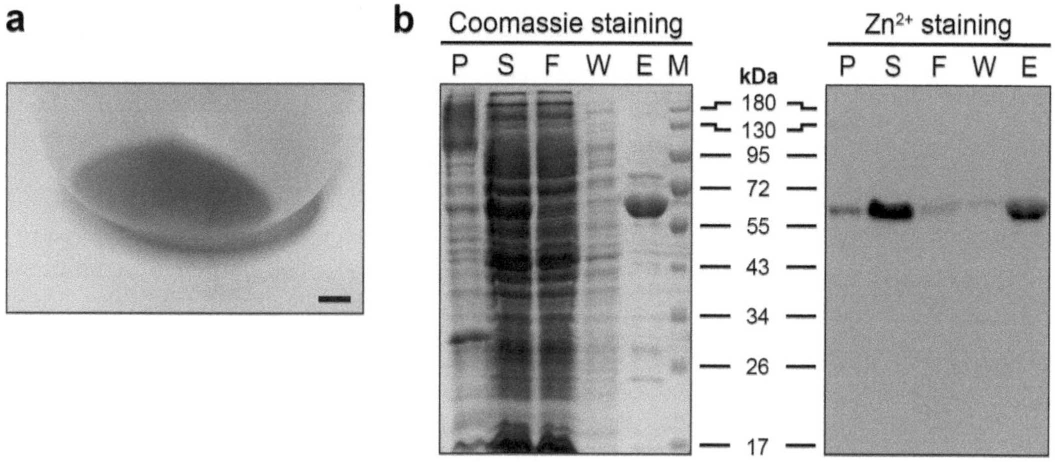

Fig. 2 Production and purification of Cph1∗. (**a**) Blue-green colored Cph1∗-containing bacterial pellet of 1 L production culture. Scale bar: ~1 cm. (**b**) Analysis of immobilized metal affinity chromatography (IMAC) purification of Cph1∗ by SDS-PAGE followed by Zn^{2+} staining of the Cph1∗ chromophore phycocyanobilin and Coomassie staining of all proteins. *P* insoluble fraction of *E. coli* lysate, *S* soluble fraction of lysate, *F* flow through of IMAC column, *W* wash of column, *E* eluate of column, *M* protein size marker. Calculated Cph1∗ molecular mass: 59.9 kDa

3.2 Purification of Cph1*

Cph1* is purified via its hexahistidine tag by IMAC. For large scale purifications, it is recommended to use an FPLC system. Small amounts (<100 mg Cph1*, equals ~8 L production culture) can be also conveniently purified using gravity flow columns.

1. Thaw frozen tubes containing the bacteria in Ni-lysis buffer in a water bath at 37 °C. Once they are completely thawed, immediately transfer them on ice.
2. Lyse bacteria by three cycles in a French press or high-pressure homogenizer at 1000–1500 bar (*see* **Note 5**). After each cycle, cool down lysate on ice below 10 °C.
3. Clarify lysate by centrifugation at 30,000 × *g* for 30 min at 4 °C (*see* **Note 6**).
4. Equilibrate Ni-NTA column with 5 column volumes (CV) of Ni-lysis buffer. Use ~2.5 mL of Ni-NTA beads per 1 L production culture (*see* **Note 7**). Next, load the clarified lysate on the Ni-NTA column, wash with 12 CV of Ni-wash buffer and elute purified Cph1* with 6 CV of Ni-elution buffer. Collect fractions with a volume of ~0.3 CV each during elution.
5. Merge the blue-green colored elution fractions containing purified Cph1* and store the protein in a closed tube or bottle at 4 °C in the dark (*see* **Note 8**).
6. The purification process and the purity of the eluate can be monitored by SDS-PAGE followed by Zn^{2+} and Coomassie staining (Fig. 2b). For Zn^{2+} staining, incubate the SDS-PAGE gel after electrophoresis for 15 min in 1 mM zinc acetate solution [17]. Visualize Zn^{2+}-induced fluorescence of the Cph1*-coupled chromophore PCB by excitation with UV-B light (*see* **Note 9**). Afterward, proteins on the same SDS-PAGE gel can be stained with Coomassie.

3.3 Synthesis of CyPhyGels

The buffer exchange of Cph1* to reaction buffer (**step 3**) should be performed on the day following protein purification. Afterward, the Cph1* protein can be frozen or it can be directly continued with the hydrogel synthesis. The synthesis should be carried out without longer interruptions and completed within 1 day.

1. At the day following protein purification, filter the Cph1* eluate with a syringe through a 0.22 μm PES filter (*see* **Note 10**).
2. Concentrate Cph1* to a concentration of 100–200 mg/mL using centrifugal concentrators (10 kDa MWCO) at 8000 × *g* (*see* **Note 11**).
3. Exchange the buffer of Cph1* to reaction buffer using a 10 mL desalting column (*see* **Note 12**). To this aim, equilibrate the column with 100 mL of reaction buffer and add up to 1 mL of concentrated Cph1* to the column. Once the protein entered the resin bed, add 1 mL of reaction buffer. When the

buffer completely entered the column, add another 1.4 mL of reaction buffer. To elute the protein, place the column over a 15 mL conical centrifugal tube and add 3 mL of reaction buffer to the column (*see* **Note 13**). If desired, Cph1∗ can be shock frozen in liquid nitrogen after elution from the desalting column (*see* **Note 14** and Fig. 3).

4. Concentrate the eluted Cph1∗ to a concentration of 100–150 mg/mL using a new centrifugal concentrator as described in **step 2**. Transfer the concentrated Cph1∗ into a microcentrifuge tube and centrifuge at 17,000 × *g* for 5 min to pellet potentially precipitated protein. Transfer supernatant into a fresh tube and homogenously mix the protein by carefully pipetting up and down (*see* **Note 15**).
5. Determine the concentration of Cph1∗ by Bradford assay or by absorbance measurement. To this aim, dilute 4 μL of the concentrated protein in 996 μL of reaction buffer (prepare at least three replicates). Determine the concentration of the diluted protein by Bradford assay using BSA as standard. Alternatively, transfer the diluted protein into a semi-micro cuvette (path length: 10 mm) and determine the absorbance of Cph1∗ at 672 nm (isosbestic point) with a spectrophotometer upon illumination of the protein for 2 min with 660 nm light (light intensity: 100 μmol/(m^2 s)). Calculate the protein concentration c [mg/mL] with the absorbance A and the dilution factor (250, 4 μL in 1 mL): $c = 1.47 \cdot 250 \cdot A$ (*see* **Note 16**). Note, that this formula is only valid for absorbance values below 0.5. If absorbance is higher, dilute sample further with reaction buffer.
6. Based on the concentration of Cph1∗, the total volume and the composition of the desired CyPhyGels, calculate the volumes of Cph1∗ protein, reaction buffer, TCEP stock solution, TEA stock solution and PEG-VS stock solution required for the hydrogel synthesis (*see* **steps 7–9** below). Typically, hydrogels with 50–70 mg/mL Cph1∗ and a molar ratio VS–Cph1∗ = 2:1 are synthesized. TCEP is used at a molar ratio Cph1∗–TCEP = 1:0.7. TEA is added to a final concentration of 100 mM. The molecular weights of 8-arm PEG-VS and Cph1∗ are 40 and 59.9 kDa, respectively. *See* **Note 17** for an exemplary calculation.
7. Mix the concentrated protein with reaction buffer and TCEP stock solution (*see* **Note 18**).
8. Following incubation for 1 h at RT, add TEA stock solution.
9. Add PEG-VS stock solution and thoroughly mix by carefully pipetting up and down.

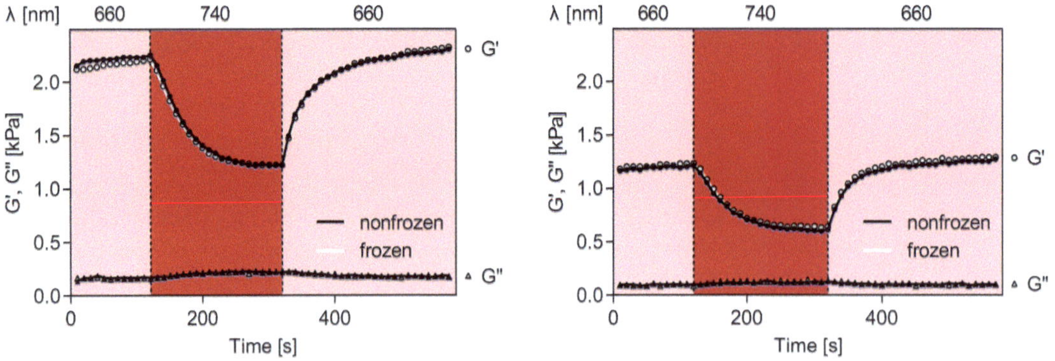

Fig. 3 Mechanical characterization of CyPhyGels synthesized from 70 mg/mL Cph1∗ with a molar vinyl sulfone (VS) to Cph1∗ ratio of 2:1. The hydrogels were synthesized from nonfrozen and frozen Cph1∗ as described in this chapter. After gelation for 22 h, the hydrogels were either incubated in cell culture medium (left panel, DMEM with 10% (v/v) fetal calf serum (FCS), 100 U/mL penicillin, 100 μg/mL streptomycin, 25 mM HEPES) or in 150 mM cysteine solution to quench unreacted vinyl sulfones (right panel) for 4 h at 37 °C, 5% CO_2. After further incubation for 20 h at 37 °C, 5% CO_2 in cell culture medium, the storage (*G*′) and loss (*G*″) moduli were measured in response to illumination with 660 and 740 nm light (both 40 μmol/(m^2 s)) by small amplitude oscillatory shear rheology (frequency: 1 Hz, deformation: 0.5%) at 37 °C as described earlier [12]

10. Transfer the mixture to the desired location(s) using a pipette (*see* **Note 19**). To synthesize disc-formed hydrogels for characterization by rheology (Fig. 3), 48 μL of the mixture are pipetted into a flat bottom well of a 96-well plate.
11. Cover the bottom of a plastic box with transparent lid with reaction buffer-soaked paper tissues, place all hydrogel reaction mixes in the box and close the lid (*see* **Note 20**).
12. Illuminate hydrogel reaction mixes through the transparent lid with 660 nm light (100 μmol/(m^2 s)) for 15 min. Then, decrease light intensity to 5 μmol/(m^2 s) and incubate the hydrogel reaction mixes at RT for 20–24 h to allow gelation (*see* **Note 21**).
13. Transfer the hydrogels into the appropriate buffer or cell culture medium for equilibration. Alternatively, to obtain softer gels, incubate the hydrogels for 4–24 h in 150 mM cysteine solution at 37 °C to quench so-far unreacted vinyl sulfones (Fig. 3).

In Fig. 3, the mechanical properties of CyPhyGels synthesized by the described protocol in response to illumination with 660 and 740 nm light are shown. The measured mechanical properties were similar to CyPhyGels synthesized earlier by a more complex protocol in absence of oxygen [12]. Moreover, freezing of Cph1∗ before hydrogel synthesis as described in **step 3** of Subheading 3.3. did not impair the measured mechanical properties and the functionality of the CyPhyGels (Fig. 3).

4 Notes

1. Before preparing the TEA stock solution, calibrate pH meter as the pH of the TEA stock solution strongly influences the pH of the coupling reaction and thus has a strong impact on the vinyl sulfone coupling chemistry. The addition of HCl will cause an increase in the temperature of the TEA solution. As pH measurement is temperature sensitive, cool down the solution to RT before checking pH again. Adjust pH until it is exactly 8.0 at RT.
2. For a high transformation efficiency of *E. coli* it is important that the volume of added DNA is small compared to the bacterial aliquot volume. Do not add more than 10% (v/v) of DNA solution to the cells. Chemically competent *E. coli* are very sensitive to shear forces before addition of the LB medium after the heat shock. Therefore, do not mix the cells by pipetting up and down or vortexing.
3. It is important to protect Cph1* (e.g., in *E. coli*, in the lysate, as purified protein or within the hydrogels) from bright light such as direct sunlight. If possible, shade the lab and only use artificial room light during sample preparation. Albeit not significantly harming Cph1*, exposure to room light should also be avoided whenever possible (e.g., by covering samples with aluminum foil). Note that, alike 660 nm light, artificial white light usually switches Cph1* into the Pfr state [12].
4. Usually, the *E. coli* cultures have an OD_{600} of ~3 before harvesting. This equals to a wet weight of the *E. coli* pellet of ~5 g per 1 L production culture. Frozen bacteria can be stored for several years at −80 °C.
5. The optimal pressure for efficient *E. coli* lysis and the number of needed cycles depend on the used French press or high-pressure homogenizer. Alternatively, *E. coli* cells can be lysed by sonication.
6. For centrifugation of large volumes of cell lysate, it is often difficult to find appropriate centrifuge beakers that allow a speed of 30,000 × *g*. Alternatively, the lysate can be centrifuged twice at 17,000 × *g* for 1 h at 4 °C.
7. Do not use a high excess of Ni-NTA beads compared to the Cph1* protein in the loaded lysate. Although the recovery of Cph1* will be higher in this case, the purity will be lower. In general, slightly overloading the Ni-NTA beads (by ~10%) will give the best trade-off between recovery and purity. As Cph1* production yield varies between different batches and binding capacity of Ni-NTA beads decreases with usage, the optimal Ni-NTA beads-to-lysate ratio has to be adjusted.

8. Do not store purified Cph1∗ longer than 1 day. If storage of Cph1∗ is necessary, the protein can be frozen after buffer exchange as described in **Note 14**. The typical yield is ~13 mg of purified Cph1∗ per 1 L of production culture.
9. GelDocs/UV chambers as typically employed for imaging of ethidium bromide-stained agarose gels can be usually used to examine Zn^{2+} stainings.
10. It is normal that some precipitates of Cph1∗ are formed after overnight incubation in Ni-elution buffer. These precipitates are effectively removed by the filtration through a 0.22 μm PES filter. If the filters are rapidly blocked, the precipitates can be pelleted by centrifugation at 8000 × *g* for 5 min before filtration of the supernatant.
11. It occurs from time to time, that new centrifugal concentrators have membrane defects that result in leakage of Cph1∗ into the lower compartment. Often, Cph1∗ precipitates in this case and is thus irreversibly lost. Therefore, when using new concentrators, only centrifuge for 1 min at the beginning and check for leakage afterward. Continue with centrifugation only if the flow through is colorless. Always centrifuge at RT and not at 4 °C to reduce protein precipitation, as solubility of Cph1∗ increases with the temperature. Throughout the concentration procedure, interrupt centrifugation every 15 min to homogenize the Cph1∗ concentration gradient by carefully inverting the concentrator. Once the Cph1∗ volume becomes very small (<500 μL) it is easier to homogenize the gradient by pipetting up and down. At such low Cph1∗ volumes, it is advised to interrupt the centrifugation more frequently to homogenize the gradient in order to prevent Cph1∗ precipitation. Note that Cph1∗ precipitation is favored at concentrations higher than ~200 mg/mL. Always handle highly concentrated Cph1∗ (>50 mg/mL) at RT and never store it at 4 °C.
12. Do not exchange the Ni-elution buffer to reaction buffer by dialysis. If we exchanged the buffer by dialysis over longer periods of time (3 days) instead of using a desalting column, hydrogel gelation did not work. We hypothesize that during dialysis cysteines get oxidized and are thus no longer available for coupling to vinyl sulfone.
13. Dependent on the actual volume of the resin material in the desalting column, the stated volumes may vary a little. The aim is to collect only the colored protein fractions (~3 mL).
14. To freeze Cph1∗ in reaction buffer after elution from the desalting column, shock freeze the protein in aliquots in liquid nitrogen and store it at −80 °C. The addition of supplements such as glycerol is not necessary. To thaw the protein, place it in

a water bath at 37 °C. After it is completely thawed, immediately transfer it to RT and continue with the hydrogel synthesis protocol. Do not freeze the protein repeatedly.

15. Never vortex or extensively shake Cph1∗ solutions as this will result in protein precipitation and foam formation.
16. The given correlation between the absorbance at 672 nm after 660 nm illumination and the protein concentration by Bradford assay was experimentally determined. Consequently, changes in the protein purity or chromophore saturation will affect this correlation. To lower the batch to batch variation of the CyPhyGels, always use the same batch of protein and the same method for determination of the protein concentration. At the isosbestic point, the absorbance of Cph1∗ in the Pfr (after 660 nm illumination) and Pr state (after 740 nm illumination) is identical and should be thus independent of the photostate. However, as the isosbestic point slightly varies in dependence of the buffer composition or the redox state of the protein, illumination with 660 nm light before measurement will help to minimize variations between different measurements.
17. In the following example, a 500 μL hydrogel with a Cph1∗ concentration of 70 mg/mL and a molar VS-to-Cph1∗ ratio of 2:1 should be synthesized. For the Cph1∗ stock solution, a concentration of 125 mg/mL is assumed. The required volumes in Subheading 3.3, **steps 7–9** would then be as follows: 280 μL Cph1∗, 107.5 μL reaction buffer, and 4.1 μL TCEP stock solution (**step** 7); 50 μL TEA stock solution (**step 8**); 58.4 μL of PEG-VS stock solution (**step 9**).
18. After addition of the TCEP stock solution to Cph1∗, immediately mix by pipetting up and down to assure homogeneous reduction of the protein.
19. Make sure that the reaction mix is homogeneous before pipetting the individual hydrogels in order to minimize variations between the different gels. For hydrogels with 50–70 mg/mL Cph1∗ and a molar VS to Cph1∗ ratio of 2:1, the reaction mix can be pipetted for at least 15 min.
20. Paper tissues soaked with reaction buffer are necessary to prevent drying of the hydrogels. Do not use water instead of reaction buffer to soak the paper tissues due to vapor diffusion.
21. Store the plastic box with the illumination panel in the dark (e.g., in a cupboard) to exclude any room light.

Acknowledgments

This work was funded by the German Research Foundation (DFG) under the Excellence Initiative (BIOSS—EXC-294) and the Excellence Strategy (CIBSS—EXC-2189—Project ID 390939984) as well as through the European Research Council under the European Community's Seventh Framework Programme (FP7/2007-2013)/ERC Grant agreement n° 259043-CompBioMat.

References

1. Watt FM, Huck WT (2013) Role of the extracellular matrix in regulating stem cell fate. Nat Rev Mol Cell Biol 14(8):467–473. https://doi.org/10.1038/nrm3620
2. Burdick JA, Murphy WL (2012) Moving from static to dynamic complexity in hydrogel design. Nat Commun 3:1269. https://doi.org/10.1038/ncomms2271
3. Rosales AM, Anseth KS (2016) The design of reversible hydrogels to capture extracellular matrix dynamics. Nat Rev Mater 1:15012. https://doi.org/10.1038/natrevmats.2015.12
4. Uto K, Tsui JH, DeForest CA, Kim DH (2017) Dynamically tunable cell culture platforms for tissue engineering and mechanobiology. Prog Polym Sci 65:53–82. https://doi.org/10.1016/j.progpolymsci.2016.09.004
5. Kloxin AM, Kasko AM, Salinas CN, Anseth KS (2009) Photodegradable hydrogels for dynamic tuning of physical and chemical properties. Science 324(5923):59–63. https://doi.org/10.1126/science.1169494
6. Guvendiren M, Burdick JA (2012) Stiffening hydrogels to probe short- and long-term cellular responses to dynamic mechanics. Nat Commun 3:792. https://doi.org/10.1038/ncomms1792
7. Zhang XL, Dong CM, Huang WY, Wang HM, Wang L, Ding D, Zhou H, Long JF, Wang TL, Yang ZM (2015) Rational design of a photo-responsive UVR8-derived protein and a self-assembling peptide-protein conjugate for responsive hydrogel formation. Nanoscale 7(40):16666–16670. https://doi.org/10.1039/c5nr05213k
8. Liu LM, Shadish JA, Arakawa CK, Shi K, Davis J, DeForest CA (2018) Cyclic stiffness modulation of cell-laden protein-polymer hydrogels in response to user-specified stimuli including light. Adv Biosyst 2(12):1800240. https://doi.org/10.1002/Adbi.201800240
9. Lyu S, Fang J, Duan T, Fu L, Liu J, Li H (2017) Optically controlled reversible protein hydrogels based on photoswitchable fluorescent protein Dronpa. Chem Commun (Camb) 53(100):13375–13378. https://doi.org/10.1039/c7cc06991j
10. Wu X, Huang WM, Wu WH, Xue B, Xiang DF, Li Y, Qin M, Sun F, Wang W, Zhang WB, Cao Y (2018) Reversible hydrogels with tunable mechanical properties for optically controlling cell migration. Nano Res 11(10):5556–5565. https://doi.org/10.1007/s12274-017-1890-y
11. Wang R, Yang Z, Luo J, Hsing IM, Sun F (2017) B12-dependent photoresponsive protein hydrogels for controlled stem cell/protein release. Proc Natl Acad Sci U S A 114(23):5912–5917. https://doi.org/10.1073/pnas.1621350114
12. Hörner M, Raute K, Hummel B, Madl J, Creusen G, Thomas OS, Christen EH, Hotz N, Gübeli RJ, Engesser R, Rebmann B, Lauer J, Rolauffs B, Timmer J, Schamel WWA, Pruszak J, Römer W, Zurbriggen MD, Friedrich C, Walther A, Minguet S, Sawarkar R, Weber W (2019) Phytochrome-based extracellular matrix with reversibly tunable mechanical properties. Adv Mater 31(12):e1806727. https://doi.org/10.1002/adma.201806727
13. Kolar K, Knobloch C, Stork H, Znidaric M, Weber W (2018) OptoBase: a web platform for molecular optogenetics. ACS Synth Biol 7(7):1825–1828. https://doi.org/10.1021/acssynbio.8b00120
14. Mailliet J, Psakis G, Feilke K, Sineshchekov V, Essen LO, Hughes J (2011) Spectroscopy and a high-resolution crystal structure of Tyr263 mutants of cyanobacterial phytochrome

Cph1. J Mol Biol 413(1):115–127. https://doi.org/10.1016/j.jmb.2011.08.023

15. Hörner M, Gerhardt K, Salavei P, Hoess P, Harrer D, Kaiser J, Tabor JJ, Weber W (2019) Production of phytochromes by high-cell-density *E. coli* fermentation. ACS Synth Biol. https://doi.org/10.1021/acssynbio.9b00267
16. https://doi.org/10.3389/fphys.2019.01526
17. Berkelman TR, Lagarias JC (1986) Visualization of Bilin-linked peptides and proteins in polyacrylamide gels. Anal Biochem 156(1):194–201

Chapter 16

Design and Application of Light-Regulated Receptor Tyrosine Kinases

Stephanie Kainrath and Harald Janovjak

Abstract

Understanding how the activity of membrane receptors and cellular signaling pathways shapes cell behavior is of fundamental interest in basic and applied research. Reengineering receptors to react to light instead of their cognate ligands allows for generating defined signaling inputs with high spatial and temporal precision and facilitates the dissection of complex signaling networks. Here, we describe fundamental considerations in the design of light-regulated receptor tyrosine kinases (Opto-RTKs) and appropriate control experiments. We also introduce methods for transient receptor expression in HEK293 cells, quantitative assessment of signaling activity in reporter gene assays, semiquantitative assessment of (in)activation time courses through Western blot (WB) analysis, and easy to implement light stimulation hardware.

Key words Optogenetics, Photoreceptor, Receptor tyrosine kinase, Transient transfection, Reporter gene assay, Luciferase assay, Western blot

1 Introduction

Receptor tyrosine kinases (RTKs) are cell surface receptors that are activated by a wide variety of growth factors, cytokines and hormones. They couple to important downstream signaling pathways, such as the mitogen activated protein kinase/extracellular signal-regulated kinase (MAPK/ERK) pathway, the phosphatidylinositol 3-kinase pathway, and phospholipase C. These pathways are tightly linked to cell survival, proliferation and differentiation, making RTKs important players in development, tissue homeostasis and disease [1]. The signaling networks downstream of RTKs are orchestrated by a variety of mechanisms, such as variability in receptor-ligand binding affinity and complex lifetime, or intracellular feedback loops [2, 3]. The complexities in RTK signaling become apparent when they are disrupted in degenerative diseases and cancer [4, 5] and can be challenging to delineate using conventional pharmacological methods [1, 6, 7].

Dominik Niopek (ed.), *Photoswitching Proteins: Methods and Protocols*, Methods in Molecular Biology, vol. 2173, https://doi.org/10.1007/978-1-0716-0755-8_16,

Light-regulated RTKs (termed Opto-RTKs) are RTKs that were re-engineered for applications in optogenetics and can be activated or inactivated by light instead of their cognate biological or synthetic ligands [8–10]. Light is an orthogonal cue that allows for the manipulation of cellular processes with high spatial and temporal precision, and is thus uniquely suited for the dissection of complex and dynamic signaling networks [6, 7]. The concurrent seminal studies of the group of Won Do Heo [9, 10] and our group [8] have resulted in a flurry of approaches in which different RTKs [9, 11–14] and also enzyme receptors of other families [15–17] have been controlled by light. In these studies, the receptors have been combined with light-sensing domains (LSDs) that alter their oligomerization state in a light-dependent manner, for example, through dimerization or oligomerization reactions, thereby mimicking the endogenous activation mechanism of many RTKs. In the first engineering approach, an LSD is directly fused to the receptor (Fig. 1a, I and II), whereas in the second approach, an LSD is fused to a protein that binds to the RTK (Fig. 1a, III). Full-length receptors modified with LSDs in either of these approaches will retain ligand binding capabilities and will therefore respond to both ligands and light (Fig. 1a). To endow more stringent control, the extracellular domain (ECD) and transmembrane domain (TMD) of a receptor can be removed, truncating it to the kinase domain (KD) and C-terminus (CT) so that it is exclusively controlled by light. In this case, the KD of the receptor is commonly anchored directly to the membrane *via* a short myristoylation (myr) sequence [8, 9, 12–15, 17]. Alternatively, the TMD and ECD can be replaced with those of the p75 receptor (a low affinity nerve growth factor receptor) to serve as an orthogonal transmembrane anchor [8, 18]. In other examples, the intracellular portion of a receptor (KD and CT) was expressed in soluble form, and LSDs were employed to achieve membrane recruitment and subsequent clustering or heterodimerization between two different receptor types. Table 1 and Fig. 1b–d summarize the LSDs employed in our group and the corresponding Opto-RTKs.

Here, we present methods for the optical control of canonical cell signaling pathways ex vivo using Opto-RTKs. The protocol covers the assessment of activity of prototypical Opto-RTKs following transient transfection in HEK293 cells. We describe appropriate controls that allow for benchmarking the light-induced signal of an Opto-RTK and serve to exclude unspecific effects of expression or light, and to verify if receptor dimerization by light is sufficient to activate the signal (Table 2). Transfecting a range of DNA amounts for Opto-RTKs and controls then allows for determining ideal conditions for minimum unspecific background signaling and a maximum signal-to-background ratio.

We further describe a quantitative reporter gene assay, as well as preparation of samples for semiquantitative assessment of

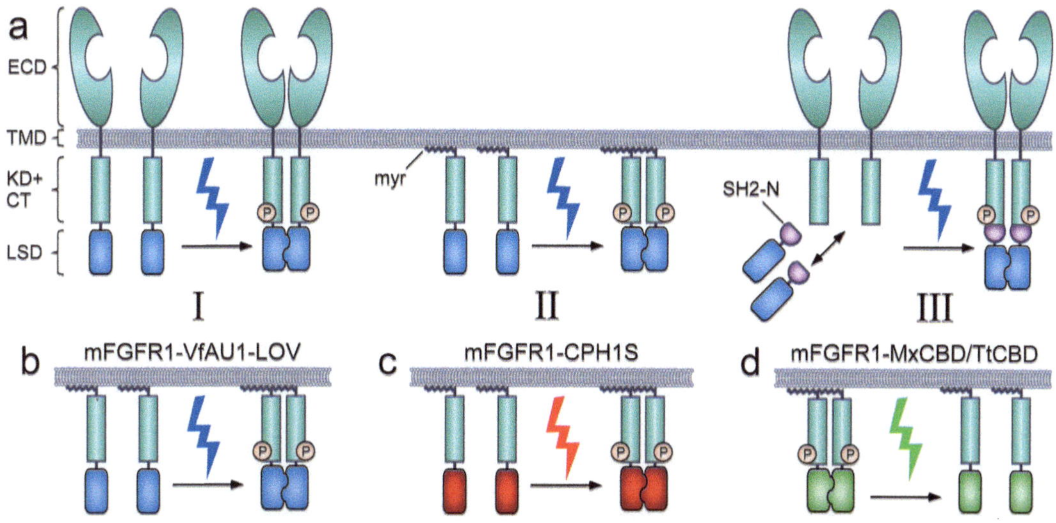

Fig. 1 Approaches to engineer Opto-RTKs. (**a**) An LSD can be directly fused to the CT of the full-length receptor or of a truncated receptor lacking the ECD and TMD. Alternatively, the LSD can be tethered to a protein that interacts with the receptor CT. (**b–d**) In these examples, the KD and CT of the murine fibroblast growth factor receptor1 (mFGFR1) is linked to the membrane via an myr anchor. (**b**) The LOV domain of Aureochrome 1 of *V. frigida* (VfAU1-LOV) dimerizes upon blue light stimulation. (**c**) The sensory domain of phytochrome 1 (CPH1S) of *Synechocystis sp.* dimerizes upon red light stimulation. (**d**) The cobalamin-binding domains (CBDs) of *M. xanthus* (MxCBD) or *T. thermophilus* (TtCBD) form a constitutive receptor complex that breaks upon green light stimulation

Table 1
LSDs employed in Opto-RTKs that were generated in our group

	VfAU1-LOV	CPH1S	MxCBD/TtCBD
λ_{max} (nm)	450	660/730	545
λ_{safe} (nm)	>520	500	>600
Cofactor	Flavin mononucleotide (FMN)	Phycocyanobilin (PCB)*	Adenosyl-cobalamin (AdoCbl)*
References	[8, 24]	[7, 14, 25]	[13, 26]

λ_{max} designates the optimal wavelength for activation (and reversal in the case of CPH1s), λ_{safe} refers to (a) wavelength (s) that will not activate the LSD and thus is safe for handling of cells expressing the domain or for imaging. LOV domains (e.g., VfAU1-LOV) and CBDs (e.g., MxCBD or TtCBD) do not absorb light above the indicated λ_{safe}, whereas the λ_{safe} of phytochromes (e.g., CPH1S) lies between two absorbance peaks and thus high intensity light from a broad band source may elicit some response. Asterisks designate exogenous cofactors that have to be supplied to cells

Opto-RTK temporal properties using the WB technique. The methods presented here are geared toward the use of receptors developed in our group, but we reason that similar considerations should be taken into account when using other light-regulated enzyme receptors.

Table 2
Control vectors and technical controls to evaluate signaling activity of an Opto-RTK in cell-based experiments

	Type of Control	Description
Transgenes	FKBP domain fusion	• Fusion of the KD and CT to an engineered FKBP (FK506 binding protein) domain, which can be forced to homodimerize by addition of the small chemical ligand AP20187 [27] • To test if the receptor can be activated by forced dimerization and to benchmark signaling levels of an induced receptor
	Fc (IgG) domain fusion	• Fusion of the KD and CT to the Fc fragment of an IgG antibody, which forces constitutive dimerization [13] • To benchmark signaling levels of a constitutively dimeric receptor
	KD and CT only	• Monomeric, membrane anchored isolated KD and CT • To benchmark background signaling levels of the transfected receptor as a "baseline"
	Dimerization-deficient receptor	• Mutation in the KD that blocks dimerization, for example, R577E in full length mFGFR1 [23] • To show that the observed signal relies on receptor dimerization and not an unspecific, effect of the LSD or light
	KD mutation	• Mutation in the KD that eliminates kinase activity, for example, Y653F/Y654F in mFGFR1 [28] • To show that the observed signal requires catalytic activity in the Opto-RTK and is not an unspecific, effect of the LSD or light
	Photoinsensitive LSD	• Mutations in the LSD or absence of exogenous cofactor to prohibit photocycle • To show that the observed signal requires photoactivation of the LSD in the Opto-RTK and is not an unspecific, effect of the LSD or light
Other	Change in stimulation light	• Stimulation with a wavelength that the LSD domain does not respond to (e.g., *see* Table 1) • To show that the observed signal requires photoactivation of the LSD in the Opto-RTK and is not an unspecific, effect of the LSD or light
	Mock transfection	• Transfection with empty plasmid vector (not containing Opto-RTK gene) • To show that the observed is not an unspecific effect of light

All controls should be performed under light and dark conditions

2 Materials

2.1 Cell Culture and Transfection

1. Purified plasmid DNA for transfection: reporter plasmid (*see* **Note 1**) and optogenetic receptor construct (*see* **Note 2**).
2. Cell culture facility equipped with a laminar flow hood, water bath (37 °C), incubator (37 °C, 5% CO_2), and cell counting chamber.

3. HEK293 cells cultured according to the provider's recommendation in 75 cm^2 tissue culture flasks.
4. D10 medium: Dulbecco's modified essential medium (DMEM) supplemented with 10% fetal bovine serum (FBS) and 1% penicillin/streptomycin (P/S), sterile filtered and stored at 4 °C.
5. D5 medium: DMEM supplemented with 5% FBS, sterile filtered and stored at 4 °C.
6. COI0.5 medium: CO_2 independent medium (COI) supplemented with 0.5% FBS and 1% P/S, sterile filtered and stored at 4 °C.
7. Opti-MEM I reduced serum medium (Thermo Fisher Scientific).
8. Phosphate-buffered saline (PBS) without Ca^{2+} and Mg^{2+}.
9. 0.25% Trypsin–EDTA solution.
10. Polyethyleneimine (PEI) transfection reagent: Completely dissolve PEI to a final concentration of 1 mg/mL (typically, using a total volume of 100 mL) in ultrapure H_2O (adjusted to pH 2.0 with HCl) then bring pH to 7.0 with NaOH. Aliquot and store at −80 °C for up to 1 year. Once thawed, keep at 4 °C for up to 1 month [19].
11. 4 mg/mL poly-L-ornithine hydrobromide (PLO) solution in ultrapure H_2O.
12. Black or white walled, clear flat-bottom polystyrene 96-well plates for reporter gene assay or 35 mm polystyrene dishes for analysis by WB (*see* **Note 3**).

2.2 Light Activation and Reporter Readout

1. Small illumination incubator (*see* **Note 4**) (Fig. 2a, b).
2. RGB LED strips (e.g., SMD 5050 LEDs, 5 m length, 60 LEDs/ with spray water shielding) with power supply, remote control and dimmer (*see* **Note 5**).
3. Digital thermometer with probe.
4. Optical power meter (e.g., Sanwa LP-1) capable of measuring light in the 400–900 nm range.
5. Plate reader capable of measuring luminescence and/or fluorescence.

2.3 Luciferase Assay Buffers and Stock Solutions [20]

1. Lysis buffer: 100 mM Tris–Cl, 40 mM Tris-base, 75 mM NaCl, 3 mM $MgCl_2$, 0.25% Triton X-100. Store at room temperature (RT).
2. *Renilla* salts buffer: 45 mM Na_2EDTA, 30 mM sodium pyrophosphate ($Na_4O_7P_2$), 1.425 M NaCl. Store at RT.

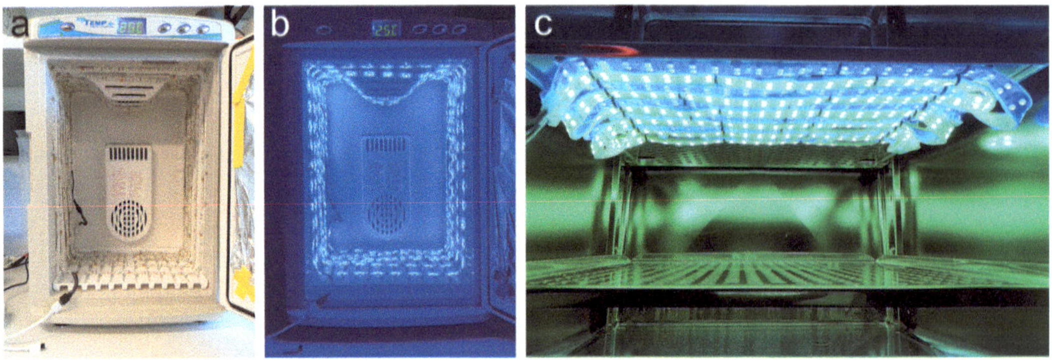

Fig. 2 Set-up of customized incubators for illumination. (**a**) A Benchmark Scientific MyTemp Mini incubator was repurposed for optogenetic experiments by equipping it with two strips of 3528 SMD LEDs (for a total of 250 LEDs). (**b**) The maximum light intensities achieved with this setup are ~330 μW/cm^2 for blue, 130 μW/cm^2 for red, and 170 μW/cm^2 for green light. (**c**) A cell incubator with CO_2 supply and humidification was equipped with waterproof 5050 SMD LEDs. A 5 m strip (600 LEDs) was cut in two 2.5 m halves which can be connected to power supplies individually. The strips were attached to the bottom of the metal shelf and affixed with cable ties. The setup enables dual-color optogenetic experiments, for example, simultaneous blue and green illumination. The maximum intensity when using both strips with the same color is ~2.1 mW/cm^2 for blue, 760 μW/cm^2 for red, and 820 μW/cm^2 for green light

3. Firefly assay stock solutions: Dissolve reagents in ultrapure H_2O unless otherwise indicated. Prepare aliquots of 500 mM dithiothreitol (DTT, 300 μL aliquots), 10 mM coenzyme A (CoA, 600 μL aliquots), 100 mM adenosine 5′-triphosphate disodium salt (ATP, 45 μL aliquots), and 80 mg/mL luciferin free acid (in lysis buffer, 52.5 μL aliquots). Store aliquots at −80 °C for up to several months.
4. *Renilla* assay stock solutions: Dissolve and prepare aliquots of 10 mM Ataluren (PTC124, in DMSO, 60 μL aliquots), 2 mM h-Coelenterazine (h-CTZ, in EtOH containing 1% HCl, 50 μL aliquots). Store aliquots at −80 °C for up to several months.
5. 3× Firefly assay reagent (Firefly-AR): Combine one of each stock aliquot (DTT, CoA, ATP, luciferin free acid) with 9 mL lysis buffer. Store at −20 °C for up to several weeks.
6. 3× *Renilla* assay reagent (*Renilla*-AR): Combine one of each stock aliquot (PTC124, h-CTZ) with 9.9 mL *Renilla* salts buffer. Store at −20 °C for up to several weeks.

2.4 WB Sample Preparation

1. RIPA buffer: 150 mM NaCl, 1% Triton X-100, 0.1% SDS, 50 mM Tris–Cl pH 8.0, 0.5% sodium deoxycholate, protease inhibitor cocktail. Keep at 4 °C for up to several weeks, or in aliquots at −20 °C for up to 1 year.
2. Cell scrapers.
3. Microcentrifuge, chilled (4 °C).

4. 4x Laemmli loading buffer: 40% glycerol, 240 mM Tris–Cl pH 6.8, 8% SDS, 0.04% bromophenol blue. Store aliquots at −20 °C and add 5% β-mercaptoethanol fresh before use.

3 Methods

3.1 Preparation of Plates and Dishes

1. Perform cell culture under sterile conditions in a laminar flow hood. Disinfect tools and containers with media and reagents by spraying their outside with 70% EtOH followed by wiping them down with a clean paper towel and placing them in the hood.
2. Coat the wells of a 96-well plate or 35 mm dishes with PLO using 70 μL PLO solution per well or 1 mL PLO solution per dish and incubate for 3 h at 37 °C or overnight (O/N) at 4 °C.
3. Remove PLO solution and wash once with 100 μL PBS per well or 2 mL PBS per dish and allow to air-dry for a few minutes.

3.2 Transient Reverse Transfection

This section describes the reverse transfection of HEK293 cells in clear-bottom 96-well plates for reporter gene assays or in 35 mm dishes for WB. In reverse transfections, cells are seeded and transfected in a single step by adding the cells to wells containing the transfection mix (*see* **Note 6**).

1. Warm trypsin solution, PBS, D5, D10, and Opti-MEM I in the water bath to 37 °C.
2. Prepare a master mix by combining sufficient Opti-MEM I (24 μL per well or 237.5 μL per dish) and PEI (1 μL per well or 12.5 μL per dish) for each condition that will be tested. Let stand at RT for 5 min.
3. In individual microcentrifugation tubes, combine Opti-MEM I and DNA for each condition to a total volume of 25 μL per well or 250 μL per dish (*see* **Note 7** for further information about DNA amounts).
4. Add Opti-MEM I-PEI mix to each Opti-MEM I-DNA mix 1:1. Mix by pipetting up and down a few times. Incubate 20 min at RT.
5. In the meantime, remove the tissue culture flask with HEK293 cells from the incubator and aspirate medium.
6. Add 10 mL of PBS and rock the flask back and forth to gently wash the cells.
7. Aspirate PBS and add 1 mL of trypsin solution. Let stand at RT for 1 min or until cells are fully detached. Gently tap against the side of the flask a few times to facilitate detachment.

8. Add 5 mL of D5 to stop trypsinization, gently pipet up and down with a serological pipet five times to resuspend cells and detach clumps.
9. Transfer cells to a 15 mL centrifugation tube and spin down at 500 × *g* for 3 min.
10. Aspirate medium and gently resuspend the cell pellet in 5 mL fresh D5.
11. Transfer 1 mL cell suspension to a new culture flask with 9 mL D10 for future passaging after 2–3 days.
12. Count the cells using a counting chamber and dilute to 500,000 cells/mL in D5.
13. Add 50 μL of Opti-MEM I-DNA-PEI mix to each well or 500 μL to each dish.
14. Add 100 μL of the cell suspension for a final 50,000 cells to each well, or 2 mL for a final 10^6 cells to each dish. Place the plates or dishes in the cell culture incubator.
15. After 6 h, change medium to prewarmed COI0.5 (*see* **Note 8** for further instructions on cofactor supplementation). Proceed to Subheading 3.3 for reporter gene assays, or Subheading 3.4 for WB sample preparation.

3.3 Light Stimulation and Reporter Gene Assay

1. Adjust the temperature inside the light incubator to ensure that it is at or slightly below 37 °C while LEDs are on, and adjust light intensity using the dimmer and power meter (*see* **Note 9**).
2. The steps during light stimulation depend on whether light-activated or light-*in*activated receptors are studied. For light-activated receptors (e.g., mFGFR-VfAU1-LOV), wrap both the "light" and "dark" plate in foil to keep them in the dark during cell starving overnight. The next morning, unwrap one plate and place both plates in an incubator equipped with LEDs for 6–8 h depending on the reporter of choice (*see* **Note 10**). For light-inactivated receptors (e.g., mFGFR-MxCBD or -TtCBD), wrap only the "dark" plate in foil and immediately place both plates in the light incubator for 16 h/O/N illumination.
3. After light stimulation, aspirate medium from all wells.
4. For fluorescent reporters, add 50–100 μL of PBS and read fluorescence in a plate reader (*see* **Note 11**).
5. For luciferase reporters, dilute 3× Firefly-AR to 1× with H_2O. Add 60 μL 1× Firefly-AR per well and immediately read luminescence in a plate reader. Representative results are shown in Fig. 3a.
6. If in addition a *Renilla* reporter is used, add 30 μL 3× Renilla-AR (for a final 1× concentration) per well on top of the luciferase assay reagent and read luminescence in a plate reader.

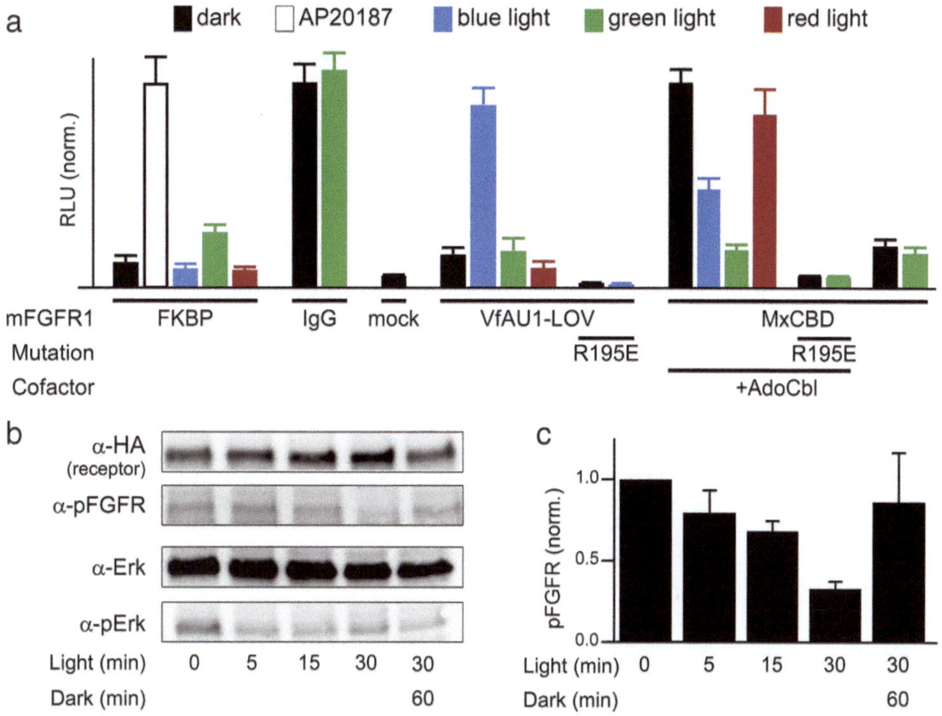

Fig. 3 Representative reporter gene assay and WB results. (**a**) Reporter assays with mFGFR1 vectors and controls. From left to right: mFGFR1-FKBP can be activated through forced dimerization by addition of the orthogonal dimerizing ligand AP20187. mFGFR1-IgG produces a constitutive dimer in the dark and under green light (green light was tested because it reduces the activity of mFGFR1-MxCBD), whereas mock-transfected cells establish a baseline signal. mFGFR1-VfAU1-LOV is activated by blue light but not by other wavelengths and the signal corresponds to that of mFGFR1-FKBP. The R195E mutation (corresponding to R577E in full-length mFGFR1) prevents dimerization and thus receptor activation [23]. mFGFR1-MxCBD forms a signaling state in the dark that is unaffected by red light and can be disrupted by blue light and green light. In absence of the cofactor AdoCbl, no signaling is observed. Results are taken from Grusch et al. [8] (normalized to mFGFR1-FKBP) and Kainrath et al. [13] (normalized to mFGFR1-IgG, dark). (**b**) WB analysis of an illumination time course. mFGFR1-MxCBD expressing cells were illuminated for different durations before they were harvested and levels of phosphorylated mFGFR1, total mFGFR1 (via a C-terminal HA-tag), phosphorylated Erk and total Erk were determined. (**c**) Band intensity was quantified in Image Studio Lite and the pErk/Erk ratio was determined. (Figures (**b**) and (**c**) are reproduced and adapted from Kainrath et al. 2017 [13] under CC. In the above shown examples, light intensities of ~200 μW/cm^2 (blue), ~170 μW/cm^2 (green), and ~14 μW/cm^2 (red light) were applied for 8–16 h

3.4 Light Stimulation and Sample Preparation for WB Analysis

This section covers the sample preparation for SDS-PAGE and WB analysis with a focus on the special considerations when handling samples from cells that were transfected with Opto-RTKs. The analysis of the samples can then be performed according to any general WB protocol, depending on available equipment and the specific signaling pathway of interest.

1. Following medium change after transfection, wrap the dishes in foil individually and keep in the dark O/N. Both light-activated and light-inactivated receptors are treated the same because no transcriptional reporter is employed here.

2. The next day, stimulate cells in the dishes for desired amounts of time in a light incubator (*see* **Note 12**).
3. Immediately after stimulation, place dishes on ice.
4. Wash cells with 1 mL ice-cold PBS by pipetting against the wall of the dish, gently tilting the dish and aspirating from the edge without disturbing the cell layer.
5. Add 180 μL ice-cold RIPA buffer, scrape cells and transfer cells to pre-chilled microcentrifugation tubes. From here on, keep cells on ice.
6. Spin down at 4 °C at 20,000 × *g* for 20 min.
7. Transfer supernatant to fresh, pre-chilled tubes. Discard the pellet.
8. Mix 90 μL sample with 30 μL 4× Laemmli buffer.
9. Denature proteins at 95 °C for 5 min. After this step, samples can be frozen and stored at −20 °C for several days.
10. Spin down at 10,000 × *g* for 1 min.
11. Load duplicate gels with 30 μL of each sample per well (*see* **Note 13**) and proceed to perform SDS-PAGE and WB. Representative results are shown in Fig. 3b, c. The remainder of the samples can be frozen as backup and stored at −20 °C for several days.

4 Notes

1. Transcriptional reporters are available commercially, from researchers or through nonprofit plasmid depositories e.g., Addgene.org. The choice of reporter plasmid will depend on the interrogated signaling pathway and preferred readout. For instance, a serum response element reporter is suited for analysis of the MAPK/ERK pathway. Luciferase reporters provide robust luminescence readouts. For fluorescent reporter proteins, the wavelength used to excite the Opto-RTK should not overlap with the excitation of the reporter fluorophore, otherwise illumination may bleach the reporter and reduce signal. Careful choice of LSD and reporter allows for the design of dual-color fluorescent assays with minimal wavelength overlap [14]. Choosing two different reporter systems can allow multiplexing to investigate more than one pathway or to use one signaling and one viability reporter, for example, a pathway-activated firefly luciferase and a constitutively expressed *Renilla* luciferase.
2. A strong promoter, such as the cytomegalovirus (CMV) promoter, can potentially lead to overexpression of the receptor. A consequence of overexpression is high baseline activity, also for

corresponding control vectors (e.g., receptor without the LSD), resulting in a requirement for low vector amounts in transfections. A truncated version of the CMV promoter (CMVtr) can be useful to titrate expression levels [13, 14, 21]. As a rule of thumb, shorter, myristoylation-anchored constructs are at higher risk of being overexpressed, making CMVtr a suitable choice. P75-anchored or full-length constructs are larger proteins that require shuttling through the ER and Golgi network to the secretory pathway, and generally benefit from expression under the stronger full-length CMV promoter.

3. For luminescence assays, opaque/white bottom well plates can be used. Cell-culture grade plates for adherent cell culture are often pre-treated to facilitate cell attachment and thus do not require coating with PLO.

4. To stimulate cells in dishes, plates, as well as whole animals (e.g., zebrafish and *Drosophila*) with light, we repurpose consumer grade incubators, for example, designed for reptile breeding or as mini-fridges, which feature cooling/heating (but no CO_2 control). The LEDs can then simply be taped to the walls in parallel turns. LED strips can also be mounted in conventional cell line incubators with CO_2 control and humidification (Fig. 2c).

5. To control light intensity, the remote control provided with the LED strips can be used. However, many commercial LED strips achieve reduced intensity through frequency modulation (rapid flickering not visible to the human eye) rather than by reducing the power output. To lower intensity with continuous illumination, a dimmer should be used, or alternatively light absorption filters (e.g., neutral density filter foils obtained from the photography industry).

6. Alternatively, cells can be forward transfected by seeding at a slightly lower density (e.g., 30,000 cells per well for 96-well plates) in D5, incubating O/N and adding the transfection mix to them on the next day. In some cases, this can improve cell viability and transfection efficiency.

7. DNA amount and cell number can be scaled by surface area for larger assay formats. For reporter assays in the 96-well plate format, we use a maximum of 250 ng DNA with 1 μg PEI (1 μL of 1 mg/mL stock) per well. Our standard transfection to detect MAPK/ERK pathway consists of 150 ng pFR-LUC, 10 ng Elk trans-reporter (PathDetect System, Agilent Technologies), 10 ng *Renilla* viability reporter, and up to 30 ng receptor DNA per well. In initial studies with a new receptor, we test DNA amounts ranging from 0.1 ng to 30 ng, also in combination with strong and weak promoters (*see* **Note 2**).

DNA amounts for transfection and duration of stimulation can vary depending on the reporter gene assay. Generally, it is recommended to follow the plasmid supplier's requirements and protocols. For reporter gene assays, prepare all transfections in duplicate on two plates (one sample plate will be illuminated, the other will be the dark control). For WB, transfect cells in 35 mm dishes with a maximum of 2.5 μg receptor DNA and 12.5 μg PEI. Prepare one dish for each sample or point in a time course (e.g., dark control, 1 min illumination, 5 min illumination, 5 min illumination followed by 5 min dark recovery).

8. Depending on LSD, supplementation with an exogenous cofactor may be necessary. For CBDs, cells are grown in the presence of 10 μM AdoCbl for 1–2 passages before transfection for reporter gene assays. For WB the cofactor is added to the starve medium after transfection. The AdoCbl cofactor and supplemented cells should be handled under red light (650 nm) to minimize bleaching. For CPH1s, 10 μM PCB are added to the starve medium. PCB is extremely light sensitive and both the cofactor and supplemented cells should be handled exclusively under green light (550 nm) to avoid bleaching of the isolated cofactor and activation of the receptor. Not all suppliers of PCB provide material of high quality and testing of PCB from multiple suppliers is recommended.
9. We generally utilize 200 $\mu W/cm^2$ as an initial starting value for reporter gene assays employing an SRE luciferase reporter. Of note, CPH1s has been shown to be activated by intensities as low as 6.2 $\mu W/cm^2$ [14]. For CBDs, varying light intensity allows to titrate the receptor signal [13]. Light intensities can be measured with power meters, some of which may not be equipped with spectral analysis properties or filters to allow for the direct measurement of different colors of light. In such cases, a conversion factor may have to be applied. Refer to the manufacturer's specifications for details.
10. In our experience, 6–8 h stimulation is sufficient to obtain a readout for most reporter plasmids. Continuous illumination for up to 16 h (O/N) with red, green or blue light of 200 $\mu W/cm^2$ intensity has no significant impact on HEK293 cell viability. Other cell types might be sensitive to light exposure. In such cases, a pulsing regime (e.g., 15 s on, 45 s off) might be preferable. A short pulse of light is sufficient to activate light sensing domains and they will remain dimerized for up to several minutes (VfAU1-LOV, CPH1s).
11. For fluorescent readouts, note that many commercial well plates exhibit autofluorescence, for example, 384-well plates that we previously used (Cat Nr. #3712, Corning) upon

excitation with light <500 nm. The excitation and emission wavelengths for fluorescent reporter readout may have to be optimized to detect signals [22].

12. An example for a time course of light activation would be to place one dish each in the light for 0 min (dark control), 5 min, 15 min, 30 min, and 30 min light followed by 60 min dark. After WB analysis, these time points give an indication over signaling activity on the minute to hour timescale, as shown in the representative WB data in Fig. 3b.
13. We recommend performing WB analysis in duplicate blots to be able to stain against total target protein and against phosphorylated protein (e.g., pErk and total Erk when testing for the MAPK/ERK pathway). In both cases, include an antibody against a housekeeping protein (e.g., vinculin) as an internal standard. For quantification, determine the intensity of each band relative to internal standard for its respective lane. Then determine the ratio of phosphoprotein to total protein for each individual sample as shown in the representative WB data in Fig. 3c.

References

1. Lemmon MA, Schlessinger J (2010) Cell signaling by receptor tyrosine kinases. Cell 141 (7):1117–1134. https://doi.org/10.1016/j.cell.2010.06.011
2. Neben CL, Lo M, Jura N, Klein OD (2019) Feedback regulation of RTK signaling in development. Dev Biol 447(1):71–89. https://doi.org/10.1016/j.ydbio.2017.10.017
3. Simi A, Ibáñez CF (2010) Assembly and activation of neurotrophic factor receptor complexes. Dev Neurobiol 70(5):323–331. https://doi.org/10.1002/dneu.20773
4. Casaletto JB, McClatchey AI (2012) Spatial regulation of receptor tyrosine kinases in development and cancer. Nat Rev Cancer 12 (6):387–400. https://doi.org/10.1038/nrc3277
5. Robertson SC, Tynan JA, Donoghue DJ (2000) RTK mutations and human syndromes: when good receptors turn bad. Trends Genet 16(6):265–271. https://doi.org/10.1016/S0168-9525(00)02021-7
6. Goglia AG, Toettcher JE (2019) A bright future: optogenetics to dissect the spatiotemporal control of cell behavior. Curr Opin Chem Biol 48:106–113. https://doi.org/10.1016/j.cbpa.2018.11.010
7. Guglielmi G, Falk HJ, De Renzis S (2016) Optogenetic control of protein function: from intracellular processes to tissue morphogenesis. Trends Cell Biol 26(11):864–874. https://doi.org/10.1016/j.tcb.2016.09.006
8. Grusch M, Schelch K, Riedler R, Reichhart E, Differ C, Berger W, Ingles-Prieto A, Janovjak H (2014) Spatio-temporally precise activation of engineered receptor tyrosine kinases by light. EMBO J 33(15):1713–1726. https://doi.org/10.15252/embj.201387695
9. Kim N, Kim Jin M, Lee M, Kim Cha Y, Chang K-Y, Heo Won D (2014) Spatiotemporal control of fibroblast growth factor receptor signals by blue light. Chem Biol 21(7):903–912. https://doi.org/10.1016/j.chembiol.2014.05.013
10. Chang KY, Woo D, Jung H, Lee S, Kim S, Won J, Kyung T, Park H, Kim N, Yang HW, Park JY, Hwang EM, Kim D, Heo WD (2014) Light-inducible receptor tyrosine kinases that regulate neurotrophin signalling. Nat Commun 5:4057. https://doi.org/10.1038/ncomms5057
11. Bugaj LJ, Spelke DP, Mesuda CK, Varedi M, Kane RS, Schaffer DV (2015) Regulation of endogenous transmembrane receptors through optogenetic Cry2 clustering. Nat Commun 6:6898. https://doi.org/10.1038/ncomms7898
12. Duan L, Hope JM, Guo S, Ong Q, François A, Kaplan L, Scherrer G, Cui B (2018) Optical activation of TrkA signaling. ACS Synth Biol

7(7):1685–1693. https://doi.org/10.1021/acssynbio.8b00126

13. Kainrath S, Stadler M, Reichhart E, Distel M, Janovjak H (2017) Green-light-induced inactivation of receptor signaling using cobalamin-binding domains. Angew Chem Int Ed 56 (16):4608–4611. https://doi.org/10.1002/anie.201611998
14. Reichhart E, Ingles-Prieto A, Tichy A-M, McKenzie C, Janovjak H (2016) A phytochrome sensory domain permits receptor activation by red light. Angew Chem Int Ed 55 (21):6339–6342. https://doi.org/10.1002/anie.201601736
15. Li Y, Lee M, Kim N, Wu G, Deng D, Kim JM, Liu X, Heo WD, Zi Z (2018) Spatiotemporal control of TGF-β signaling with light. ACS Synth Biol 7(2):443–451. https://doi.org/10.1021/acssynbio.7b00225
16. Ramachandran A, Vizán P, Das D, Chakravarty P, Vogt J, Rogers KW, Müller P, Hinck AP, Sapkota GP, Hill CS (2018) TGF-β uses a novel mode of receptor activation to phosphorylate SMAD1/5 and induce epithelial-to-mesenchymal transition. ELife 7: e31756. https://doi.org/10.7554/eLife.31756
17. Sako K, Pradhan SJ, Barone V, Inglés-Prieto Á, Müller P, Ruprecht V, Čapek D, Galande S, Janovjak H, Heisenberg C-P (2016) Optogenetic control of nodal signaling reveals a temporal pattern of nodal signaling regulating cell fate specification during gastrulation. Cell Rep 16(3):866–877. https://doi.org/10.1016/j.celrep.2016.06.036
18. Muthuswamy SK, Gilman M, Brugge JS (1999) Controlled dimerization of ErbB receptors provides evidence for differential signaling by homo- and heterodimers. Mol Cell Biol 19 (10):6845–6857. https://doi.org/10.1128/MCB.19.10.6845
19. Polyethylenimine (PEI), linear (1 mg/mL) (2008). Cold Spring Harb Protoc 2008 (3): pdb.rec11323. doi:https://doi.org/10.1101/pdb.rec11323
20. Baker JM, Boyce FM (2014) High-throughput functional screening using a homemade dual-glow luciferase assay. J Vis Exp 88. https://doi.org/10.3791/50282
21. Watanabe N, Mitchison TJ (2002) Single-molecule speckle analysis of actin filament turnover in lamellipodia. Science 295 (5557):1083–1086. https://doi.org/10.1126/science.1067470
22. Ingles-Prieto A, Reichhart E, Muellner MK, Nowak M, Nijman SM, Grusch M, Janovjak H (2015) Light-assisted small-molecule screening against protein kinases. Nat Chem Biol 11(12):952–954. https://doi.org/10.1038/nchembio.1933
23. Bae JH, Boggon TJ, Tome F, Mandiyan V, Lax I, Schlessinger J (2010) Asymmetric receptor contact is required for tyrosine autophosphorylation of fibroblast growth factor receptor in living cells. Proc Natl Acad Sci U S A 107(7):2866–2871. https://doi.org/10.1073/pnas.0914157107
24. Takahashi F, Yamagata D, Ishikawa M, Fukamatsu Y, Ogura Y, Kasahara M, Kiyosue T, Kikuyama M, Wada M, Kataoka H (2007) AUREOCHROME, a photoreceptor required for photomorphogenesis in stramenopiles. Proc Natl Acad Sci U S A 104 (49):19625–19630. https://doi.org/10.1073/pnas.0707692104
25. Strauss HM, Schmieder P, Hughes J (2005) Light-dependent dimerisation in the N-terminal sensory module of cyanobacterial phytochrome 1. FEBS Lett 579 (18):3970–3974. https://doi.org/10.1016/j.febslet.2005.06.025
26. Jost M, Fernandez-Zapata J, Polanco MC, Ortiz-Guerrero JM, Chen PY, Kang G, Padmanabhan S, Elias-Arnanz M, Drennan CL (2015) Structural basis for gene regulation by a B12-dependent photoreceptor. Nature 526(7574):536–541. https://doi.org/10.1038/nature14950
27. Welm BE, Freeman KW, Chen M, Contreras A, Spencer DM, Rosen JM (2002) Inducible dimerization of FGFR1: development of a mouse model to analyze progressive transformation of the mammary gland. J Cell Biol 157 (4):703–714. https://doi.org/10.1083/jcb.200107119
28. Mohammadi M, Dikic I, Sorokin A, Burgess WH, Jaye M, Schlessinger J (1996) Identification of six novel autophosphorylation sites on fibroblast growth factor receptor 1 and elucidation of their importance in receptor activation and signal transduction. Mol Cell Biol 16 (3):977–989

Chapter 17

All-Optical Miniaturized Co-culture Assay of Voltage-Gated Ca^{2+} Channels

Viviana Agus and Harald Janovjak

Abstract

Light-activated proteins enable the reversible and spatiotemporal control of cellular events in optogenetics. Optogenetics is also rapidly expanding into the field of drug discovery where it provides cost-effective and noninvasive approaches for cell manipulation in high-throughput screens. Here, we present a prototypical cell-based assay that applies Channelrhodopsin2 (ChR2) to recapitulate physiological membrane potential changes and test for voltage-gated ion channel (VGIC) blockade. ChR2 and the voltage-gated Ca^{2+} channel 1.2 ($Ca_V1.2$) are expressed in individual HEK293 cell lines that are then co-cultured for formation of gap junctions and an electrical syncytium. This co-culture allows identification of blockers using parallel fluorescence plate readers in the 384-well plate format in an all-optical mode of operation. The assay is transferable to other VGICs by modularly combining new and existing cell lines and potentially also to other drug targets.

Key words Optogenetics, High-throughput screening, 384-well plate, Voltage-gated ion channel, Syncytium, $Ca_V1.2$, Channelrhodopsin, All-optical, FLIPR

1 Introduction

Ion channels are pore-forming membrane proteins that act as biological gates and control the movement of ions across cell membranes. Ion channels play a key role in tuning cell excitability, for example, by establishing cell type-specific resting membrane potentials, and in shaping induced electrical activity, for example, during intercellular synaptic communication or intracellular action potential propagation [1, 2]. Latter process is catalyzed by VGICs that open upon a transient increase in cell membrane potential and, not surprisingly, are the targets of drugs in several neuronal, cardiac and muscular dysfunction [3, 4]. Nevertheless, VGICs still remain largely underexplored in therapy, in part due to a lack of physiological and cost-effective interrogation methods that are suited for large-scale compound testing in early drug discovery [5]. In principle, the fast and transient membrane potential changes that open

Dominik Niopek (ed.), *Photoswitching Proteins: Methods and Protocols*, Methods in Molecular Biology, vol. 2173, https://doi.org/10.1007/978-1-0716-0755-8_17,

VGICs can be delivered by electrophysiology (e.g., the patch-clamp technique) in a reversible manner, but high experimental cost and low-throughput render this approach not suitable for large-scale screening campaigns (e.g., for testing hundreds of thousands of compounds). On the other hand, cost-effective fluorescent ion flux assays utilize irreversible chemical stimulants to elicit depolarization or hyperpolarization (e.g., K^+ challenges or ionophores) but therefore poorly recapitulate some physiological mechanisms of VGIC activation.

Recently, the field of optogenetics has provided light-activated and genetically encoded means to apply transient voltage changes in living cells as a powerful biological substitute of direct electrical stimulation [6, 7]. Light-induced depolarization is achieved by cation-permeable ion channels [8, 9] that are well-suited to mimic physiological voltage changes: they can be switched on rapidly (e.g., the ChR2 L132C and T159C variants [10, 11] within milliseconds) and their activation does not require any physical contact or cytoplasm dialysis, in contrast to electrophysiological or chemical techniques. Because of these properties, optogenetic tools have been recently integrated in drug discovery pipelines that target VGICs and also other membrane receptor drug targets, such as kinase-linked receptors [12–14]. In their most sophisticated implementation, these methods allow all-optical modes of operation in which light is used as the stimulus and the readout [15–21] thereby obviating the use of chemicals and limiting the sample handling and intervention steps.

In this section, we present a method to activate a VGIC by ChR2 in a co-culture cell system. This method uses HEK293 cells that endogenously express connexin-43 (Cx43) hemichannels, which in dense cultures will form intercellular gap junction connections in syncytia permitting electrical transfer from a light-activated "driver" cell population to a voltage-sensitive "target" cell population. This syncytium concept has been used to study in vivo neuronal networks [22] as well as to control in vitro excitability [23], either based on neurotransmitter-mediated synapses or gap junctions [24]. Specifically, we here describe how to obtain syncytial cell populations in 384-well plates from modularly combined cell lines, ultimately achieving simultaneous light activation and imaging of multiple wells using parallel plate readers such as the Molecular Devices Fluorescence Imaging Plate Reader (FLIPR). Two distinct stable clonal cell lines are generated one of which expresses ChR2 and the other $K_{IR}2.3$ and $Ca_V1.2$ subunits (Fig. 1). The ChR2 cell line serves for light-induced depolarization, whereas the $K_{IR}2.3/Ca_V1.2$ line is the voltage-sensitive target cell line. In this cell line, the inward rectifier K^+ channel $K_{IR}2.3$ lowers the resting membrane potential from the ~−25 mV typical for wild type HEK293 cells [25] toward the K^+ equilibrium membrane potential (typically −80 to −90 mV). This relatively negative

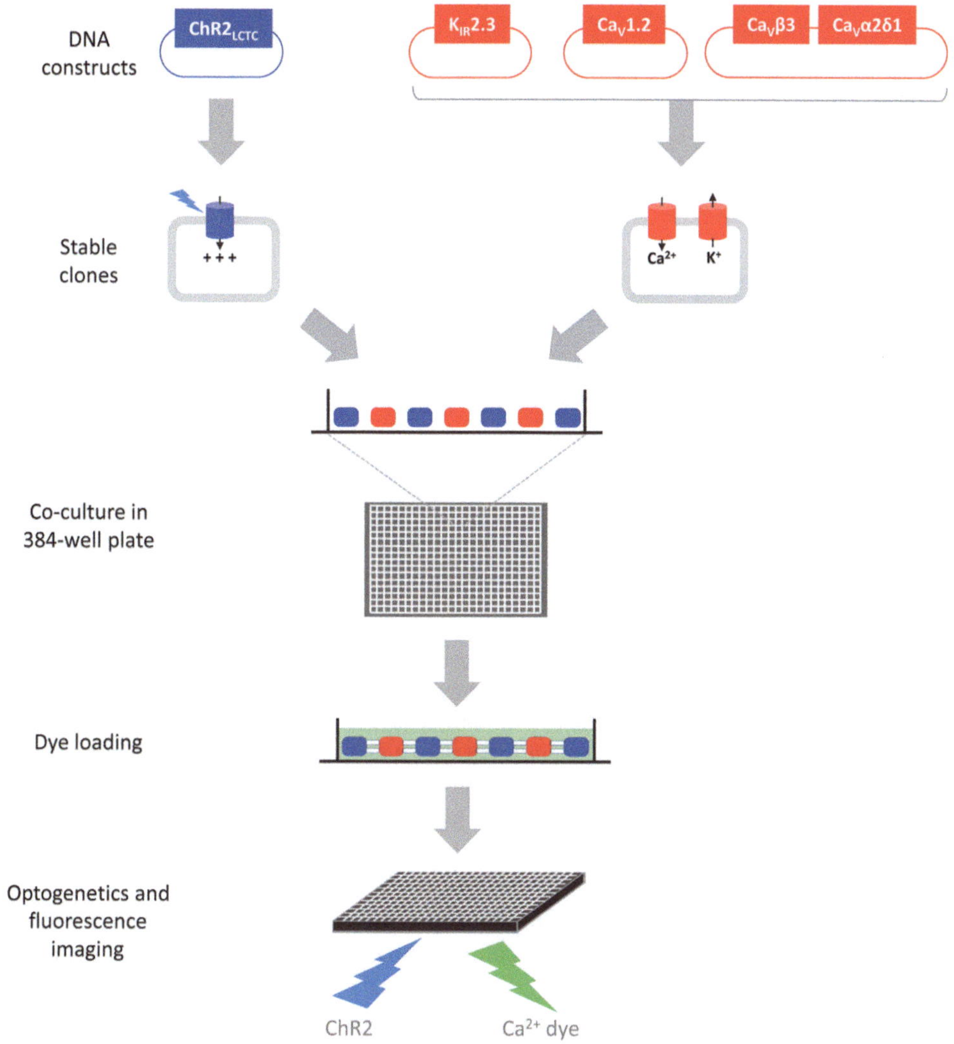

Fig. 1 Schematic representation of the all-optical co-culture assay in multiwell plates. Genes encoding the optogenetic tool and VGIC with accessory subunits are stably expressed in distinct HEK293 cell populations. Pure clones are isolated and cultivated separately, then mixed together during seeding into 384-well plates. A 24 h co-culture step enables the formation of gap junction connections between the light-induced depolarization cells and the VGIC target cells. The co-culture is then prepared for signal detection (i.e., loaded with a green-fluorescent Ca^{2+} sensitive organic indicator). ChR2 photocurrent is triggered by the blue light flashes applied for imaging of the green fluorescent Ca^{2+} indicator, resulting in simultaneous stimulation and detection of the voltage-gated Ca^{2+} current in an all-optical experiment

membrane potential maintains the Ca_V channel in the closed state, avoiding cytotoxic effects due to Ca^{2+} overload and allowing the opening upon the rapid light-induced depolarization stimulus. We will discuss how to configure light stimulation and recording parameters in the FLIPR in order to activate ChR2 and to monitor Ca_V activity through a fluorescent ion indicator, providing large-scale drug screening capabilities with physiological and low-cost optical

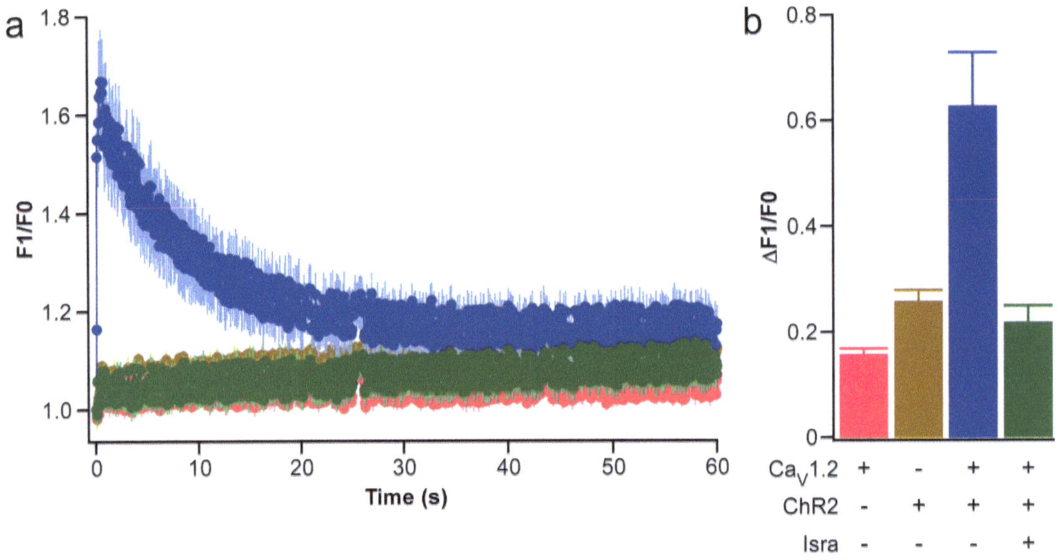

Fig. 2 Optical control of Ca$_V$1.2 Ca^{2+} flux in a ChR2 co-culture. (**a**) HEK293 cells expressing ChR2 LCTC were seeded in 384-well plates with HEK293 cells expressing Ca$_V$1.2. After 24 h of co-culture, cells were loaded with Fluo-8,NW Ca^{2+} sensitive dye in the absence (blue traces) or presence (green traces) of 10 μM isra. Mono-culture of HEK293 cells expressing ChR2 LCTC (brown traces) or Ca$_V$1.2 (red traces) were also included as controls (these cells were not treated with isra). Fluorescence changes are monitored upon blue light illumination. Average kinetics (*F*1/*F*0 baseline-corrected values) from six replicate wells for each condition are shown with standard deviation shown as shaded sticks (*see* **Note 19**). (**b**) Analysis of maximal fluorescence changes for the experiment shown in **a** (maximal − minimal *F*1/*F*0 values over the 60 s measurement, averages ± standard deviations for six replicate wells)

control. We demonstrate the performance of the approach using isradipine (isra) [26], a prototypical dihydropyridine class blocker of Ca$_V$s (Fig. 2). The target channel Ca$_V$1.2 serves as an example, and the method can be extended to other VGICs, drug targets and optogenetic tools [12–14].

2 Materials

2.1 Molecular Biology

1. Chemically competent *E. coli* cells.
2. Glass rod or beads and sterile loops.
3. Clear glass tubes (autoclaved and capped).
4. LB culture medium: 0.5% yeast extract, 1% tryptone, 1% NaCl, pH 7.2. Medium is sterile filtered and autoclaved prior to the addition of antibiotics.
5. Block heater or water bath for cell heat shock (42 °C).
6. Shaker for incubation (225 rpm, 37 °C).
7. S.O.C. culture medium: 0.5% yeast extract, 2% tryptone, 10 mM NaCl, 2.5 mM KCl, 10 mM $MgCl_2$, and 10 mM

$MgSO_4$. The medium is autoclaved prior to the addition of glucose to a final concentration of 20 mM, followed by sterile filtering.

8. 10 cm diameter LB agar plates supplemented with appropriate antibiotic selection (kanamycin for pIRES2-EGFP vector, zeocin for pBudCE4.1, and ampicillin for all other vectors).
9. Mammalian codon usage optimized gene encoding ChR2 (*Chlamydomonas reinhardtii* ChR2/chlamyopsin4 light-gated ion channel; XM_001701673.1; amino acids 1 to 309, with L132C and T159C substitutions (ChR2 LCTC); *see* **Note 1**), and human genes encoding $K_{IR}2.3$ (NM_004981.1) (*see* **Note 2**), $Ca_V1.2$ α-1 subunit (NM_001129843.1) and Ca_V accessory subunits ($Ca_v\beta3$, CACNB3, NM_000725.3 and $Ca_V\alpha2\delta1$, CACNA2D1, NM_000722.2) cloned in mammalian constitutive expression vectors (pcDNA3.1Hygro, pcDNA3, pIRES2-EGFP and pBudCE4.1, respectively; *see* **Note 3**) and sequence-verified.
10. Plasmid DNA purification kit.
11. Spectrophotometer for DNA quantification and assessment of purity (absorbance measurements at 260 and 280 nm).
12. Autoclaved water.

2.2 Cell Culture and Transfection

1. Cell culture facility equipped with laminar flow hood (in which the light can be switched off), water bath (37 °C), incubator (37 °C, 5% CO_2), and benchtop centrifuge (up to 180 × *g*).
2. HEK293 cells (ATCC#CRL-1573) or derivatives cultured according to the provider's recommendations (*see* **Note 4**).
3. Dulbecco's phosphate buffered saline (PBS) without Ca^{2+} and Mg^{2+}, stored at 4 °C and equilibrated at 37 °C before use.
4. Trypsin–EDTA solution, stored at 4 °C and briefly equilibrated at 37 °C before use.
5. 500 mL 0.22 μm cellulose nitrate filter system.
6. T75 (75 cm^2) polystyrene cell culture flasks with vent cap.
7. 96-well polystyrene cell culture plates.
8. Cell counting chamber (e.g., BECKMAN COULTER Z1 Particle Counter).
9. Electroporation cuvettes (Gene Pulser, 0.4 cm electrode gap, Bio-Rad).
10. Opti-MEM I reduced serum medium.
11. HEK293 complete medium: 500 mL of minimum essential medium eagle with Earle's salts (EMEM) supplemented with 50 mL fetal bovine serum, 1% penicillin-streptomycin (P/S) and 2 mM ultraglutamine (Lonza), sterile filtered, stored at 4 °C and equilibrated at 37 °C before use.

12. G418, hygromycin and zeocin antibiotic stock solutions as supplied by commercial vendors (e.g., Invivogen).
13. Test for mycoplasma contamination using a kit or a commercial service.

2.3 Cell Seeding and Dye Loading

1. 384-well black wall clear bottom poly-D-lysine–coated polystyrene assay plates (Thermo Scientific Matrix, GR 4332 CPL-10).
2. Electronic multichannel pipette with 125 μL sterile filter tips.
3. Black microplate lids (Greiner Bio-One).
4. Aluminum foil or opaque Eppendorf tubes for light-sensitive solutions.
5. All-trans retinal (ATR; Sigma-Aldrich, 100 mM stock solution prepared in DMSO) stored in aliquots at −20 °C protected from light (*see* **Note 5**).
6. DMSO stored at room temperature (RT).
7. K0Na150 buffer: 150 mM NaCl, 2 mM $CaCl_2$, 10 mM HEPES, 10 mM Glucose, pH 7.4; sterile filtered and stored at 4 °C (*see* **Note 6**).
8. ScreenQuest Fluo-8 No Wash Ca^{2+} dye (Fluo-8,NW; AAT Bioquest) dissolved in DMSO following the manufacturer's instructions (200 μL DMSO for one vial of component A) and stored at −20 °C protected from light (*see* **Note 7**).
9. Isra (10 mM stock solution prepared in DMSO) stored in aliquots at −20 °C.

2.4 Assays Using FLIPR

1. FLIPR Tetra (Molecular Devices) equipped with aequorin option ICCD camera, blue LED excitation banks (470–495 nm), 515–575 nm emission filter and ScreenWorks (version 4.0.0.30) analysis software (*see* **Note 8**).
2. Laboratory with reduced ambient light (shaded windows and Na^+-vapor yellow lamps).

3 Methods

3.1 Plasmid Transformation, Amplification, and Fresh DNA Isolation

1. On the day of transformation, warm up LB agar plates (one for each plasmid) at 37 °C for 30 min.
2. Transform each plasmid into one sample of chemically competent *E. coli* cells, following the manufacturer's instructions.
3. Spread bacteria evenly on plates with a sterile glass rod or beads, then place plates at 37 °C upside down overnight.
4. On the following day, pick one colony from each plate using sterile disposable loops and inoculate cultures in glass tubes

containing 3 mL of LB medium with proper selective antibiotic (kanamycin, zeocin or ampicillin). Grow cultures overnight at 37 °C with 225 rpm shaking.

5. On the following day, extract plasmid from bacteria using the plasmid DNA purification kit following the manufacturer's instructions and validate DNA purity by 260/280 nm absorbance measurement, aiming for a ratio of at least 1.8.

3.2 Cell Culture, Transfection and Clone Generation

Perform the following procedures in a laminar flow hood under normal working light conditions.

1. Culture cells in a humidified incubator at 37 °C in 5% CO_2 atmosphere with a sub-cultivation ratio of 1:6 twice a week in T75 flasks.
2. Transfect cells when they reach 85–90% confluency. On the day of transfection, wash cells with 5 mL PBS, detach the cells with 1 mL trypsin (5-10 min at 37 °C in 5% CO_2) and dilute the cells by adding 5 mL Opti-MEM I medium.
3. Count cells and resuspended 3×10^6 cells in 800 μL of Opti-MEM I medium for centrifugation (5 min, $180 \times g$).
4. Transfer cells to electroporation cuvettes and add 10 μg of plasmid DNA. Incubate at RT for 5 min.
5. Electroporate cells with BIORAD Gene Pulser II (300 V, 950 μF pulse).
6. Seed transfected cells in a T75 flask with 10 mL of medium.
7. Forty-eight hours after transfection, exchange culture medium with medium containing selective antibiotics; refresh antibiotic selective medium every 2–3 days for about 10–14 days, until pools of stably transfected cells are obtained (*see* **Note 9**).
8. Maintain the ChR2 cell line stable pool in hygromycin (20 μg/mL) and the $K_{IR}2.3$ cell line stable pool in G418 (0.4 mg/mL).
9. Following selection, transfect $Ca_V1.2$ and the Ca_V accessory subunits (using the above described procedures) and select with zeocin (20 μg/mL). Maintain the obtained $K_{IR}2.3/Ca_V$ stable pool in G418 (0.2 mg/mL) and zeocin (10 μg/mL).
10. Clonal cell lines can be obtained by seeding stable pools in 96-well plates at 3 cells/well and selecting clones with functional FLIPR tests.

3.3 Cell Seeding

Perform the following procedures in a laminar flow hood under normal working light conditions unless indicated otherwise.

1. Harvest the ChR2 cell line and the $K_{IR}2.3/Ca_V$ cell line from T75 flasks that reached 85–90% confluency.
2. Wash each cell line with 5 mL of PBS, detach cells with 1 mL trypsin (5–10 min at 37 °C in 5% CO_2) and then dilute cells by adding 5 mL medium containing P/S but no selection antibiotics.

3. Determine the cell number using the counting chamber.
4. Before proceeding with the next steps, turn the light in the laminar flow hood off.
5. Dilute the ChR2 cell line to a density of 10,000 cells in 20 μL and add ATR to final concentration of 5 μM.
6. Dilute the $K_{IR}2.3/Ca_V$ cell line to a density of 10,000 cells in 20 μL and add ATR to final concentration of 5 μM.
7. Use the sixteen-channel automatic pipet to dispense cells (10,000 ChR2 cells and 10,000 $K_{IR}2.3/Ca_V$ cells in each well, 12 replicate wells) in poly-D-lysine–coated black-clear 384-well plates (*see* **Note 10**).
8. As negative controls, seed 20,000 ChR2 or $K_{IR}2.3/Ca_V$ cells as mono-cultures (12 replicate wells each).
9. Wrap the plates with aluminum foil or black plate lids to avoid light entering from both above or below the plates.
10. Culture the cells for 24 h at 37 °C in 5% CO_2 protected from light.

3.4 Loading with Ca^{2+}-Sensitive Fluorescent Dye

Perform the following procedures with the light in the laminar flow hood off.

1. On the day of experiment, prepare a 0.5× loading solution of Fluo-8,NW by dissolving 10 μL of component A with 9.5 mL of K0Na150 buffer and adding 500 μL of component B (*see* **Note 11**). Also, add ATR to a final concentration of 5 μM. As a test compound, include isra to a final concentration of 10 μM in the loading solution of 6 replicate wells for each cell clone and the co-culture (*see* **Note 12**).
2. With the room light switched off and only dim ambient light in the room, remove the culture medium from the 384-well plate by manual overthrow and tap on a paper towel.
3. Add 20 μL of loading solution to each well using the 16-channel pipette.
4. Wrap plates with aluminum foil or use black plate lids.
5. Incubate cells for 1 h at 37 °C in 5% CO_2 (*see* **Note 13**).

3.5 All-Optical Test of a Ca^{2+} Channel Blocker

The FLIPR instrument should be located in a room with reduced light (Na^+-vapor yellow lamps and shaded windows). Plates are kept wrapped in aluminum foil or black lids when they are handled outside the FLIPR. Aluminum foil or black lids are removed only when the plate is inside the instrument, that is, immediately before closing the instrument door and starting the measurement. Settings and items in the software are capitalized in the following section.

1. Using the ScreenWorks software, create a NEW PROTOCOL FILE.
2. In the protocol SETTINGS, set a FLUORESCENCE READING MODE using EXCITATION WAVELENGHT 470–495 nm, EMISSION WAVELENGHT 515–575 nm, GAIN 2000, EXPOSURE TIME 0.01 s, EXCITATION INTENSITY 100%, GATE OPEN 20% (*see* **Note 14**). This setting will achieve both the ChR2 light stimulation and the Fluo-8,NW fluorescent dye imaging. In particular, ChR2 will be repetitively activated throughout the measurement, while $Ca_V1.2$ will rapidly open upon first light depolarization and then shift into inactivated refractory state (*see* **Notes 15** and **16**).
3. In the protocol SETTINGS, choose ASSIGN 384-well assay plate to READ PLATE POSITION and configure DATA FILE NAME and location.
4. In the ANALYSIS window, choose CONFIGURE GROUPS, then ADD NEW GROUP and in the GROUP window assign GROUP NAME and color, then click on the well positions to assign group features to the corresponding replicate wells.
5. In the ANALYSIS window, configure RESPONSE OVER BASELINE correction (BASELINE 1 START=read 1; BASELINE END=read 2). This will display fluorescence changes as $Fl/F0$ values (as shown in Fig. 2a), where Fl is the RELATIVE FLUORESCENCE UNITS value of each read and $F0$ is the mean of RELATIVE FLUORESCENCE UNITS value from BASELINE 1 START to BASELINE END reads (*see* **Note 17**).
6. Add a READ PROCESS with a FIRST INTERVAL of 0.066 s READ TIME INTERVAL and 900 NUMBER OF READS. This will result in about 60 s total reading time (*see* **Note 18**).
7. Select RUN EXPERIMENT and save the data file in the *.fmd format.
8. In the ANALYSIS window, activate GROUP SELECTION MODE. From the SELECT GROUPS function choose the groups to be visualized and then select AVERAGE and SHOW STANDARD DEVIATION in DETAIL GRAPH (*see* **Note 19**).
9. In the FILE menu choose EXPORT FILE, define FILE EXPORT DIRECTORY, then in STATISTICS menu check ENABLE and select MAX-MIN as COMPUTE, from START READ 1 to END READ 900. Typical analyzed values are shown in Fig. 2b expressed as averages over six replicate wells.

4 Notes

1. Other ChR2 variants can be used in place of ChR2 LCTC. For instance, from our experience both the T159C fast kinetics variant and the D156A step-function opsin can efficiently induce light depolarization in in vitro ion flux assays.
2. $Ca_V1.2$ belongs to "high voltage-gated" Ca^{2+} channels that open at around −40 mV [27]. Thus the resting membrane potential of nonexcitable cells, such as the HEK293 cells, should be lowered to keep these channels closed and to study their complete activation window. A suited hyperpolarizing agent is the $K_{IR}2.3$ inward rectifier K^+ channel used in this example, or other K^+ channels with similar hyperpolarizing properties (e.g., $K_{IR}2.1$ or other members of the K_{IR} family). Expression of this additional agent is likely generally required when studying most human VGICs but not for voltage-insensitive drug targets.
3. pBudCE4.1 is an expression vector that permits constitutive expression of two genes (here, $Ca_v\beta3$, CACNB3, NM_000725.3, and $Ca_V\alpha2\delta1$). To further reduce possible cytotoxicity of Ca^{2+} influx through $Ca_V1.2$ channel, an inducible expression system could be used for the expression of $Ca_V1.2$ α-1 subunit.
4. Through endogenous expression of Cx43 hemichannels [28], HEK293 cells spontaneously form gap-junction connection when cultured at sufficient density to enable cell contact (e.g., 20,000 cells in a well of a 384-well plate with 10 mm^2 growth area). In case that cell types that do not express endogenous connexins are used (e.g., HeLa cells [29]), Cx43 or other subtypes could be recombinantly expressed to enable gap junction formation.
5. ATR is the ChR2 chromophore cofactor [30] and we found that 5 μM concentration provided the optimal light depolarizing activity of ChR2 and no cytotoxicity. We also tested 2 and 10 μM in HEK293 cells with similarly good results. The concentration of ATR should be adjusted depending on the cell line used, as different cell lines may require different supplementation or might be differently sensitive to ATR or to the DMSO used as solvent. Other optogenetic tools may require other cofactors, such as linear tetrapyrroles in the case of phytochromes [21].
6. The K0Na150 buffer used in this example was designed to have a minimal extracellular concentration of K^+. Because of $K_{IR}2.3$ expression, the low extracellular K^+ concentration will drive the membrane voltage near to the K^+ equilibrium potential. In our experience, this buffer showed good functionality of both

ChR2 light-induced depolarization and $Ca_V1.2$ Ca^{2+} current. Other buffers such as the Tyrode's buffer can also be used with similar results. For each combination of optogenetic tool and VGIC different assay buffers should be tested in order to find the optimal condition for both tool and target function.

7. Alternative fluorescent Ca^{2+} indicators can be used (e.g., Fluo-4 or Cal-520) or also genetically encoded Ca^{2+} indicators (e.g., GCaMP). These indicators have spectral properties similar to the Fluo-8,NW indicator used in our example and can be measured with the same excitation wavelength and emission wavelength used in this example. For each indicator, the specific parameters GAIN and GATE OPEN % might be set in order to obtain relative fluorescence values within the dynamic range of FLIPR instrument. In case Ca^{2+} indicators with different spectral properties are used (e.g., Fura-2 dyes or RCaMP indicators), a DUAL READ Fluorescence protocol could be designed where Read_Mode_1 will activate the ChR2 as in this example and Read_Mode_2 will monitor indicator fluorescence.
8. The choice of optogenetic tool can be based on the intensity of the light source used for excitation (in the case of FLIPR, the intensity is ~0.07 mW/mm^2 [13]). Channelrhodopsins with low light sensitivity, such as wild type ChR2 [30], might be not efficiently activated by FLIPR, and this may also apply to other optogenetic tools [13].
9. The optogenetic tool and the voltage-gated target can be either stably expressed as in the described example or transiently transfected in separate cell lines [29]. Transient expression has been previously applied in other all-optical drug screens [18, 19].
10. For each combination of optogenetic tool and VGIC, the response obtained with this co-culture approach can be compared with the response obtained with a standard coexpression approach (i.e., transfecting the optogenetic tool and VGIC into the same cell line). In some cases, we found that the co-culture approach results in lower signal-to-noise ratio, whereas in other cases targeting of multiple genes into distinct cell lines resulted in much more efficient expression and increased functionality.
11. We found that Fluo-8,NW dye and component B can be used at half concentration (0.5× instead of the 1× suggested by manufacturer) with no significant reduction of signal amplitude and reproducibility.
12. In the described example, the $Ca_V1.2$ blocker was included in the Ca^{2+} dye loading solution, thus its effect is tested after 1 h incubation. Shorter incubation times can be tested by adding

the blocker using manual pipetting or using a 384-tip dispensing head (such as the one included in the FLIPR). For instance, 5 μL/well of blocker (prepared 5× concentrated in K0Na150 buffer) could be added to the 20 μL/well of dye loading solution just 5 or 10 min prior to running the FLIPR optical stimulation and Ca^{2+} imaging. It should be noticed that this step should be done protected from light (e.g., in a room with reduced illumination and/or with Na^{+}-vapor lights, or inside the FLIPR).

13. Dye loading can be performed either at 37 °C or RT for 45–90 min without significantly affecting the assay performance.
14. The optimal values for EXCITATION INTENSITY % and GATE OPEN % should be determined for each cell line, optogenetic tool, and fluorescent indicator used. In particular, EXCITATION INTENSITY % will affect both the optogenetic tool and the fluorescent indicator activation, while GATE OPEN % will affect only the detection of the fluorescent indicator response.
15. ChR2 is a blue light activated cation channel, mainly permeable to Na^{+} and K^{+} [31], whereas $Ca_V1.2$ is selectively permeable to Ca^{2+}. This means that the Ca^{2+} influx measured in the ChR2-Ca_V syncytium will be mediated almost exclusively by $Ca_V1.2$ conductance.
16. For combinations of optogenetic tools and voltage-gated targets different from the ones in this example, it is advisable to validate the response of the individual cell lines to establish the contribution of each protein to the signal of the co-culture assay.
17. For a better comparison between different experiments and/or adjacent wells, we suggest to apply the RESPONSE OVER BASELINE correction yielding $F1/F0$ values (Fig. 2), in order to correct the data for slight differences in cell density or dye uptake that can affect the absolute fluorescence value measured by the instrument. This correction also enables comparing the fluorescence fold change obtained at the FLIPR with other fluorescence readers with different dynamic ranges. If this correction is not desired, data file can be also processed with SUBTRACT BIAS ON SAMPLE 1 or no corrections, in order to visualize the RELATIVE FLUORESCENCE UNITS counted by the FLIPR camera.
18. We recommend to keep the sampling rate of the imaging as high as possible (frequency of at least 10 Hz) to track the fast Ca^{2+} rise upon light stimulation.

19. Fluorescence kinetics can be also inspected as overlay of single replicate wells by choosing the WELL SELECTION MODE in the ANALYSIS WINDOW and by selecting all the wells to be visualized in the DETAIL GRAPH. This visualization allows for identifying possible bad replicates where an outlier response is obtained due to technical effects. These outlier wells will be not selected when assigning wells to correspondent GROUPS.

References

1. Mayer ML (2005) Glutamate receptor ion channels. Curr Opin Neurobiol 15:282–288
2. Bezanilla F (2005) Voltage-gated ion channels. IEEE Trans Nanobioscience 4:34–48
3. Imbrici P, Nicolotti O, Leonetti F, Conte D, Liantonio A (2018) Ion channels in drug discovery and safety pharmacology. Methods Mol Biol 1800:313–326
4. Zamponi GW (2016) Targeting voltage-gated calcium channels in neurological and psychiatric diseases. Nat Rev Drug Discov 15:19–34
5. Garcia ML, Kaczorowski GJ (2016) Ion channels find a pathway for therapeutic success. Proc Natl Acad Sci U S A 113:5472–5474
6. Deisseroth K (2015) Optogenetics: 10 years of microbial opsins in neuroscience. Nat Neurosci 18:1213–1225
7. Reiner A, Isacoff EY (2013) The brain prize 2013: the optogenetics revolution. Trends Neurosci 36:557–560
8. Fenno L, Yizhar O, Deisseroth K (2011) The development and application of optogenetics. Annu Rev Neurosci 34:389–412
9. Zhang F, Aravanis AM, Adamantidis A, de Lecea L, Deisseroth K (2007) Circuit-breakers: optical technologies for probing neural signals and systems. Nat Rev Neurosci 8:577–581
10. Pan ZH, Ganjawala TH, Lu Q, Ivanova E, Zhang Z (2014) ChR2 mutants at L132 and T159 with improved operational light sensitivity for vision restoration. PLoS One 9:e98924
11. Berndt A et al (2011) High-efficiency channelrhodopsins for fast neuronal stimulation at low light levels. Proc Natl Acad Sci U S A 108:7595–7600
12. Zhang H, Cohen AE (2017) Optogenetic approaches to drug discovery in neuroscience and beyond. Trends Biotechnol 35:625–639
13. Agus V, Janovjak H (2017) Optogenetic methods in drug screening: technologies and applications. Curr Opin Biotechnol 48:8–14
14. Song C, Knopfel T (2016) Optogenetics enlightens neuroscience drug discovery. Nat Rev Drug Discov 15:97–109
15. Agus V, Picardi P, Redaelli L, Scarabottolo L, Lohmer S (2018) Three-dimensional control of ion channel function through optogenetics and co-culture. SLAS Discov 23:102–108
16. Zhang H, Reichert E, Cohen AE (2016) Optical electrophysiology for probing function and pharmacology of voltage-gated ion channels. Elife 5:e15202
17. Klimas A et al (2016) OptoDyCE as an automated system for high-throughput all-optical dynamic cardiac electrophysiology. Nat Commun 7:11542
18. Prigge M, Rosler A, Hegemann P (2010) Fast, repetitive light-activation of CaV3.2 using channelrhodopsin 2. Channels (Austin) 4:241–247
19. Ingles-Prieto A et al (2015) Light-assisted small-molecule screening against protein kinases. Nat Chem Biol 11:952–954
20. Streit J, Kleinlogel S (2018) Dynamic all-optical drug screening on cardiac voltage-gated ion channels. Sci Rep 8:1153
21. Reichhart E, Ingles-Prieto A, Tichy AM, McKenzie C, Janovjak H (2016) A phytochrome sensory domain permits receptor activation by red light. Angew Chem Int Ed Engl 55:6339–6342
22. Kim CK, Adhikari A, Deisseroth K (2017) Integration of optogenetics with complementary methodologies in systems neuroscience. Nat Rev Neurosci 18:222–235
23. Jia Z et al (2011) Stimulating cardiac muscle by light: cardiac optogenetics by cell delivery. Circ Arrhythm Electrophysiol 4:753–760
24. Saez JC, Retamal MA, Basilio D, Bukauskas FF, Bennett MV (2005) Connexin-based gap junction hemichannels: gating mechanisms. Biochim Biophys Acta 1711:215–224
25. Hsu K, Han J, Shinlapawittayatorn K, Deschenes I, Marban E (2010) Membrane potential depolarization as a triggering mechanism for Vpu-mediated HIV-1 release. Biophys J 99:1718–1725
26. Catterall WA, Perez-Reyes E, Snutch TP, Striessnig J (2005) International Union of

Pharmacology. XLVIII. Nomenclature and structure-function relationships of voltage-gated calcium channels. Pharmacol Rev 57:411–425

27. Ravindran A, Kobrinsky E, Lao QZ, Soldatov NM (2009) Functional properties of the CaV1.2 calcium channel activated by calmodulin in the absence of alpha2delta subunits. Channels (Austin) 3:25–31
28. Gemel J, Valiunas V, Brink PR, Beyer EC (2004) Connexin43 and connexin26 form gap junctions, but not heteromeric channels in co-expressing cells. J Cell Sci 117:2469–2480
29. Haq N et al (2013) A high-throughput assay for connexin 43 (Cx43, GJA1) gap junctions using codon-optimized aequorin. Assay Drug Dev Technol 11:93–100
30. Nagel G et al (2003) Channelrhodopsin-2, a directly light-gated cation-selective membrane channel. Proc Natl Acad Sci U S A 100:13940–13945
31. Deisseroth K, Hegemann P (2017) The form and function of channelrhodopsin. Science 357:6356

Chapter 18

Optogenetics and CRISPR: A New Relationship Built to Last

Jan Mathony, Mareike D. Hoffmann, and Dominik Niopek

Abstract

Since the breakthrough discoveries that CRISPR-Cas9 nucleases can be easily programmed and employed to induce targeted double-strand breaks in mammalian cells, the gene editing field has grown exponentially. Today, CRISPR technologies based on engineered class II CRISPR effectors facilitate targeted modification of genes and RNA transcripts. Moreover, catalytically impaired CRISPR-Cas variants can be employed as programmable DNA binding domains and used to recruit effector proteins, such as transcriptional regulators, epigenetic modifiers or base-modifying enzymes, to selected genomic loci. The juxtaposition of CRISPR and optogenetics enables spatiotemporally confined and highly dynamic genome perturbations in living cells and animals and holds unprecedented potential for biology and biomedicine.

Here, we provide an overview of the state-of-the-art methods for light-control of CRISPR effectors. We will detail the plethora of exciting applications enabled by these systems, including spatially confined genome editing, timed activation of endogenous genes, as well as remote control of chromatin–chromatin interactions. Finally, we will discuss limitations of current optogenetic CRISPR tools and point out routes for future innovation in this emerging field.

Key words CRISPR-Cas, Optogenetics, Photoreceptor, Genome editing, anti-CRISPR

1 Introduction

When Emmanuelle Charpentier, Jennifer Doudna, and colleagues demonstrated in 2012 that CRISPR-Cas9 is an easily programmable, RNA-guided DNA endonuclease [1], they kicked off a revolution in the biosciences. Subsequent reports published by the Feng Zhang and George Church labs showed that CRISPR-Cas9 can be adapted for genome editing in human cells [2, 3] and numerous research groups around the world soon transferred this technology to other cell types and species. The power of CRISPR lies in its simplicity, as it requires only two components: A protein, the CRISPR-Cas nuclease, most frequently Cas9, and a single guide (sg)RNA, which targets the nuclease to selected, genomic loci by means of sequence-complementary. While Cas9 can, in principle, target any custom DNA sequence, it requires the presence of a protospacer adjacent motif (PAM) next to the ~20 nucleotide

Dominik Niopek (ed.), *Photoswitching Proteins: Methods and Protocols*, Methods in Molecular Biology, vol. 2173,
https://doi.org/10.1007/978-1-0716-0755-8_18, © Springer Science+Business Media, LLC, part of Springer Nature 2020

sgRNA-complementary sequence. In case of the most commonly applied Cas9 ortholog from *Streptococcus pyogenes* (*Spy*Cas9), the PAM sequence comprises NGG. This constraint can, however, be overcome by using engineered, PAM-relaxed *Spy*Cas9 [4–6] (or Cas12 [7]) mutants or by selecting natural Cas9 orthologs that show different PAM requirements [8, 9].

In its native configuration, Cas9 induces double-strand breaks at the target site, which can trigger two major repair pathways in mammalian cells. The first, named nonhomologous end-joining (NHEJ) is highly error-prone. During NHEJ, cells often insert or delete single or multiple base pairs around the break point of the chromosome before ligating the ends back together. When this process happens within protein-encoding regions of a gene, frame-shift mutations may occur that often result in a nonfunctional protein variant. The second pathway, called homology-directed repair (HDR), employs the genetic information of the sister chromatid to precisely repair the break. Importantly, the HDR pathway can be hijacked by supplying an exogenous repair template in trans, that is, a single or double-stranded DNA fragment flanked by sequences with homology to the break-adjacent regions. Thereby, precise changes such as knockins of longer sequence stretches at selected positions within the genome can be triggered. In contrast to Cas9, which cuts both DNA strands at the same position (blunt) or maximum one base pair shifted, Cas12-mediated cleavage results in four base staggered ends [10]. Moreover, Cas12 can process RNA guides from longer transcripts, which simplifies multiplexing, that is, simultaneous targeting of various different loci at once [11, 12]. Finally, Cas13 is an RNA-guided, RNA-targeting CRISPR effector facilitating the cleavage (knockdown) or editing of selected RNAs [13–15].

Apart from employing catalytically active Cas9/−12 and Cas13 for sequence-specific DNA and RNA cleavage, endonuclease activity deficient variants of these have been created by mutagenesis (referred to as d(ead)Cas) [10, 13, 15, 16]. These can be used as programmable DNA or RNA binding domains allowing for sequence-specific recruitment of custom effector domains. The latter can be either genetically fused to dCas or localized to it by other means (e.g., via aptamers integrated into the RNA guide). In case of Cas9 and Cas12, these effectors can, for instance, be transcriptional activators, repressors or epigenetic modifiers [16–24] that are used to activate or repress selected genes and/or perturb chromatin states.

Typically, when delivering CRISPR-Cas effectors into cells and organisms, for example, by transient transfection of DNA, RNA, Cas:RNA ribonucleoprotein (RNP) complexes or by viral delivery, the onset of an induced perturbation cannot be precisely controlled in time and space. Moreover, once present inside a cell, the CRISPR-Cas effector will remain active until its protein and/or

RNA components have been degraded. As a result, the spatiotemporal resolution at which CRISPR perturbations can be made is limited. For end-point measurements (e.g., genome-wide knockout screens), this may not be an issue. When studying dynamic processes, such as control of gene expression networks, genome 3D organization or cellular adaptation to changes in the genome and transcriptome, the ability to precisely control the activity of CRISPR-Cas effectors in time and space from the outside is, however, extremely relevant. Moreover, it is known that persistent activity of CRISPR effector increases the chance of off-target effects and/or toxicity [25–27].

Combining CRISPR-Cas with optogenetics has the potential to overcome these limitations by permitting spatiotemporally confined activity of CRISPR effectors and facilitating highly dynamic (epi)genome or transcriptome perturbations. Available optogenetic CRISPR tools can roughly be categorized into five different groups: (1) split systems in which light is employed to reconstitute an active CRISPR-Cas effector from two separate, inactive parts; (2) single-component CRISPR-Cas9 switches, that is, Cas-photoreceptor fusions the function of which is dependent on light; (3) photocaging approaches using UV-labile chemicals; (4) light-dependent expression or delivery of CRISPR; and (5) light-dependent Cas9 inhibition via engineered, photoswitchable anti-CRISPR proteins. Of note, these categories are not entirely exclusive as some tools might fall into more than one category. In the following, we will introduce the available optogenetic tools for CRISPR control and point out their particularly notable features and potential limitations.

2 Optogenetic Split CRISPR-Cas9 Systems

A powerful way to control CRISPR perturbations by optogenetics is using light-inducible dimerizers to reconstitute CRISPR effectors from split parts. In 2015, the groups of Moritoshi Sato and Charles Gersbach independently reported the application of this principle to control CRISPR-based transcriptional activators, thereby facilitating light-inducible expression of endogenous genes in mammalian cells [28, 29] (Fig. 1a and b). Their systems are based on a p65 [28] or VP64 [29] transactivator domain fused to the blue light receptor cryptochrome 2 (CRY2) from *Arabidopsis thaliana* and *Spy*dCas9 fused to CRY2s interaction partner CIB1. Upon blue light excitation, CRY2 and CIB1 bind to one another, resulting in the assembly of dCas9-p65 [28] or dCas9-VP64 [29] complexes. When targeted to endogenous promoters, these complexes activate transcription. The authors demonstrated that their system can induce expression of various genes such as *ASCL1*, *MYOD1*, *NANOG* , and *IL1RN* in human cells.

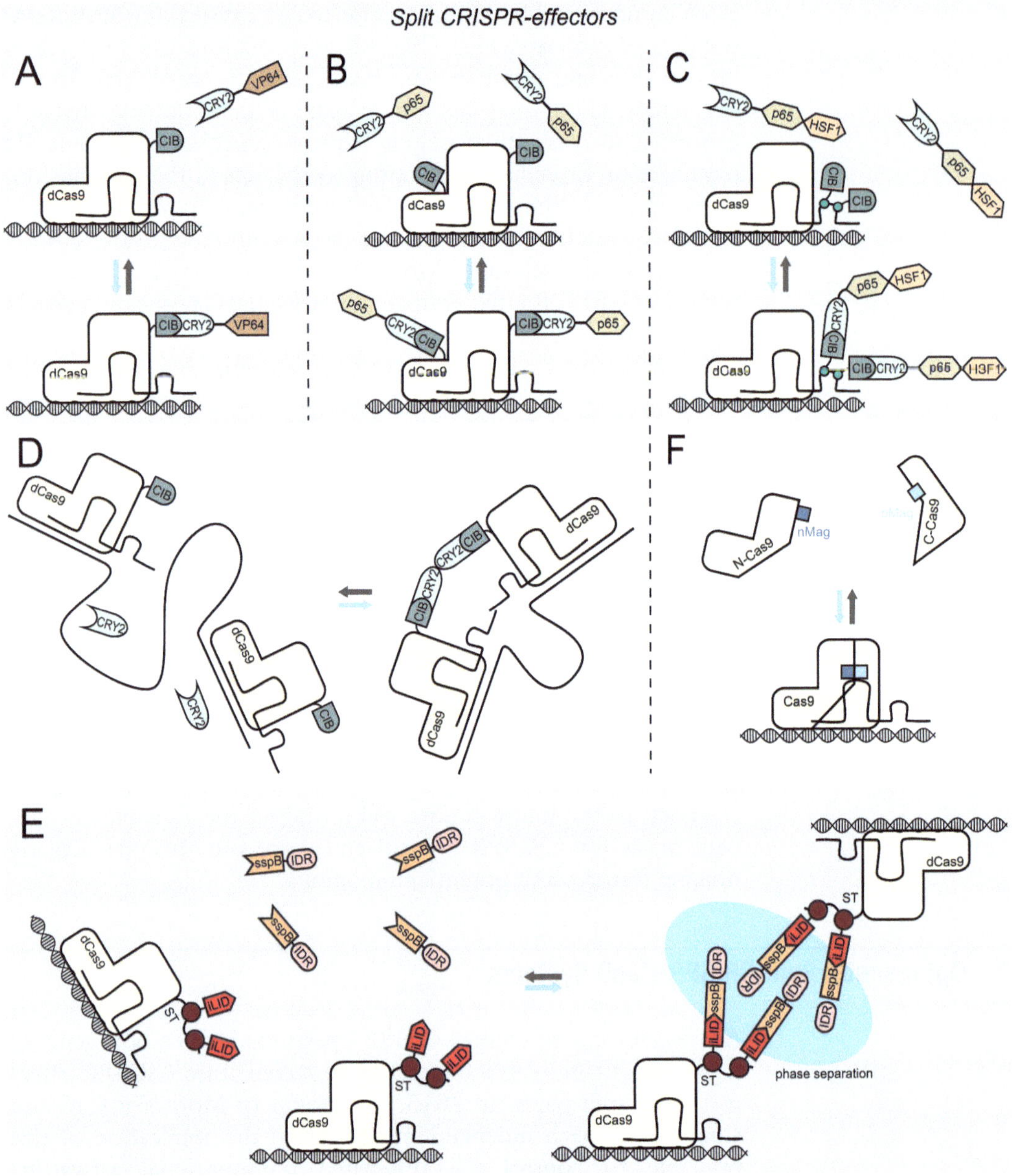

Fig. 1 Split CRISPR effectors dependent on blue light. (**a**–**c**) Light-inducible gene activation via CRY2-CIB mediated heterodimerization of *Spy*dCas9 and transcriptional activators VP64 [29] (**a**), p65 [28] (**b**) and p65-HSF1 [31] (**c**). In **a** and **b**, CIB is genetically fused to Cas9. In **c**, CIB is recruited to MS2 aptamer repeats integrated into the sgRNA via genetic fusion to MS2 aptamer-binding protein. (**d**) Light-activated dynamic looping (LADL) can bring two otherwise distal genomic loci into spatial proximity [32]. (**e**) The CasDrop system enables light-dependent liquid–liquid phase transitions [35]. (**f**) Photoactivatable genome editing with paCas9. paCas9 comprises a N- and C-terminal split Cas9 fragments fused to nMag and pMag, respectively. Irradiation triggers reconstitution of a catalytically active Cas9 that can perform targeted double-strand breaks

Importantly, these systems can be multiplexed simply by coexpressing multiple sgRNAs. Thereby, the same promoter can be targeted at different positions to increase the overall expression that can be triggered. Furthermore, using sgRNAs targeting two different promoters enabled simultaneous activation of two different genes. Of note, both systems are reversible: Alternating between phases of blue light irradiation and incubation in the dark resulted in oscillatory behavior of mRNA abundance [29]. Finally, the tools are not limited to human cells, but can be adapted to other cell types, for example, from zebrafish [30].

In the aforementioned configuration, one dCas9 molecule is employed to recruit a single transactivator domain to a given genomic site. While sufficient to activate expression from endogenous genes, these systems show a low overall potency, that is, the amount of mRNA produced from the targeted gene is limited. The strategy's efficiency can be dramatically improved, however, by targeting multiple transactivators to a single Cas9 molecule. The CPTS2.0 (for CRISPR–Cas9-based photoactivatable transcription system 2.0) employs multiple MS2 protein-binding aptamer repeats integrated into the sgRNA as scaffold for transactivator binding [31]. Light-dependent recruitment is achieved by fusing CIB1 to MS2 (also called MCP for MS2 cop protein) and CRY2 to a p65-HSF1 tandem transactivator. Upon photoexcitation, multiple activation domains are recruited to a single Cas9 molecule, thus highly improving the efficacy of light-induced gene expression [31] (Fig. 1c).

Apart from recruiting effectors to selected genomic sites, clever twists to the aforementioned strategies enable perturbation of the 3D genome conformation or the nucleus biophysical properties. Light-activated dynamic looping (LADL) [32], for example, an optogenetic technique developed by Jennifer E. Phillips-Cremins group at the University of Pennsylvania, can be used to bring into spatial proximity two otherwise distal genomic sites (Fig. 1d). LADL employs dCas9-CIBN fusions targeted to two selected regions of a chromosome via different sgRNAs. It further requires CRY2, which not only binds CIBN upon blue-light irradiation, but also oligomerizes [33, 34], thereby forming a bridge between the dCas9-CIBN molecules [32]. The LADL system was applied, for instance, to redirect a stretch enhancer to the *Zfp462* promoter, thus triggering an increase in *Zfp462* expression.

CasDrop, on the other hand, facilitates the induction of liquid–liquid phase transitions in the nucleus at selected chromatin regions and thereby enables studying biophysical features that underly genome regulation [35] (Fig. 1e). CasDrop comprises three parts. First, a dCas9 fused to the SunTag, a repeating peptide array that can recruit multiple copies of an antibody-fusion protein [36]. Second, the SunTag-binding antibody fused to iLID, an *Avena sativa* light-oxygen-voltage 2 (LOV2) domain-based optogenetic heterodimerizer. Third, the iLID interaction partner SspB fused to an intrinsically disordered region of a protein. Multivalent interactions of the latter are known to facilitate intracellular phase transitions that result in local condensation and thus formation of liquid droplets [35]. By coexpressing these three proteins in HEK293 (human embryonic kidney cells) or U2OS (human osteosarcoma cells) together with sgRNAs targeting selected loci, Shin et al. were able to induce droplet formation adjacent to the Cas9-targeted regions such as telomeres [35]. They further showed that condensates grow preferably at regions of low chromatin density (euchromatin) and that droplet formation can spatially rearrange the chromatin [35].

It has to be noted that in all of the aforementioned tools, dCas9 is always bound to the genomic target site(s) irrespective of light, while photoactivation triggers recruitment of an additional protein to the DNA-bound dCas9. This has important implications, since dCas9 itself can cause perturbation of genome regulation, for example, via CRISPR interference (CRISPRi) [16]. One way to address this issue is to split Cas9 into two parts, which are unable to bind DNA on their own, but reconstitute a functional Cas9 upon dimerization. This concept was first put into practice by the group of Feng Zhang at the Broad Institute. The authors split Cas9 at several positions and fused the N- and C-terminal split fragments to FK506 binding protein 12 (FKBP) and FKBP rapamycin binding (FRB) [37]. Rapamycin-induced dimerization of FKBP and FRB then lead to reconstitution of a functional Cas9 [37]. Nihongaki et al. applied the identical approach, but combined it with light induction instead of a chemical trigger. Their photoactivatable Cas9 (paCas9) is based on *Spy*Cas9, split between residues 713/714 [38] (Fig. 2). The N- and C-terminal Cas9 split fragments were fused to pMag and nMag, which stands for positively and negatively charged magnets [38]. These are engineered derivatives of the Vivid (VVD) LOV domain from *Neurospora crassa*, which naturally homodimerizes upon blue light excitation. Due to the artificially introduced charges at the protein–protein interface, pMag and nMag can only form heterodimers (i.e., pMag can bind nMag), but no more homodimers (pMag is repelled by pMag; nMag is repelled by nMag). In the absence of blue light, the magnet-bearing split Cas9 fragments do not dimerize and are thus unable to bind and cleave DNA. Upon photoactivation, however, the pMag and nMag parts heterodimerize, thereby reconstituting a DNA binding and endonuclease-competent Cas9 protein (Fig. 1f). Nihongaki et al. observed blue light-dependent formation of insertions and deletions, which are indicative of Cas9 induced double-strand breaks, at genomic sites targeted with paCas9 in HEK293T cells [38]. They could also spatially confine paCas9 activity by using a photomask to shield parts of a culture dish from the light trigger. As paCas9 binds DNA in a light-

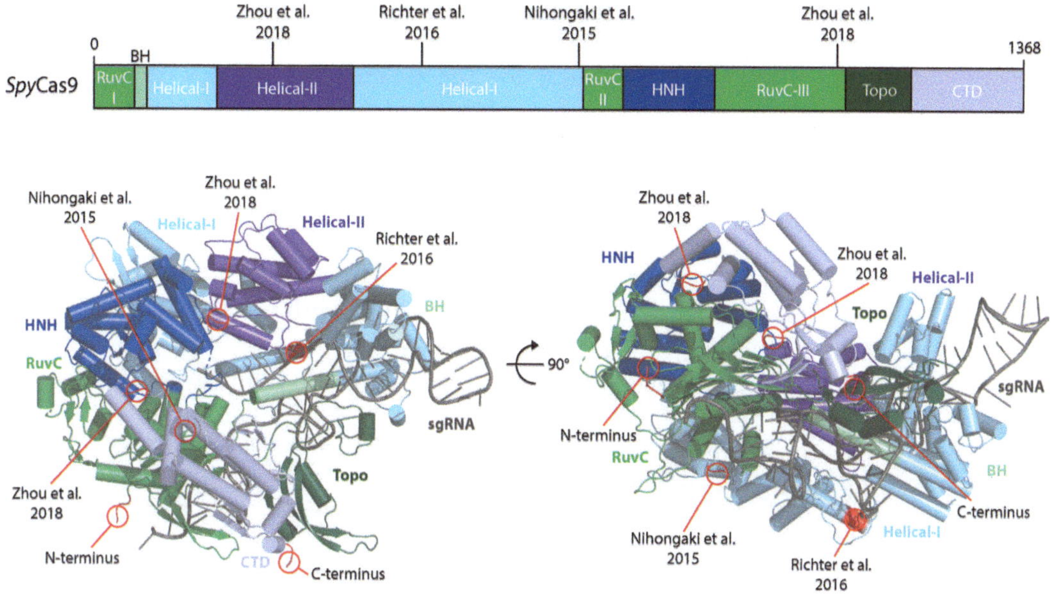

Fig. 2 Domain organization (top) and structure (bottom) of *Spy*Cas9. Sites that have been harnessed for engineering of optogenetic CRISPR effectors are indicated (PDB 5VW1)

dependent manner, this concept is compatible with both, catalytically active Cas9 as well as dCas9. The researchers demonstrated this by using a catalytically impaired paCas9 to silence a luciferase reporter gene via CRISPRi in a light-dependent manner [38]. In a subsequent paper [31], Nihongaki et al. further modified their catalytically inactive paCas9 variant to facilitate light-inducible transcription by adapting the SAM (synergistic activator mediator) strategy [20]. SAM comprises three parts: (1) A dCas9 fused to VP64, (2) a sgRNA incorporating two MS2 RNA aptamers and (3) an MS2-fused P65-HSF1 tandem transactivator mentioned above. By splitting dCas9-VP64 into an N- and C-terminal fragment fused to pMag and nMag, respectively, the formation of a DNA-binding competent Cas9-SAM complex could be controlled with light. This system, which is referred to as Split-CPTS2.0, exceeds the efficiency of the CPTS and CPTS2.0 systems mentioned before. It was also the first of its kind shown to be capable of inducing a cellular phenotype, namely, neuronal differentiation of induced pluripotent stem cells (iPSC) triggered by light-dependent activation of the *NEUROD1* gene. Finally, magnets-induced dimerization of a split-CRISPR effector has also been successfully adapted to Cas12 (Cpf1) [39].

3 Single-Component CRISPR-Cas9 Switches

While optogenetic split CRISPR-Cas systems are relatively easy to design and customize, they demand coexpression of two separate proteins (the split fragments), which then reassociate in a light-dependent manner. The absolute and relative expression of the split fragments is critical for the functionality of this approach: An overall low and/or very heterogeneous expression of the two components will likely result in suboptimal light-regulation or low efficacy.

This limitation can be addressed, at least in parts, by engineering single-component optogenetic CRISPR effectors, in which the light-dependent conformational change of a photoreceptor directly controls the function of a fused full-length Cas9 or dCas9. Toward this goal, the group of Andreas Möglich created fusions between the *Rhodobacter sphaeroides* (*Rs*)LOV domain and *Spy*Cas9 [40]. The *Rs*LOV is a homodimer, which dissociates upon blue-light exposure. The design hypothesis was that—provided *Rs*LOV was inserted at a suitable position into Cas9—the *Rs*LOV-Cas9 fusion would be DNA-binding incompetent when in a homodimeric state (i.e., *Rs*LOV2 dark state), but Cas9 function would be regained upon light-mediated dissociation (Fig. 3a). Based on the Cas9 crystal structure, the authors picked >200 Cas9 candidate surface sites and then created a library of *Rs*LOV-dCas9 fusions bearing the photoreceptor domain at these chosen sites. They also included flexible linkers of 15–29 residues appended at either N- or C-terminus of the *Rs*LOV. The resulting library was then screened for candidates that showed light-dependent DNA binding. To do so, the researchers coexpressed the dCas9-*Rs*LOV fusions together with an RFP reporter and an RFP-targeting sgRNA in *E. coli*. The authors then performed an iterative fluorescence-activated cell sorting (FACS)-based screen. They alternated between incubation of

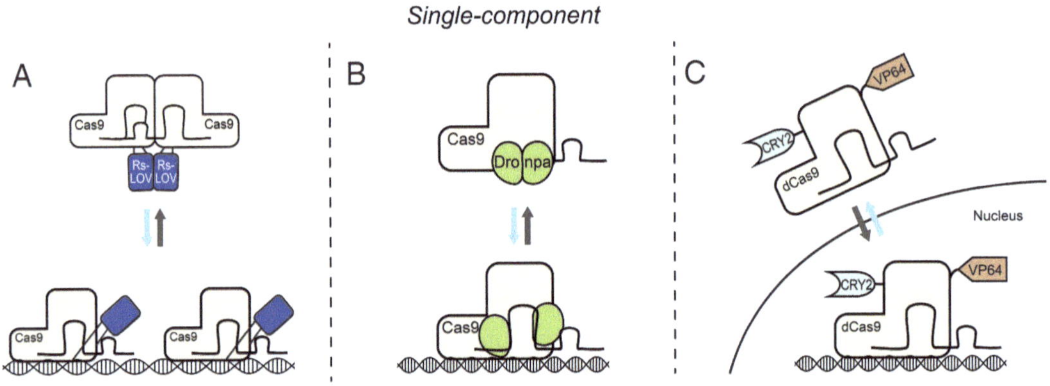

Fig. 3 Single-component CRISPR-Cas9 switches. (**a**) PaRC9 is a fusion of *Spy*Cas9 to the *Rs*LOV domain. Blue light irradiation triggers dissociation of the *Rs*LOV parts and thereby releases the activity of Cas9 [40]. (**b**) Ps-Cas9 harnesses the light-mediated dissociation of paDronpa for optogenetic control of Cas9 DNA binding [42]. The ps-Cas9 system is compatible with *Spy*Cas9 and *Sau*Cas9. (**c**) CRY2 is cleared from the nucleus within several hours of light induction. This property can be harnessed to block gene activation via dCas9-VP64 in a light-depenent manner [44]

cultures in the blue light or dark followed by sorting out the populations that show weak or strong RFP expression, respectively. Of note, sorting was performed at 29 °C, as the library showed an RFP expression comparable to the negative control irrespective of light at higher temperatures. After several rounds of screening and further optimization, Richter et al. had identified a variant named photoactivatable *Rs*LOV-Cas9 (paRC9) that showed noticeable light-dependent activity. In this variant, the *Rs*LOV is located between Cas9 residues F478 and V481 (E479 and E480 are deleted), which—actually—corresponds to a site that had already been identified in an independent domain insertion study by the Doudna lab [41] (Fig. 2). Albeit conceptually very interesting, its modest dynamic range (3- to 6-fold difference between light and dark Cas9 activity) and requirement of rather low temperatures (29 °C) limit the application spectrum of paRC9.

An alternative to paCas9 is the single-chain photoswitchable Cas9 (ps-Cas9) developed by the Michael Z. Lin lab at Stanford [42] (Fig. 3b). Zhou and Lin et al. had previously reported an interesting strategy for optogenetic control of protein activity using an engineered, dimeric Dronpa variant (paDronpa) [43]. To adapt this strategy to Cas9, paDronpa was inserted at two sites flanking the DNA binding cleft of Cas9. More precisely, one copy was introduced into the protospacer-associated motif domain of the nuclease lobe (behind K1246), while the other was placed into the recognition domain (REC2; behind A259) (Fig. 2). In the dark, the intramolecular interaction of both paDronpa copies masks the DNA binding cleft, thus blocking Cas9 DNA binding. Upon photoexcitation with cyan light, however, the two intramolecular paDronpa molecules dissociate, thereby releasing Cas9 activity. Importantly, this strategy is compatible with catalytically active Cas9 as well as dCas9 and can be applied to regulate different Cas9 orthologs, namely, *Spy*Cas9 and the smaller *Staphylococcus aureus* (*Sau*)Cas9 [8]. The major downside of this approach is that even in the fully induced state, ps-Cas9 is considerably less active (>2-fold lower indel frequencies) as compared to wild-type Cas9 [8], indicating that the presence of the paDronpa domains interferes—to some extent—with Cas9 function. Moreover, the insertion of two paDronpa copies results in a size of the Cas9-fusion of ~5.6 kb, which can present a limitation with respect to delivery. Adeno-associated virus vectors, for instance, have a packaging capacity of only ~4.8 kb.

Apart from activating Cas9 upon photostimulation, the opposite mode of regulation, that is, light-induced blockage of Cas9 function, has also been explored (Fig. 3c). The group of Chandra Tucker at the University of Colorado found that proteins genetically fused to CRY2 are cleared from the nucleus within several hours of light induction [44]. They fused CRY2 to Cas9-VP64, coexpressed the construct together with sgRNAs targeting the

IL1RN locus in HEK293T cells and assessed the abundance of IL1RN transcript upon light stimulation by qPCR. Tucker and colleagues found that IL1RN mRNA counts were reduced by fourfold in the dark as compared to the light condition [44]. However, the block of Cas9 activity upon irradiation was incomplete, that is, IL1RN expression in the light was still increased relative to control samples expressing CRY2-dCas9-VP64 together with a nontargeting sgRNA [44].

4 Photochemical Approaches

Apart from using photoreceptors to conditionally perturb Cas9 conformation, optical control can also be achieved via photochemical switches. These systems are not fully genetically encoded, that is, they depend on exogenous supply of a photolabile chemical. Provided this is not an issue in context of the envisaged application or experimental system, photochemical approaches present a powerful alternative to photoreceptor-based CRISPR tools. Based on the published data, photochemical systems tend to be tighter than photoreceptor-based tools, meaning that they exhibit lower CRISPR-Cas9 background activity in the noninduced state. Their irreversible nature and—with few exceptions—dependency on UV light for activation, which is highly energetic and shows poor tissue penetration, present limitations for photochemical tools.

Hemphill et al. used genetic code expansion to incorporate photocaged amino acids at functionally relevant positions into Cas9 [45] (Fig. 4a). Cas9 K866 undergoes a conformational change upon sgRNA binding, possibly required to correctly position the target DNA strand prior to cleavage. Hemphill et al. introduced an amber stop codon (TAG) at position K866 into Cas9 [45]. They then coexpressed this Cas9 construct with an engineered pyrrolysyl-tRNA (PylT)/tRNA synthetase pair in mammalian cells. The authors also supplied photocaged lysine (PCK), that is, a lysine derivative bearing a UV-labile moiety attached to the amino acid's functional group, to the medium. PCK is recognized by the engineered tRNA synthetase, loaded onto the UAG-targeting PylT and eventually incorporated at position 866 of Cas9 during translation. Hemphill et al. showed that the presence of the PCK blocks Cas9 activity, while exposure of cells to 365 nm light for a few minutes removes the photolabile group on K866 and thereby restores Cas9 function. The system shows low background activity and is compatible with *Spy*Cas9 and -dCas9. The requirement of various components, that is, the Cas9 transgene with amber stop codon, the engineered PylT, tRNA synthetase as well as PCK, needs to be taken into account when applying this approach.

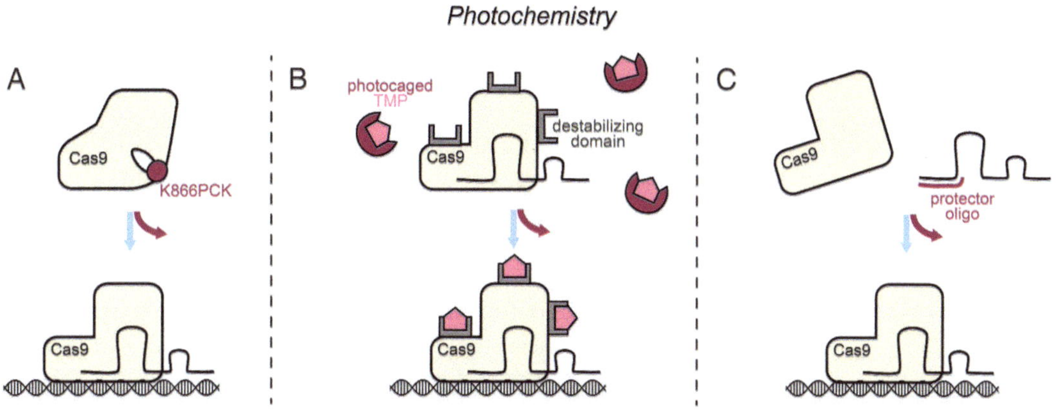

Fig. 4 Photochemical approaches. (**a**) Incorporating a photoacaged lysine at position 866 into *Spy*Cas9 results in UV-light-dependent Cas9 activity [45]. (**b**) Fusion of Cas9 to DHFR targets Cas9 for degradation; TMP stabilizes DHFR. Coupling TMP to a photocleavable moeity results in light-dependent stabilization of Cas9 [46]. (**c**) CRISPR-plus employs photocleavable oligonucleotides to unmask sgRNAs in a light-dependent manner [47]

An alternative strategy has been invented by Manna et al., who used a photochemical strategy independent of genetic code expansion [46]. They constructed a Cas9 fused to destabilizing domains of *E. coli* dihydrofolate reductase (DHFR) (Fig. 4b). The presence of these domains targets Cas9 to proteasomal degradation. Trimethoprim (TMP) addition prevents degradation by stabilizing the DHFR domains. To render TMP-mediated stabilization of Cas9 light-inducible, Manna et al. appended photocleavable moieties to TMP, namely, 2-(2-nitrophenyl)-propoxycarbonyl (NPPOC) or thiocoumarine, which are removed upon irradiation with 385 or 470 nm light, respectively. Thus, when expressing the Cas9-DHFR fusion and supplying the TMP-derivative, Cas9 activity can be restored in a light-dependent manner. Its compatibility with two different light activation wavelengths, the possibility of multimodal fine-tuning by using different light doses as well as different concentrations of the photolabile chemical render this method particularly versatile.

On top of controlling the function or abundance of the Cas9 protein, the sgRNA is also an interesting target for photochemical control of CRISPR. The CRISPR-plus (CRISPR-precise light-mediated unveiling of sgRNAs) method employs sgRNAs-complementary ssDNA oligonucleotides as "protectors" of sgRNAs [47] (Fig. 4c). The protector ssDNA contains photocleavable groups at every sixth position. Upon UV irradiation, the protector is cleaved into short 6-mer fragments that have a reduced melting temperature (T_m) and thus a diminished affinity to the sgRNA target as compared to the full-length protector. As result, the sgRNA is released from the protector and thus able to associate with Cas9 and initiate target DNA binding. This method was tested extensively in in vitro DNA cleavage assays and was shown to also function in HeLa (human cervix carcinoma) cells [47]. In the latter case, some background cleavage was observed after several days of incubation, possibly due to spontaneous unbinding of the protector group in the cellular environment [47]. The merit of this method lies in its simplicity and versatility, that is, it is, in principle, independent of the used Cas9 ortholog and should be compatible with catalytically active and inactive Cas9s. The spontaneous release of the protector from sgRNAs observed in mammalian cells after several days of incubation might present a caveat, in particular in cases in which leakiness, that is, Cas9 activity in absence of the light trigger, cannot be tolerated [47].

5 Light-Dependent Expression or Delivery

All of the aforementioned strategies employ modified Cas9 or sgRNA variants, the function of which directly depends on the state of a fused receptor or bound chemical. However, it is also possible to achieve optical control of CRISPR without altering Cas9 or the sgRNA itself. One way to do so is via light-dependent transgene expression or delivery. A particular advantage of these approaches is that they often are compatible with near-infrared (NIR) activation wavelengths, thus causing low toxicity and facilitating deep tissue penetration, the latter of which is of particular importance for in vivo applications. Also, these strategies are, in principle, compatible with any CRISPR-Cas system or ortholog.

The FACE system (for far-red light activated CRISPR-dCas9 effector) by Shao et al. [48] is based on BphS, a far-red light-activated bacteriophytochrome diguanylate cyclase and a c-di-GMP-responsive transcription factor, BldD [49]. Upon light-mediated activation of BphS, intracellular c-di-GMP levels increase, thereby resulting in BldD dimerization and thus expression from a BldD-dependent promoter. Shao et al. placed FGTA4, a fusion of MS2 aptamer-binding protein to p65 and HSF1, under control of the BldD-dependent promoter. Hence, upon far-red light stimulation (730 nm light), FGTA4 is expressed and recruited to dCas9 via MS2 aptamers integrated into the sgRNA (Fig. 5a). According to the data presented by Shao et al., the FACE system shows very low Cas9 background activity in the dark and can yield a several hundred fold increase in target gene expression upon light activation. In addition, the authors demonstrated that their system (1) can be easily tuned by adjusting the illumination time or light intensity, (2) is reversible and (3) functions in different cell types, including HEK, HeLa and hMSC-TERT (transformed human mesenchymal stem cells). Shao et al. also compared the performance of their system to the blue light-inducible CPTS 2.0 tool (see above) in vivo. They used electroporation for delivery of the corresponding

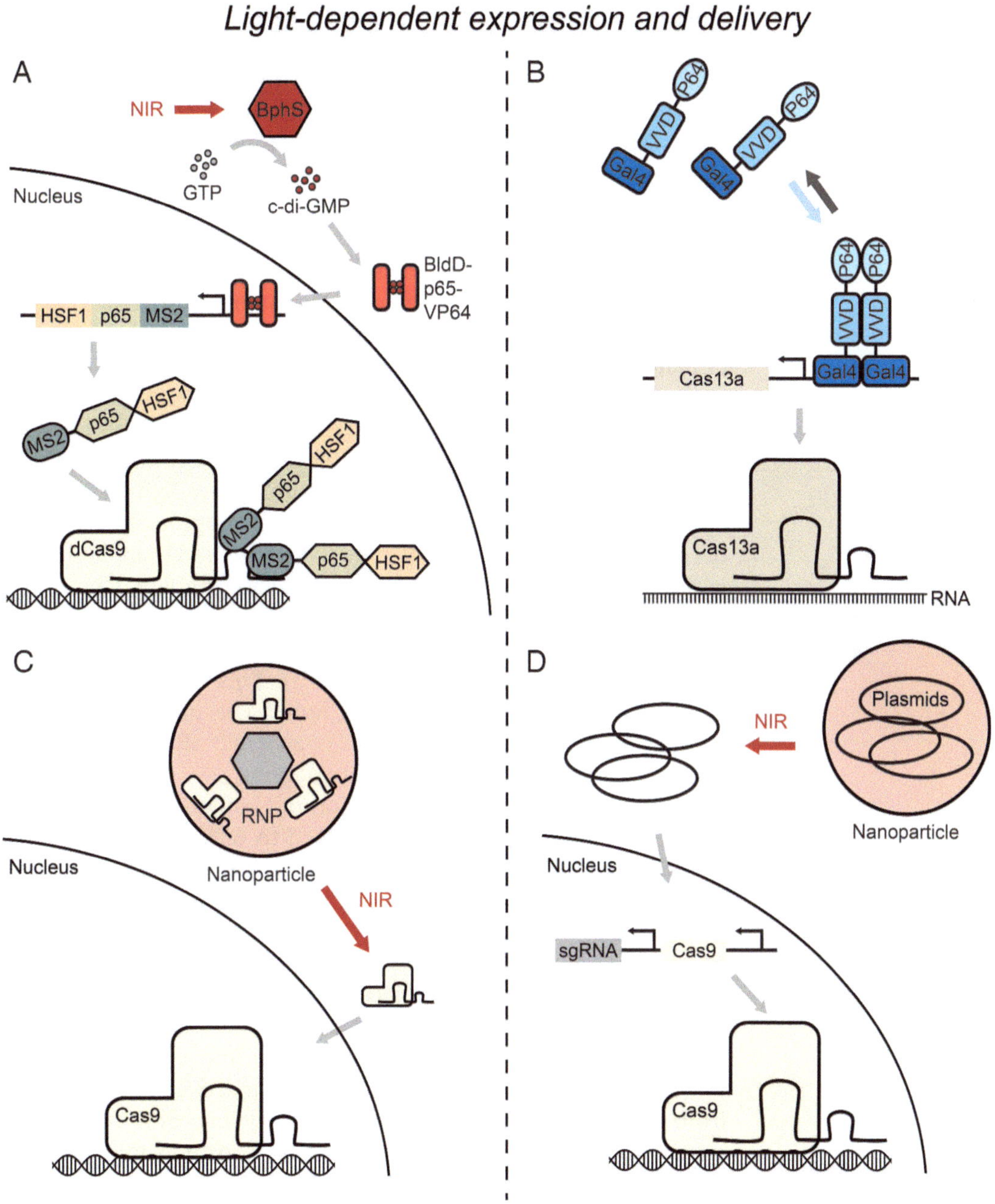

Fig. 5 CRISPR-Cas control by light-dependent expression and delivery. (**a**) The FACE system uses a far-red light (730 nm)-dependent transgene system to control the abundance of FGTA4, a fusion of MS2 aptamer-binding protein to p65 and HSF1. Upon far-red light induction, FGTA4 is expressed and binds to MS2 aptamers incorporated into the sgRNA [48]. Cas9 and the sgRNA are constitutively expressed. (**b**) Blue light-activated expression of Cas13a allows optogenetic control of RNA targeting [52]. (**c**, **d**) NIR-mediated release of Cas9 RNPs (**c**) or Cas9-encoding plasmids (**d**). See main text for details

constructs into the *tibialis posterior* muscles of mice and targeted the genes of laminin subunit alpha 1 (*Lama1*) or follistatin (*Fst*), which are possible candidate genes for treatment of myopathies. Remarkably, the authors found that only the FACE, but not the CPTS 2.0, enabled significant activation of these target genes under the used experimental conditions. The authors also demonstrated far-red light-mediated differentiation of iPSCs in culture upon induction of endogenous *Neurog2* [48].

Apart from the particularly notable example above, a variety of light-dependent transgene expression systems exist, that can—in principle—be harnessed for CRISPR-Cas control [50]. These are, in many cases, easy to implement, as they usually only require 2 3 components, that is, a transgene driven from an engineered promoter and either a split transcription factor (TF) or TF-photoreceptor fusion (in some cases, an exogenous chromophore further has to be added). Importantly, light-dependent transgene systems cannot only be adapted for optogenetic control of (*Spy*)Cas9. On the contrary, they should be readily applicable to, in principle, any CRISPR-Cas system and Cas ortholog. Cheng and colleagues, for example, used the LightON system [51] based on an engineered, blue light-inducible Gal4 transcription factor, to control expression of Cas13a [52] (Fig. 5b). This way, targeted RNA cleavage could be controlled with light. The downside of such approaches is that gene expression is generally slow as compared to, for instance, protein–protein interactions.

Besides controlling the expression of Cas9 or Cas9-dependent effector domains, light can also be used to release Cas9-encoding plasmids or Cas9-sgRNA RNPs from carrier molecules.

Following this working principle, Pan et al. used upconversion nanoparticles (UCNPs) to convert NIR radiation (800 or 980 nm) to high-energy UV light [53]. They coupled Cas9 RNPs to the UCNPs via a UV-sensitive moiety (4-(hydroxymethyl)-3-nitrobenzoic acid; ONA) and then coated the UCNP-Cas9 particles with polyethylenimine (PEI). Upon transfection into mammalian cells, the Cas9 RNPs remain UCNP bound and inactive, unless released from the UCNPs by NIR stimulation [53] (Fig. 5c). To demonstrate the power of this strategy, Pan et al. treated mice bearing tumor xenografts with the UCNP-Cas9-RNP complexes. They used a sgRNA targeting the gene encoding protein kinase1 (*PLK-1*), a key regulator of mitosis. Remarkably, repeated intratumoral injection of UCNP-Cas9-RNPs followed by irradiation with NIR light over the course of several weeks resulted in a considerable delay in tumor progression as compared to the nonirradiated control group. This is likely due to cytotoxicity induced upon disruption of the *PLK-1* gene. Together, these data indicate that, beyond applications in basic research, optogenetic CRISPR tools could provide exciting opportunities for spatially confined treatment of malignancies. Lyu et al., on the other hand, employed a photolabile semiconducting polymer nanotransducer (pSPN) to release Cas9-encoding plasmids in a light-dependent manner [54] (Fig. 5d). Upon NIR (680 nm) irradiation, the pSPN generates singlet-oxygen ($^1O_2$), thereby releasing plasmids coupled to the polymer via $^1O_2$-cleavable linkers [54]. Both of these systems work in mammalian cells as well as in mice. A potential disadvantage as compared to photoreceptor-based approaches is the lack of reversibility and, possibly, the specialized chemical knowledge required for their implementation.

6 Light-Dependent Cas9 Inhibition Via Engineered Anti-CRISPR Proteins

Anti-CRISPR proteins (Acrs) are small, potent inhibitors of CRISPR systems [55–57]. They originate from the evolutionary war between bacteria and bacteriophages: The bacteria use CRISPR to destroy invading nucleic acids, while phages express Acrs to impair the CRISPR immunity. A plethora of Acrs has been discovered that efficiently block the function of various Cas9 and Cas12 orthologs [58–63].

Different inhibitory mechanisms have been reported for different Acrs, including masking of Cas9 catalytic sites, blocking of DNA binding, covalent modification of the Cas protein or enzymatic cleavage of the Cas-bound RNA guide [60, 64–67]. Naturally, Acrs are constitutive inhibitors that block the function of their cognate Cas ortholog(s) whenever present in the cell. Our lab has recently shown that by engineering Acr-photoreceptor chimeric proteins, Cas9 inhibition can be placed under optogenetic regulation (Fig. 6a) [68]. More precisely, we embedded the *Avena sativa* LOV2 domain into selected surface sites of AcrIIA4 (Fig. 6b). When coexpressed with *Spy*Cas9, the AcrIIA4-LOV2 chimeras bind *Spy*Cas9 and thereby block its activity in the dark. Upon irradiation with blue light, however, Cas9 is released and available for DNA targeting and genome editing (Fig. 6a). We named the AcrIIA4-LOV2 fusion CASANOVA, an acronym for **C**RISPR–Cas9 **a**ctivity **s**witching via **a** **n**ovel **o**ptogenetic **v**ariant of **A**crIIA4. Importantly, AcrIIA4 blocks *Spy*Cas9 function by impairing its DNA binding and so does CASANOVA [58]. In consequence, CASANOVA can be used not only to control genome editing via catalytically active *Spy*Cas9, but also target locus binding of *Spy*dCas9-effector fusions [68]. It was also the first optogenetic tool to be harnessed for light-mediated CRISPR labeling (see the chapter "Light-inducible CRISPR labeling" by Hoffmann et al. in this book). Importantly, the CASANOVA concept is not limited to AcrIIA4 and *Spy*Cas9, but can be transferred to other Acrs and Cas orthologs (preprint by Hoffmann and Mathony et al., bioRxiv, https://doi.org/10.1101/858589 and unpublished data). The merit of the CASANOVA approach is that users can stick to their

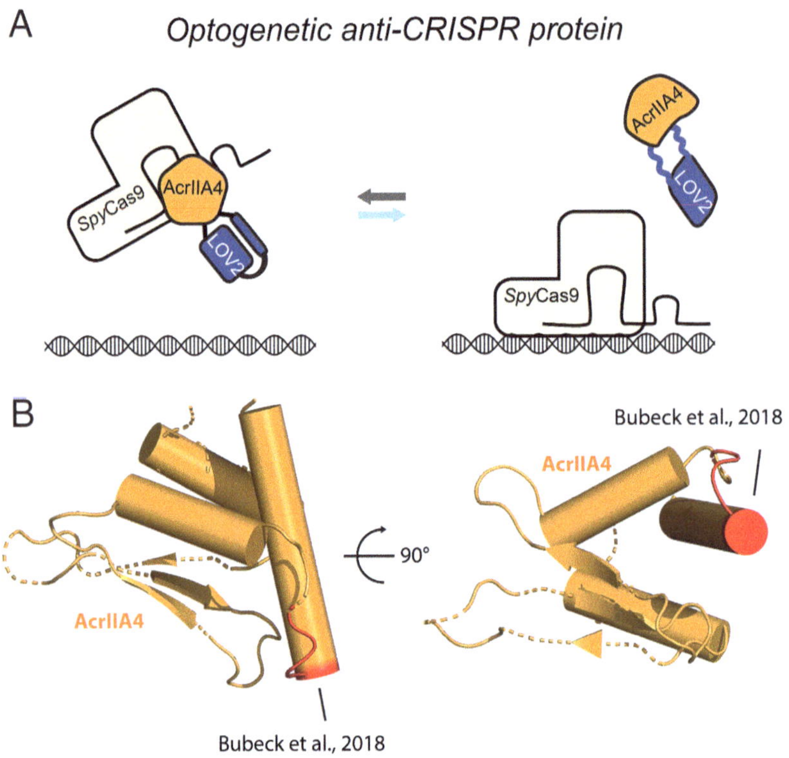

Fig. 6 Light-dependent Cas9 inhibition by CASANOVA, an engineered, switchable anti-CRISPR protein. (**a**) Schematic of CASANOVA function. In the dark, CASANOVA binds *Spy*Cas9, thereby blocking its activity. Upon blue light irradiation, CASANOVA dissociates from Cas9, hence releasing Cas9 function [68]. (**b**) Structure of AcrIIA4. The surface site at which the LOV2 domain was incorporated into the AcrIIA4 structure is indicated in red. (PDB 5VW1)

established Cas9 and sgRNA variants and only need to add CASANOVA to their experimental system to achieve optogenetic control. This renders CASANOVA highly versatile. Its limitations are the requirement of a blue-light activation wavelength as well as the need to efficiently overexpress CASANOVA, as otherwise, Cas9 activity will not be fully blocked in the dark.

7 Conclusions and Future Perspectives

Optogenetics and CRISPR are two powerful, highly complementary technologies. Combining these technologies facilitates targeted genome perturbations in mammalian cells with high spatiotemporal precision. Here, we reviewed the available methods for optogenetic CRISPR-Cas control via engineered Cas9-photoreceptor fusions, functionalized sgRNAs, light-inducible expression and delivery as well as light-dependent inhibition. We note that the available tools all come along with specific advantages and disadvantages with respect to their complexity and number of required components, delivery, light wavelength used for activation, kinetics and dynamic range of light regulation, compatibility with different CRISPR-Cas variants/orthologs and experimental systems as well as potential reversibility. In consequence, there is no one-fits-all solution suitable for any experimental context or anticipated application. Rather, users interested in applying optogenetic control to CRISPR should weigh the pros and cons of different methods in light of the application they envisage. The unprecedented speed of innovation in the CRISPR and optogenetics fields suggests that the available repertoire of light-switchable CRISPR systems will further expand in the near future.

With respect to reversibility, that is, the ability to shut off CRISPR-Cas activity following a preceding light-activation phase, it should be noted that the DNA-bound Cas9-sgRNA state has a half-live of several hours, at least when the sgRNA sequence perfectly matches the DNA target [69]. Optogenetic approaches that permit DNA binding of Cas upon light activation [38, 43, 68] do usually not enforce its dissociation in the absence of the stimulus. Thus, currently the reversibility of such tools is dependent on the kinetics of Cas9 unbinding from DNA and overall Cas9 turnover. This limitation might be at least partially overcome, for instance, by placing Cas9 degradation under light control [70–72].

Thus far, most optogenetic CRISPR tools have been centered around Cas9, most predominantly *Spy*Cas9. The accelerating speed of discovery, adaptation and rational improvement of CRISPR systems [4–6, 8, 10, 12–14, 73, 74] paired with advances in optogenetics and protein design [68, 75–78], however, suggest enormous potential for future innovation at the intersection of optogenetics and CRISPR. Looking further ahead, we believe that optogenetic control of CRISPR-mediated perturbations will facilitate dissection of highly dynamic genome regulatory processes such as transcription, genome 3D conformation and epigenetics. Optogenetic CRISPR control might also find application in gene therapy, where it could enable precise dosing of CRISPR-mediated interventions in patients [79] and aid confining therapeutic genome editing to selected tissues. This will be of particular interest in context of diseases for which light can be easily delivery to the therapeutically relevant tissue (e.g., skin, mouth or eyes). In conclusion, CRISPR and optogenetics present a case of mutual attraction and their combination will provide outstanding opportunities for basic research and therapy in the years to come.

Acknowledgments

We thank Katharina Niopek for feedback on the manuscript. M.D.H. was supported by a Helmholtz International Graduate School for Cancer Research scholarship (DKFZ, Heidelberg).

References

1. Jinek M, Chylinski K, Fonfara I et al (2012) A programmable dual-RNA-guided DNA endonuclease in adaptive bacterial immunity. Science 337(6096):816–821. https://doi.org/10.1126/science.1225829
2. Mali P, Yang L, Esvelt KM et al (2013) RNA-guided human genome engineering via Cas9. Science 339(6121):823–826. https://doi.org/10.1126/science.1232033
3. Cong L, Ran FA, Cox D et al (2013) Multiplex genome engineering using CRISPR/Cas systems. Science 339(6121):819–823. https://doi.org/10.1126/science.1231143
4. Hu JH, Miller SM, Geurts MH et al (2018) Evolved Cas9 variants with broad PAM compatibility and high DNA specificity. Nature 556(7699):57–63. https://doi.org/10.1038/nature26155
5. Kleinstiver BP, Prew MS, Tsai SQ et al (2015) Broadening the targeting range of *Staphylococcus aureus* CRISPR-Cas9 by modifying PAM recognition. Nat Biotechnol 33(12):1293–1298. https://doi.org/10.1038/nbt.3404
6. Kleinstiver BP, Prew MS, Tsai SQ et al (2015) Engineered CRISPR-Cas9 nucleases with altered PAM specificities. Nature 523(7561):481–485. https://doi.org/10.1038/nature14592
7. Kleinstiver BP, Sousa AA, Walton RT et al (2019) Engineered CRISPR-Cas12a variants with increased activities and improved targeting ranges for gene, epigenetic and base editing. Nat Biotechnol 37(3):276–282. https://doi.org/10.1038/s41587-018-0011-0
8. Ran FA, Cong L, Yan WX et al (2015) In vivo genome editing using *Staphylococcus aureus* Cas9. Nature 520(7546):186–191. https://doi.org/10.1038/nature14299
9. Esvelt KM, Mali P, Braff JL et al (2013) Orthogonal Cas9 proteins for RNA-guided gene regulation and editing. Nat Methods 10:1116. https://doi.org/10.1038/nmeth.2681
10. Zetsche B, Gootenberg JS, Abudayyeh OO et al (2015) Cpf1 is a single RNA-guided endonuclease of a class 2 CRISPR-Cas system. Cell 163(3):759–771. https://doi.org/10.1016/j.cell.2015.09.038
11. Campa CC, Weisbach NR, Santinha AJ et al (2019) Multiplexed genome engineering by Cas12a and CRISPR arrays encoded on single transcripts. Nat Methods 16(9):887–893. https://doi.org/10.1038/s41592-019-0508-6
12. Zetsche B, Heidenreich M, Mohanraju P et al (2017) Multiplex gene editing by CRISPR-Cpf1 using a single crRNA array. Nat Biotechnol 35(1):31–34. https://doi.org/10.1038/nbt.3737
13. Cox DBT, Gootenberg JS, Abudayyeh OO et al (2017) RNA editing with CRISPR-Cas13. Science 358(6366):1019–1027. https://doi.org/10.1126/science.aaq0180
14. Abudayyeh OO, Gootenberg JS, Essletzbichler P et al (2017) RNA targeting with CRISPR-Cas13. Nature 550(7675):280–284. https://doi.org/10.1038/nature24049
15. Abudayyeh OO, Gootenberg JS, Konermann S et al (2016) C2c2 is a single-component programmable RNA-guided RNA-targeting CRISPR effector. Science 353(6299):aaf5573. https://doi.org/10.1126/science.aaf5573
16. Qi LS, Larson MH, Gilbert LA et al (2013) Repurposing CRISPR as an RNA-guided platform for sequence-specific control of gene expression. Cell 152(5):1173–1183. https://doi.org/10.1016/j.cell.2013.02.022
17. Maeder ML, Linder SJ, Cascio VM et al (2013) CRISPR RNA-guided activation of endogenous human genes. Nat Methods 10(10):977–979. https://doi.org/10.1038/nmeth.2598
18. Perez-Pinera P, Kocak DD, Vockley CM et al (2013) RNA-guided gene activation by CRISPR-Cas9-based transcription factors. Nat Methods 10(10):973–976. https://doi.org/10.1038/nmeth.2600
19. Mali P, Aach J, Stranges PB et al (2013) CAS9 transcriptional activators for target specificity screening and paired nickases for cooperative genome engineering. Nat Biotechnol 31(9):833–838. https://doi.org/10.1038/nbt.2675

20. Konermann S, Brigham MD, Trevino AE et al (2015) Genome-scale transcriptional activation by an engineered CRISPR-Cas9 complex. Nature 517(7536):583–588. https://doi.org/10.1038/nature14136
21. Hilton IB, D'Ippolito AM, Vockley CM et al (2015) Epigenome editing by a CRISPR-Cas9-based acetyltransferase activates genes from promoters and enhancers. Nat Biotechnol 33(5):510–517. https://doi.org/10.1038/nbt.3199
22. Thakore PI, D'Ippolito AM, Song L et al (2015) Highly specific epigenome editing by CRISPR-Cas9 repressors for silencing of distal regulatory elements. Nat Methods 12 (12):1143–1149. https://doi.org/10.1038/nmeth.3630
23. Vojta A, Dobrinic P, Tadic V et al (2016) Repurposing the CRISPR-Cas9 system for targeted DNA methylation. Nucleic Acids Res 44 (12):5615–5628. https://doi.org/10.1093/nar/gkw159
24. Josipovic G, Zoldos V, Vojta A (2019) Active fusions of Cas9 orthologs. J Biotechnol 301:18–23. https://doi.org/10.1016/j.jbiotec.2019.05.306
25. Bisaria N, Jarmoskaite I, Herschlag D (2017) Lessons from enzyme kinetics reveal specificity principles for RNA-guided nucleases in RNA interference and CRISPR-based genome editing. Cell Syst 4(1):21–29. https://doi.org/10.1016/j.cels.2016.12.010
26. Shen CC, Hsu MN, Chang CW et al (2018) Synthetic switch to minimize CRISPR off-target effects by self-restricting Cas9 transcription and translation. Nucleic Acids Res 47 (3):e13. https://doi.org/10.1093/nar/gky1165
27. Shin J, Jiang F, Liu JJ et al (2017) Disabling Cas9 by an anti-CRISPR DNA mimic. Sci Adv 3(7):e1701620. https://doi.org/10.1126/sciadv.1701620
28. Nihongaki Y, Yamamoto S, Kawano F et al (2015) CRISPR-Cas9-based photoactivatable transcription system. Chem Biol 22 (2):169–174. https://doi.org/10.1016/j.chembiol.2014.12.011
29. Polstein LR, Gersbach CA (2015) A light-inducible CRISPR-Cas9 system for control of endogenous gene activation. Nat Chem Biol 11(3):198–200. https://doi.org/10.1038/nchembio.1753
30. Putri RR, Chen L (2018) Spatiotemporal control of zebrafish (*Danio rerio*) gene expression using a light-activated CRISPR activation system. Gene 677:273–279. https://doi.org/10.1016/j.gene.2018.07.077
31. Nihongaki Y, Furuhata Y, Otabe T et al (2017) CRISPR-Cas9-based photoactivatable transcription systems to induce neuronal differentiation. Nat Methods 14(10):963–966. https://doi.org/10.1038/nmeth.4430
32. Kim JH, Rege M, Valeri J et al (2019) LADL: light-activated dynamic looping for endogenous gene expression control. Nat Methods 16(7):633–639. https://doi.org/10.1038/s41592-019-0436-5
33. Che DL, Duan L, Zhang K et al (2015) The dual characteristics of light-induced cryptochrome 2, homo-oligomerization and Heterodimerization, for optogenetic manipulation in mammalian cells. ACS Synth Biol 4 (10):1124–1135. https://doi.org/10.1021/acssynbio.5b00048
34. Bugaj LJ, Choksi AT, Mesuda CK et al (2013) Optogenetic protein clustering and signaling activation in mammalian cells. Nat Methods 10(3):249–252. https://doi.org/10.1038/nmeth.2360
35. Shin Y, Chang YC, Lee DSW et al (2018) Liquid nuclear condensates mechanically sense and restructure the genome. Cell 175 (6):1481–1491. e1413. https://doi.org/10.1016/j.cell.2018.10.057
36. Tanenbaum ME, Gilbert LA, Qi LS et al (2014) A protein-tagging system for signal amplification in gene expression and fluorescence imaging. Cell 159(3):635–646. https://doi.org/10.1016/j.cell.2014.09.039
37. Zetsche B, Volz SE, Zhang F (2015) A split-Cas9 architecture for inducible genome editing and transcription modulation. Nat Biotechnol 33(2):139–142. https://doi.org/10.1038/nbt.3149
38. Nihongaki Y, Kawano F, Nakajima T et al (2015) Photoactivatable CRISPR-Cas9 for optogenetic genome editing. Nat Biotechnol 33(7):755–760. https://doi.org/10.1038/nbt.3245
39. Nihongaki Y, Otabe T, Ueda Y et al (2019) A split CRISPR-Cpf1 platform for inducible genome editing and gene activation. Nat Chem Biol 15(9):882–888. https://doi.org/10.1038/s41589-019-0338-y
40. Richter F, Fonfara I, Bouazza B et al (2016) Engineering of temperature- and light-switchable Cas9 variants. Nucleic Acids Res 44(20):10003–10014. https://doi.org/10.1093/nar/gkw930
41. Oakes BL, Nadler DC, Flamholz A et al (2016) Profiling of engineering hotspots identifies an allosteric CRISPR-Cas9 switch. Nat Biotechnol 34(6):646–651. https://doi.org/10.1038/nbt.3528

42. Zhou XX, Zou X, Chung HK et al (2018) A single-chain Photoswitchable CRISPR-Cas9 architecture for light-inducible gene editing and transcription. ACS Chem Biol 13 (2):443–448. https://doi.org/10.1021/acschembio.7b00603
43. Zhou XX, Fan LZ, Li P et al (2017) Optical control of cell signaling by single-chain photo-switchable kinases. Science 355 (6327):836–842. https://doi.org/10.1126/science.aah3605
44. Pathak GP, Spiltoir JI, Hoglund C et al (2017) Bidirectional approaches for optogenetic regulation of gene expression in mammalian cells using Arabidopsis cryptochrome 2. Nucleic Acids Res 45(20):e167. https://doi.org/10.1093/nar/gkx260
45. Hemphill J, Borchardt EK, Brown K et al (2015) Optical control of CRISPR/Cas9 gene editing. J Am Chem Soc 137 (17):5642–5645. https://doi.org/10.1021/ja512664v
46. Manna D, Maji B, Gangopadhyay SA et al (2019) A singular system with precise dosing and spatiotemporal control of CRISPR-Cas9. Angew Chem Int Ed Engl 58(19):6285–6289. https://doi.org/10.1002/anie.201900788
47. Jain PK, Ramanan V, Schepers AG et al (2016) Development of light-activated CRISPR using guide RNAs with Photocleavable protectors. Angew Chem Int Ed Engl 55 (40):12440–12444. https://doi.org/10.1002/anie.201606123
48. Shao J, Wang M, Yu G et al (2018) Synthetic far-red light-mediated CRISPR-dCas9 device for inducing functional neuronal differentiation. Proc Natl Acad Sci U S A 115(29): E6722–E6730. https://doi.org/10.1073/pnas.1802448115
49. Ryu MH, Gomelsky M (2014) Near-infrared light responsive synthetic c-di-GMP module for optogenetic applications. ACS Synth Biol 3(11):802–810. https://doi.org/10.1021/sb400182x
50. Horner M, Muller K, Weber W (2017) Light-responsive promoters. Methods Mol Biol 1651:173–186. https://doi.org/10.1007/978-1-4939-7223-4_13
51. Wang X, Chen X, Yang Y (2012) Spatiotemporal control of gene expression by a light-switchable transgene system. Nat Methods 9 (3):266–269. https://doi.org/10.1038/nmeth.1892
52. Qi F, Tan B, Ma F et al (2019) A synthetic light-switchable system based on CRISPR Cas13a regulates the expression of LncRNA MALAT1 and affects the malignant phenotype of bladder cancer cells. Int J Biol Sci 15 (8):1630–1636. https://doi.org/10.7150/ijbs.33772
53. Pan Y, Yang J, Luan X et al (2019) Near-infrared upconversion-activated CRISPR-Cas9 system: a remote-controlled gene editing platform. Sci Adv 5(4):eaav7199. https://doi.org/10.1126/sciadv.aav7199
54. Lyu Y, He S, Li J et al (2019) A photolabile semiconducting polymer nanotransducer for near-infrared regulation of CRISPR/Cas9 gene editing. Angew Chem Int Ed Engl 58 (50):18197–18201. https://doi.org/10.1002/anie.201909264
55. Bondy-Denomy J, Pawluk A, Maxwell KL et al (2013) Bacteriophage genes that inactivate the CRISPR/Cas bacterial immune system. Nature 493(7432):429–432. https://doi.org/10.1038/nature11723
56. Pawluk A, Bondy-Denomy J, Cheung VH et al (2014) A new group of phage anti-CRISPR genes inhibits the type I-E CRISPR-Cas system of *Pseudomonas aeruginosa*. MBio 5(2): e00896. https://doi.org/10.1128/mBio.00896-14
57. Bondy-Denomy J, Garcia B, Strum S et al (2015) Multiple mechanisms for CRISPR-Cas inhibition by anti-CRISPR proteins. Nature 526(7571):136–139. https://doi.org/10.1038/nature15254
58. Rauch BJ, Silvis MR, Hultquist JF et al (2016) Inhibition of CRISPR-Cas9 with bacteriophage proteins. Cell 168(1–2):150–158.e110. https://doi.org/10.1016/j.cell.2016.12.009
59. Marino ND, Zhang JY, Borges AL et al (2018) Discovery of widespread type I and type V CRISPR-Cas inhibitors. Science 362 (6411):240–242. https://doi.org/10.1126/science.aau5174
60. Knott GJ, Thornton BW, Lobba MJ et al (2019) Broad-spectrum enzymatic inhibition of CRISPR-Cas12a. Nat Struct Mol Biol 26 (4):315–321. https://doi.org/10.1038/s41594-019-0208-z
61. Pawluk A, Amrani N, Zhang Y et al (2016) Naturally occurring off-switches for CRISPR-Cas9. Cell 167(7):1829–1838. e1829. https://doi.org/10.1016/j.cell.2016.11.017
62. Pawluk A, Staals RH, Taylor C et al (2016) Inactivation of CRISPR-Cas systems by anti-CRISPR proteins in diverse bacterial species. Nat Microbiol 1(8):16085. https://doi.org/10.1038/nmicrobiol.2016.85
63. Lee J, Mir A, Edraki A et al (2018) Potent Cas9 inhibition in bacterial and human cells by AcrIIC4 and AcrIIC5 anti-CRISPR proteins.

MBio 9(6):e02321–e02318. https://doi.org/10.1128/mBio.02321-18
64. Dong GM, Wang S et al (2017) Structural basis of CRISPR-SpyCas9 inhibition by an anti-CRISPR protein. Nature 546 (7658):436–439. https://doi.org/10.1038/nature22377
65. Garcia B, Lee J, Edraki A et al (2019) Anti-CRISPR AcrIIA5 potently inhibits all Cas9 homologs used for genome editing. Cell Rep 29(7):1739–1746. e1735. https://doi.org/10.1016/j.celrep.2019.10.017
66. Harrington LB, Doxzen KW, Ma E et al (2017) A broad-Spectrum inhibitor of CRISPR-Cas9. Cell 170(6):1224–1233. e1215. https://doi.org/10.1016/j.cell.2017.07.037
67. Zhu YW, Zhang F, Huang ZW (2018) Structural insights into the inactivation of CRISPR-Cas systems by diverse anti-CRISPR proteins. BMC Biol 16:32. https://doi.org/10.1186/s12915-018-0504-9
68. Bubeck F, Hoffmann MD, Harteveld Z et al (2018) Engineered anti-CRISPR proteins for optogenetic control of CRISPR-Cas9. Nat Methods 15(11):924–927. https://doi.org/10.1038/s41592-018-0178-9
69. Ma H, Tu LC, Naseri A et al (2016) CRISPR-Cas9 nuclear dynamics and target recognition in living cells. J Cell Biol 214(5):529–537. https://doi.org/10.1083/jcb.201604115
70. Renicke C, Schuster D, Usherenko S et al (2013) A LOV2 domain-based optogenetic tool to control protein degradation and cellular function. Chem Biol 20(4):619–626. https://doi.org/10.1016/j.chembiol.2013.03.005
71. Sun W, Zhang W, Zhang C et al (2017) Light-induced protein degradation in human-derived cells. Biochem Biophys Res Commun 487 (2):241–246. https://doi.org/10.1016/j.bbrc.2017.04.041
72. Bonger KM, Rakhit R, Payumo AY et al (2014) General method for regulating protein stability with light. ACS Chem Biol 9(1):111–115. https://doi.org/10.1021/cb400755b
73. Liu JJ, Orlova N, Oakes BL et al (2019) CasX enzymes comprise a distinct family of RNA-guided genome editors. Nature 566 (7743):218–223. https://doi.org/10.1038/s41586-019-0908-x
74. Kleinstiver BP, Pattanayak V, Prew MS et al (2016) High-fidelity CRISPR-Cas9 nucleases with no detectable genome-wide off-target effects. Nature 529(7587):490–495. https://doi.org/10.1038/nature16526
75. Dagliyan O, Tarnawski M, Chu PH et al (2016) Engineering extrinsic disorder to control protein activity in living cells. Science 354 (6318):1441–1444. https://doi.org/10.1126/science.aah3404
76. Bedbrook CN, Yang KK, Robinson JE et al (2019) Machine learning-guided channelrhodopsin engineering enables minimally invasive optogenetics. Nat Methods 16 (11):1176–1184. https://doi.org/10.1038/s41592-019-0583-8
77. Hoffmann MD, Bubeck F, Eils R et al (2018) Controlling cells with light and LOV. Adv Biosystems 2(9):1800098. https://doi.org/10.1002/adbi.201800098
78. Gainza P, Sverrisson F, Monti F et al (2019) Deciphering interaction fingerprints from protein molecular surfaces using geometric deep learning. Nat Methods 17(2):184–192. https://doi.org/10.1038/s41592-019-0666-6
79. Kemaladewi DU, Bassi PS, Erwood S et al (2019) A mutation-independent approach for muscular dystrophy via upregulation of a modifier gene. Nature 572(7767):125–130. https://doi.org/10.1038/s41586-019-1430-x

Index

A

B

C

Dominik Niopek (ed.), *Photoswitching Proteins: Methods and Protocols*, Methods in Molecular Biology, vol. 2173, https://doi.org/10.1007/978-1-0716-0755-8, © Springer Science+Business Media, LLC, part of Springer Nature 2020

D

E

F

G

H

I

W

X

Y

Z

MIX
Papier aus verantwortungsvollen Quellen
Paper from responsible sources
FSC® C105338

If you have any concerns about our products,
you can contact us on
ProductSafety@springernature.com

In case Publisher is established outside the EU,
the EU authorized representative is:
Springer Nature Customer Service Center GmbH
Europaplatz 3, 69115 Heidelberg, Germany

Printed by Libri Plureos GmbH
in Hamburg, Germany